Science and Technical Writing

A Manual of Style

Science and Technical Writing

A Manual of Style

SECOND EDITION

Philip Rubens,
General Editor

ROUTLEDGE
New York and London

Published in 2001 by
Routledge
29 West 35th Street
New York, NY 10001

Published in Great Britain by
Routledge
11 New Fetter Lane
London EC4P 4EE

Routledge is an imprint of the Taylor & Francis Group

Printed in the United States of America on acid-free paper
Design and typography: Jack Donner

10 9 8 7 6 5 4 3 2 1

Library of Congress Cataloging-in-Publication Data

Science and technical writing : a manual of style / Philip Rubens, editor. — 2nd ed.
p. cm.
Includes bibliographical references and index.
ISBN 0–415–92550–9 (alk. paper) — ISBN 0–415–92551–7 (pbk. : alk paper)
1. Technical writing. I. Rubens, Philip.
T11. S378 2001
808'.0666—dc21 00–032837

Contents

List of Exhibits xxxi

Contributors xxxiii

Preface xxxv

1. Audience Analysis and Document Planning 1

Analyzing an Audience 1

Conducting an Audience Analysis 1

Identifying Audience Characteristics 2

Assessing Audience Objectives and Needs 2

Creating an Audience Profile 3

Analyzing the Information 3

Identifying Information Characteristics 3

Purpose and Structure 4

Assumptions about Audience Training 4

Frequency and Pattern of Use 4

Textual Features 4

Classifying Documents by Type 5
Using Information Characteristics to Define Documents 5
Defining Marketing Documents 6
Defining Conceptual Documents 6
Defining Procedural Documents 6
Defining Tutorials 7
Defining Job Aids 7
Defining Referential Documents 8
Using a Typology to Guide Document Development 8
Determining the Appropriate Medium 9
Articles 11
Booklets 11
Brochures 11
Newsletters 12
Correspondence 12
Letters 12
Memoranda 13
Electronic Mail (E-Mail) 13
Manuals (Guides) 13
Tutorial/Training Guides 14
User Guides/Operator Manuals 15
Reference Manuals 16
User Reference Manuals 16
Job Aids 16
Reports 17
Scientific Research Reports 17
Business Research Reports 18
Progress/Status Reports 18
Proposals 18
Feasibility Studies 19
Help Systems 20
Wizards 21

Websites 21
Site Elements 22
Site Types 23
Site Access 23

Planning a Document's Content 24

Collecting Information about the Subject 24
Analyzing Written Source Material 24
Interviewing Sources 25
Conducting a Hands-on Evaluation 26
Selecting an Organizational Method 26
Preparing an Outline 27

2. Writing for Non-native Audiences 31

Minimum Word Strategy 32

Controlled English 32

Global English 33

General Sentence Structure Guidelines 34
Use Short, Simple Sentences 34
Say What You Mean 34
Simplify Verb Phrases and Tenses 35
Untangle Long Noun Phrases 35
Make Sure It Is Clear What Each Pronoun Is Referring To 36
Place a One-Word Modifier in front of the Word That It Modifies 36
Make Sure It Is Obvious What Each Prepositional Phrase Is Modifying 37
Make Sure It Is Clear What Each Relative Clause Is Modifying 37
Make Sure It Is Obvious What Each Infinitive Phrase Is Modifying 38
Consider *Fronting* Prepositional Phrases and Other Sentence Constituents 38
Make Each Sentence Syntactically and Semantically Complete 39

Avoid Unusual Grammatical Constructions 39
The "Get" Passive 39
Causative "Have" And "Get" 39
"Given" And "Given That" 40
"In That" 40
"Should" as a Conditional Tense Modal Verb 40
"Need Not" 41
Inverted Sentences 41
Beware of Inherently Ambiguous Constructions 41
"Require" + an Infinitive or "Need" + an Infinitive 41
"Must Have Been" 42
Avoid Adverbial Interrupters 42
Use Standard, Conventional English 43
Use Standard Verb Complements 43
Do Not Use Transitive Verbs Intransitively, or Vice Versa 44
Do Not Violate Semantic Restrictions 45
Use Singulars and Plurals Appropriately 46
Use the Article "the" Only When It Is Clear Which Instance of the Following Noun You Are Referring To 46
Syntactic Cues Procedure Guidelines **47**
Do Not Write Telegraphically 48
Consider Expanding Past Participles into Relative Clauses 48
Consider Editing Present Participles 49
Search for All Occurrences of the Word "And" 51
Search for All Occurrences of "Or" 53
Look for Long Noun Phrases 54
Look for Specific Verb Forms 54
Look for "Give" and "Assign" as Verbs 54
Look for the Word "If" 54
Consider Expanding Adjectives into Relative Clauses 55
Terminology Guidelines **55**
Develop a Multilingual Terminology Database 55
Each Word Should Have Only One Meaning 56
Standardize your Terminology 57
Do Not Use Synonyms 57

Use Conventional Word Combinations and Phrases 57
Standardize Your Use of Prepositions 57
Avoid Using Words and Phrases That Have Multiple Common Meanings 58
Avoid Using Idiomatic Two- or Three-Word Verbs 58
Avoid Using a Phrase When There Is a Single Word That You Could Use Instead 58
Avoid Slang and Colloquialisms 59
Avoid Using Figurative Language 59
Avoid Humor, Satire, and Irony 59
Avoid Referring to Culture-Specific Ideas 59
Use Nouns as Nouns, Verbs as Verbs, and So On 60
Do Not Use Verbs as Nouns 60
Do Not Add Verb Suffixes to Nouns, Conjunctions, and Adjectives 60
Do Not Use Adjectives as Nouns 60
Beware of Words That May Be Grammatically or Semantically Incomplete 61

3. Grammar, Usage, and Revising for Publication 63

Grammar Basics 63

Phrase Types 63
Clause Types 66
Sentence Types 66

Common Problems in Grammar, Style, and Usage 67

Problems with Nouns 67
Abstract versus Concrete Nouns 67
Plural versus Possessive Nouns 67
Appositives and Parenthetical Definitions 68
Gerunds and Infinitives 68
Noun Stacks 68

Problems with Pronouns 69
The Problem with Pronoun References 69
Compound Pronoun Antecedents 69
Collective Nouns as Pronoun Antecedents 70
Problems with Second-Person Pronouns 70
Agreement of Pronouns 70
Sexist Language and Pronouns 70
Problems with Verbs 71
Verb Agreement with Collective Noun Subjects 71
Verb Agreement with Subject Complements 71
Verb Agreement with Compound Subjects 71
Problems with Modifiers 72
Restricting, Intensifying, and Absolute Modifiers 72
Adjective versus Adverb Forms 72
Misplaced Modifiers 73
Dangling Participles 73
Dangling Modifiers 74
Squinting Modifiers 74

Revision Strategies **74**

Revision Strategy 1: Examining Components 75
Revision Strategy 2: Examining Basic Grammar ("Triage") 77
Revision Strategy 3: Eliminating Sentence Faults 77
Dependent Clauses as Fragments 77
Phrases as Fragments 78
Run-on Sentences 78
Fragments as Used in Lists 78
Revision Strategy 4: Establishing Consistency 79
Establishing Logical Transitions between Grammatical Structures 79
Deleting Topical Shifts 80
Creating Parallelism 81

4. Punctuating Scientific and Technical Prose 85

Punctuation of Common Sentence Structures 85

Comma 87

Compound Sentence and Comma 87
Compound Predicate and Comma 87
Introductory Phrase or Clause Punctuated with a Comma 87
Introductory Expression after Conjunction and Comma 88
Restrictive Phrase or Clause and Comma 88
Participial Phrase as Subject and Comma 89
Items in Series and Comma 89
Phrases and Clauses in Series and Comma 89
Transitional Word or Phrase and Comma 89
Transposed Terms and Comma 90
Apposition and Comma 90
Contrasting Phrase and Comma 90
Deliberate Omissions and Comma 90
Similar or Identical Words and Comma 91
Two or More Adjectives in Series and Comma 91
Direct Quotations and Comma 91
Direct Questions and Comma 92
Dates and Comma 92
Specifying Names with Commas 92
Numbers and Commas 92
Reference Numbers and Commas 93

Semicolon 93

Semicolon and Clauses Joined without Conjunction 93
Semicolon and Clauses Joined by Conjunction 93

Transitional Adverbs and Semicolon 94
Transitional Terms and Semicolon 94
Elements in a Series and Semicolon 94

Colon 94

Elements in a Series and Colon 94
Amplification and Colon 95
Long Quotation and Colon 95
Ratios and Colon 95
Time of Day and Colon 95

Dash 96

Amplification or Illustration and Dash 96
Em Dash and Other Punctuation 96
Range of Numbers or Time and En Dash 96
Two-Noun Combinations and En Dash 97

Period 97

Vertical Lists and the Period 97
Abbreviations and Periods 98
Raised Periods 99
Ellipsis Points 99

Question Mark 99

Exclamation Point 100

Hyphen 100

Unit Modifier Containing Adjectives and Hyphen 100
Unit Modifier Containing Number and Unit of Measure 101
Unit Modifier Containing Adverb and Hyphen 101
Unit Modifier Containing a Proper Name 101
Unit Modifier Longer than Two Words 102
Single Letter and Hyphen 102

Foreign Phrase Used as Adjective 102
Predicate Adjective 102
Self- Compound and Hyphen 102
Suspended Hyphen 103
Numbers and Hyphens 103
Technical Terms and Hyphens 103
Prefixes and Suffixes and Hyphens 103

Slash 104

Apostrophe 105

Possessives and Apostrophes 105
Contractions and Apostrophes 105
Plurals and Apostrophes 105

Quotation Marks 106

Single versus Double Quotes 106
Placement of Quotes with Other Punctuation 106
Specialized Usage for Quotation Marks 107
So-Called Expressions 107

Parenthesis 108

Abbreviations and Parentheses 108
References and Parentheses (or Brackets) 108
Numbered Lists and Parentheses 108
Punctuating Parenthetical Comments 108

Brackets 109

Editorial Interjections and Brackets 109
Parentheses within Parentheses 109

Iconic Characters 109

Punctuating Equations 110

5. Using Acceptable Spelling 111

Good Spelling Practice 111

Preferred Spellings 111

Spelling Checkers 112

Plurals 113

Plurals of Foreign-Language Words 113

Singular/Plural Dilemma 114

Proper Names, Possessives, and Abbreviated Terms 114

Compound Words and Hyphenation 115

Word Division 115

General Word Division Guidelines 115

Suffixes 116

Soft Endings ("*-ce*," "*-ge*") 116

The "*-y*" Ending 116

Consonant Endings 117

The "*-ceed*," "*-sede*," "*-cede*" Rule 117

Capitalization 118

Beginning Sentence 118

Listings 118

Parentheses 118

Quotations 119

Salutations and Closings 119

Book and Article Titles 119

Proper Nouns 119

Trade Names, Trademarks, Registered, and Copyright 120

Computer Documentation 121

Troublesome Words 122

Sound (Homonyms)- and Look-Alike Words 122
Spelling Variations 122
American versus British Spelling 123
Easy Misspellings 123
Foreign Words 124

6. Incorporating Specialized Terminology 125

Abbreviations as Professional Shorthand 125

Style Guides for Abbreviated Terms 126
Abbreviated Terms in Technical and Scientific Texts 127
Initialisms 127
Acronyms 128
Indirect Articles and Acronyms 128
Symbols 129
Arbitrary Changes in Symbols 129
Creating Suitable Abbreviations 129
Abbreviating Single Words 130
Abbreviating Two Consecutive Words 130
Frequency of Use for Abbreviated Terms 130
Explaining Abbreviated Terms 130
Formatting Abbreviated Terms 131
Beginning a Sentence with Abbreviated Terms 131
Punctuating Abbreviated Forms 131
Periods in Abbreviated Forms 132
Slash in Abbreviated Forms 133
Apostrophe Plurals of Initialisms 133

Organization Names 133

Academic Organizations 133
Commercial Organizations 134
Government and Legislative Names 135
Societies and Associations 136

Military Terminology 136

Military Standards Related to Terminology 136
Aircraft Terminology 137
Armament 137
Naval Vessels 138
Surveillance Terms 138

Health and Medical Terms 139

Anatomical Terms 139
Diseases and Conditions 139
Drug Terminology 139
Infectious Organisms 140
Medical Equipment 140
Medical Procedures and Treatments 140
Vitamins and Minerals 140

Scientific Terms and Symbols 141

Astronomical Terms 141
Planets 141
Moons 142
Comets 142
Meteors 142
Constellations 142
Stars 143
Classes of Stars 144
Novas 144
Galaxies 144
Biological Terms 145
Chemistry Terminology 146
Elements 146
Chemical Compounds 148
Chemical Notation 148

Formulas 149

Geology Terms 149

Physics Terminology 150

Laws, Principles, and Theories 150

Phenomena 150

Atomic and Nuclear Particles 151

Technology Terms 151

Computer Terms 152

Trends in Terminology 152

Hardware Terminology 152

Software Terminology 153

Commercial Programs and Application Packages 153

Personal Computer Software 153

Mainframe Computer Software 153

Releases and Versions of Products 154

Computer Commands 154

Computer Messages 154

Software Development Documentation 155

Programming Languages 155

Procedures 156

Electronics 157

Circuit Technologies 157

Passive Electronic Components 157

Active Electronic Components 158

Logic Circuits 158

Integrated Circuits 159

Telecommunications 160

Telecommunication Abbreviations and Acronyms 160

Transmission and Control Characters 160

Device States 160

7. Using Numbers and Symbols 161

Using Numerals 161

Measurements 162

Mathematical Expressions 162

Decimals 162

Percentage 162

Proportion 163

Time Spans 163

Dates and Clock Time 163

Writing Fractions and Large Numbers 164

Writing Fractions and Mixed Numbers 164

Writing Large Numbers 164

Using Words to Express Numbers 165

Using Hyphens with Numbers 166

Using Scientific Notation 168

Comparing Powers of Ten 169

Numbers in Tables and Graphs 170

Reporting Significant Digits 171

Orders of Magnitude 173

Units of Measurement 173

Choosing a System of Units 173

Using SI Units and Symbols 174

Using Traditional Units 178

Choosing Appropriate Symbols 178

Mathematical Symbols 179

Arithmetic, Algebra, Number Theory 179

Trigonometric and Hyperbolic Functions 180

Elementary and Analytic Geometry 181

Calculus and Analysis 181

Logic and Set Theory 182

Statistics 182

Mathematical Typography 183

Use of Roman Type 184

Use of Italic Type 184

Use of Greek Letters 184

Use of Boldface 184

Handling Mathematical Expressions 185

Horizontal Spacing of Elements 185

Placement of Equations 186

Breaking Equations 187

Miscellaneous Typographic Details 189

Punctuating Mathematics 190

8. Using Quotations, Citations, and References 193

Legal Guidelines 194

Quotations 194

Quotation Usage 194

How to Use Quotations 195

When to Cite Quotations 195

When Not to Cite a Source 196

Types of Quotations 196

Types of Direct Quotations 196
Guidelines for Direct Quotations 197
Guidelines for Nonstandard Direct Quotations 197
Epigraphs 197
Quoting Oral Sources 198
Quoting from Correspondence 198
Quoting within Footnotes 199
Foreign-Language Quotations 199
Guidelines for Indirect Quotes 199
Properly Setting Quotations in Text 200
Guidelines for Setting a Quotation within the Text 200
Introducing Block (Set-off) Quotations 200
Informal Introduction 201
Formal Introduction 201
Introducing Indirect Quotations 202
Revising Quotations 202
Adding Material to Quotations 202
Omitting Parts of Block (Set-off) Quotations 203
Omitting Parts of Run-in Quotations 204
Capitalization 205
Capitalizing Run-in Quotations 206
Capitalizing Block Quotations 206
Punctuation 206
Quotation Marks 207
Quotation Marks in Block (Set-off) Quotations 207
Commas and Periods 207
Other Punctuation and Quotations 208
Citations, Notes, and References 208
Selecting an Appropriate Style Guide 209
Using a Supportive Database Program 210
Understanding the Basic Aspects of Documentation 210
In-Text Citations 211
Complete Reference in Text 211
Illustrations and Data Displays as In-Text Notes 213

Footnotes 213
Endnotes 214
Reference Lists 215
Bibliography 216
Citing Electronic Sources 217
Modern Language Association (MLA) 219
International Standard Organization ISO 690-2 219
American Psychological Association (APA) 219
Chicago Manual of Style 220
American Chemical Society (ACS) 220
American Medical Association (AMA) 220

9. Creating Indexes 221

Key Indexing Terms 221

Need for Indexes in Technical Material 222

The Goals of a Good Index 222

Planning an Index 223

The Audience 223
The Indexer 223
Planning the Index Length 224
Preparation Time and Indexing 224
Indexer's Style Guide and Keyword List 225

Filing Index Entries 226

Alphabetic versus Subject Indexes 226
Alphabetic Arrangements 227
General Rules for Alphabetizing 228

Index Format and Design 229

Indented or Paragraph Format 229
Indented 229
Paragraph 229

Layout Considerations 230
Type Choices 230
Punctuation Considerations 231
Format and Design Example 232

Indexing Procedure 232

An Overview of the Indexing Procedure 233
Determining What to Index 233
Matter to Omit 234
Using the Computer in Index Preparation 234
Word Processors and Alphabetized Lists 235
Word Processors and Embedded Indexing 235
Stand-Alone Index Composition Programs 236
Markup Languages and Indexes 237
Indexing with High-End Publishing Programs 237
Computer Indexing Conclusions 238
Writing and Editing Index Entries 239
Headings and Subheadings 240
Synonyms 242
Cross-References 242
Rules Governing Index Locators 244
Rules Governing Special Cases 244
Abbreviations and Acronyms 244
Accents, Apostrophes, and Diacritical Marks 245
Foreign Terms 245
Latin Names 245
Entities and Alphabetic Ordering 245
Proper Names 245
Numerals 246
Symbols 246
Including an Introductory Note 247
Final Proofreading 247

10. Creating Nontextual Information 249

Electronically Acquiring, Processing, and Printing or Displaying Illustrations 250

Acquiring Illustrations 250

Processing Illustrations 253

Printing or Displaying Illustrations 254

Selecting Illustrations 255

Photographs 256

Renderings 256

Line Drawings 257

Infographics 259

Symbols and Icons 259

Preparing Illustrations for Foreign Audiences 260

Planning for Cultural Differences 260

Providing for Text Expansion 260

Using Symbols and Icons 261

Constructing Illustrations 261

Gathering Source Information 262

Working with the Artist 262

Printing Illustrations 263

Special Considerations for Producing Illustrations 263

Type Legibility for Illustration Text 263

Positioning Illustrations 263

Selecting Views 264

Including a Hand 264

Highlighting 264

Numbering Illustrations 266
Labeling Illustrations 267
Captioning Illustrations 267
Subtitling Illustrations 268
Callouts 268
Illustration Notes 268

11. Creating Usable Data Displays 269

What Are Data Displays: Tables, Charts, and Diagrams? 269
Which Should I Use 270

Characteristics of Data Displays 271
Legible 271
Integrated with Text 271
Easy-to-Locate Information 272
Complete Data Displays 272
Simple Data Displays 272
Accessible Data Displays 272
Numbering Data Displays 272
Primary Title for Data Displays 273
Subtitle for Data Displays 273

Using Technology to Create Data Displays 273

Tables 275
When to Use Tables 275
Parts of Tables 275
Column Headings 276
Column Subheadings 277
Column Numbers 278
Column Spanners 278
Field Spanners 280

Stub 281
Stub Head 281
Row Headings 281
Field 284
Rules 285
Row Rules 286
Column Rules 286
Table Border 286
Head-Stub-Field Separators 286
Side-by-Side Table Separators 287
Table Notes 287
Specific Notes 287
Table Contents 288
Text in Tables 289
Numbers as Table Content 289
Dates in Tables 290
Times in Tables 290
Ranges in Tables 291
Formulas in Tables 291
Graphics in Tables 292
Missing or Negligible Values in Tables 292
Repeated Values in Tables 292
Special Tabular Problems 293
Continued Tables 293
Referring to Tables 294
Oversize Tables 294
Abbreviations in Tables 294
Simplifying Comparisons in Tables 294
Grouping and Summarizing Data in Tables 295
Common Types of Tables 296
Look-Up-a-Value Tables 296
Decision Tables 297
Distance Tables 297
Matrix Charts 298

Charts 298

When to Use Charts 299

Characteristics of an Effective Chart 299

Coherent 299

Simple 299

Concise 300

Honest 301

Visually Attractive Charts 302

Fit Charts to the Medium 303

Parts of Charts 304

Data Values on Charts 305

Data Point Symbols for Charts 305

Area or Volume Designation for Charts 307

Symbols or Pictures for Charts 307

Curves and Lines in Charts 308

Line Weight for Chart Curves 309

Line Style in Charts 309

Curve Labels in Charts 309

Scales for Charts 310

Placing Scale Labels in Charts 311

Scale Divisions 312

Scale Ranges for Data Plotting 313

Scaling Function 313

Linear Function 313

Logarithmic Function 314

Geometric Function 314

Grid Lines 315

Reference of Base Line 315

Posted Data Values 315

Annotations 315

Inserts 316

Notes for Charts 317

Border or Frame for Charts 317

Keys and Legends for Charts 318
Special Topics and Effects 318
Size of Chart Typography 318
Weights of Chart Lines 319
Abbreviations 319
Patterns, Textures, and Shading 319
Color Combinations 320
Three-Dimensional Effects 321
Oblique Views 322
Thickness 322
Drop Shadows 323
Pictorial Elements 323
Overlays and Combinations 323
Referring to Charts 323
Common Types of Charts 324
Scale Charts 324
Scatter Charts 325
Triangular Charts 326
Circular Charts 326
Curve Charts 327
Frequency Distribution Charts 328
Cumulative Curve Charts 328
Surface Charts 329
Band Surface Charts 330
100 Percent Surface Charts 331
Net-Difference Surface Charts 331
Column and Bar Charts 332
High-Low Charts 334
100 Percent Column Chart 334
Histogram 335
Time Lines 335
Paired-Bar Charts 336
Sliding-Bar Charts 336
Column/Bar Chart Combinations 337
Grouped Column/Bar Charts 337
Divided Column/Bar Charts 338
Deviation Column/Bar Charts 339
Pictographic Column/Bar Charts 340

Pie Charts 340
Pictorial Pie 342
Separate Segment 343

Diagrams **343**

Components of Diagrams 344
Symbols 344
Links 344
Patterns of Linkages 346
Common Types of Diagrams 347
Organization Charts 347
Flowcharts 348
Connection Maps 349
Dependency Diagrams 349

12. Designing Useful Documents **351**

Designing a Document's Format **351**

Using Technology to Support Document Design **352**

Designing the Page and Screen **353**

Page and Screen Size 354
White Space 356
Margins 357
Typefaces 357
Type Sizes 361
Leading (Line Spacing) 362
Line Width 362
Interactions of Type Size, Line Width, and Leading 362
Highlighting Techniques 363

Page and Screen Elements 364

Page Numbering 364
Running Headers and Footers 364
Specialized Notices: Warnings, Cautions, and Notes 365
Source Notes and References 365
Lists 366
Revision Notices 366

Page and Screen Grids 366

Designing Specific Information Types 368

Designing Brochures and Newsletters 368
Preparing a Paper or Article for Publication 370
- Title 371
- Abstract 371
- Introduction 371
- Body 372
- Conclusion 372
- References 372

Designing a Technical Manual 372
- Reader Access 373
- Front Cover 373
- Front Matter 373
  - Title Page 374
  - Copyright Page 374
  - Acknowledgments 374
  - Abstract or Preface 374
  - Table of Contents 374
  - List of Figures 374
  - About This Manual 374
- Chapter or Section Divisions 375
- Appendixes 375
- Glossary 375
- Indexes 375
- Back Cover 375

Preparing A4 Bilingual and Multilingual Documents 376
Separate Publications 376
Separate Joined Publications 376
Two-Column Side-by-Side Publications 377
Multicolumn Side-by-Side Publications 378
Designing On-Screen Information 379
Document Techniques and Information Class 379
Splash Page 380
Home Page 380
Section Pages 381
Linked and Archived Sources 381
Text: Static Information 381
Multimedia: Dynamic Information 382
Graphic Specification 382
Controlling Large Document Sets 383
Categorizing Information 383
Audience Analysis to Classify Documents 383
Frequency of Use to Classify Documents 384
Task Orientation to Classify Documents 384
System Configuration to Classify Documents 384
Making Information Accessible 384
Graphic Production Considerations 385
Preparing Camera-Ready Pages 385
Choosing Paper 386
Binding the Document 387
Wire Stitching 387
Mechanical Binding 387
Perfect Binding 388
13. Bibliography 389
Index 399

List of Exhibits

Exhibit	Title	Paragraph
1–1	Document Typology	1.17
1–2	Communication Media Characteristics	1.39
1–3	Document Delivery Methods	1.40
3–1	**Grammatical Functions of Parts of Speech**	**3.3**
3–2	Phrase Types	3.4
3–3	Clause Types	3.5
3–4	Sentence Types	3.6
3–5	Common Modifier Problems	3.30
3–6	Component Edits	3.39
3–7	Typical Error Types Found in Triage Edits	3.40
3–8	Common Transitional Words and Phrases	3.49
7–1	**Comparison of Estimated Masses of Several Icebergs**	**7.34**
7–2	Tabular Data Presented as Powers of Ten	7.36
7–3	International System of Units—Base Units	7.45
7–4	International System of Units—Derived Units	7.45
7–5	Acceptable International System of Units Prefixes	7.46
10–1	**Some Typical Input Device Resolutions**	**10.8**
10–2	Scanning Recommendations for Same-Size Continuous Tone Illustrations	10.10
10–3	Download Time for Sample Displayed Documents	10.15
11–2	**Parts of a Table**	**11.28**
11–2	**Parts of a Chart**	11.133

Contributors

1. Audience

Lori Anschuetz • Senior Project Manager, Tec-Ed, Rochester, New York • A.B. Journalism and German, University of Michigan • Senior member, Society for Technical Communication

Amber Kylie Clark • Technical Writer, Tec-Ed, Ann Arbor, Michigan • B.S. Mathematics, Eastern Michigan University

2. Non-Native Audience

John R. Kohl • Technical Editor, SAS Institute, Cary, North Carolina • M.A. Teaching English as a Second Language, University of Illinois at Champaign-Urbana

3. Grammar, 5. Spelling, 8. Quotations and References

Katherine Underwood • Manager of Publications and Information Design, Duke University, Durham, NC • M.A. Linguistics, University of Texas-Austin

4. Punctuation

Joseph E. Harmon • Senior Technical Editor, Chemical Technology Division, Argonne National Laboratory, Argonne, IL • B.S. Mathematics, University of Illinois at Champaign-Urbana; M.A. English, University of Illinois at Chicago

6. Special Terminology

John Goldie • Technical Writer and User Interface Designer • A.B. Fine Arts, Harvard University; Ed.M. Harvard Graduate School of Education

7. Numbers and Symbols

Russ Rowlett • Director of the Center for Mathematics and Science Education, University of North Carolina, Chapel Hill, NC • Ph.D. Mathematics, University of Virginia

8. Quotes and References

Brenda Knowles Rubens • Human Factors Engineer, IBM Corporation, Research Triangle Park, NC • Ph.D. Communication and Rhetoric, Rensselaer Polytechnic Institute

9. Index

Richard Vacca • Document Systems Analyst and Communications Consultant, Boston, Massachusetts • M.S. Technical Communication, Rensselaer Polytechnic Institute

10. Illustration, 11. Displays, 12. Layout, General Editor

Philip Rubens • Professor of English, East Carolina University • Ph.D. Northern Illinois University • Fellow, Society for Technical Communication

Additional artwork for these chapters contributed by:

Wendy Beth Jackelow • Medical & Scientific Illustration • NYC • wbjackelow@aol.com

Raymond Ore • Raymation • http://www.raymation.co.uk

Penn State Pointers • Penn State University • College of Agricultural Sciences • http://aginfo.psu.edu/psp

Preface

It is certainly a pleasure to be offering a second edition of this text after nearly a decade of helping writers and editors prepare useful documents. During that time, many aspects of communication have changed, many have remained the same. It is difficult to imagine that attention to grammar and presentation has gone out of style. But the advent of accessible electronic information has influenced this edition, as it should. Moreover, we have taken this opportunity to incorporate and emphasize aspects of communication that seem more germane to contemporary interests. Thus, for example, guidance with creating texts for non-native audiences appears earlier in this text—chapter 2: "Writing for Non-native Audiences." Similarly, we have incorporated text where it seemed to make sense: chapter 6: "Incorporating Specialized Terminology" includes abbreviations, chapter 8: "Using Quotations, Citations, and References" combines the techniques for citing any information source.

At the same time, we have recognized the importance of new techniques and technologies. Thus, chapter 10: "Creating Nontextual Information" offers an extended discussion of using computers to create texts and supporting visuals for both print and online display. That discussion is further supported by discussions of computerized aids in virtually every chapter. Chapter 9: "Creating Indexes," for example, offers guidance on selecting electronic indexing systems; even chapter 5: "Using Acceptable Spelling" discusses the use of electronic dictionaries and other spelling aids.

Most chapters have added significant new material. Chapter 1: "Audience Analysis and Document Planning," for instance, includes a new discussion of information types and their relationship to appropriate presentation media. Some parts of the text have undergone significant revision. Chapter 2: "Writing for Non-native Audiences" departs significantly from the previous edition by adding guidance on Controlled and Global English. Chapter 3: "Grammar, Usage, and Revising for Publication" offers slightly less guidance in favor of the discussion provided in chapter 2. The rationale in this instance is that good writing practices that make texts easier to translate should apply to any audience. Finally, chapter 7: "Using Numbers and Symbols" has been thoroughly edited to reflect contemporary practices.

The balance between discussions of text and nontextual elements in the first edition has been maintained here, although expanded in useful ways. Chapter 10: "Creating Nontextual Information," chapter 11: "Creating Usable Data Displays," and chapter 12: "Designing Useful Documents" provide guidance on all aspects of document development, both paper and online, as well as assistance in selecting appropriate technologies.

Finally, this text employs a number of techniques to help readers locate information. First, chapter topical summaries, which offer references to appropriate subsections, appear near the beginning of each chapter. Depending on the topic's complexity, many of these subsections have their own reference sections. Chapter 4: "Punctuating Scientific and Technical Prose" illustrates the usefulness of this technique. Second, the text provides internal cross-referencing to similar or supporting topics. Third, the table of contents offers a more direct link to topics by providing only topical page changes, and the index has been expanded. Fourth, more descriptive headers and footers provide improved location and searching information. We hope these techniques prove useful.

Given the history of this text, the editor and contributor wish to express their thanks to those who contributed to the previous edition: Lilly Babits, Wallace Clements, Alberta Cox, Eva Dukes, Valerie Haus, William Horton, William J. Hosier, Barry Jereb, John Kirsch, Jean Murphy, David A. T. Peterson, James Prekeges, and R. Dennis Walters. We are certainly indebted to their efforts! We would also like to thank those readers who offered useful criticism of that edition: Diana Meyers, JPL; John Hagge, Iowa State University; Carolyn D. Rude, Texas Tech University; Joseph M. Williams, University of Chicago. We thank Lori Lathrop, past president of the American Society of Indexers, for her close reading of chapter 9: "Creating Indexes." Finally, we thank both the Society for Technical Communication for granting permission to use some of the material that appears in chapter 2 and Penn State University, College of Agricultural Sciences, for permission to use one of their infographics in chapter 10.

The editor would like to acknowledge the assistance of his editors at Routledge, Kevin Ohe and Krister Swartz, of Brenda Knowles Rubens in reviewing the entire manuscript, and of Smoke and Ripley, who insisted on balancing fun and work.

1.

Audience Analysis and Document Planning

Analyzing an Audience

1.1 Before writing anything, describe an audience by

- Conducting an audience analysis (1.2),
- Identifying audience characteristics (1.5),
- Assessing audience objectives and needs (1.6),
- Creating an audience profile (1.10).

Conducting an Audience Analysis

1.2 Conduct either a formal or an informal audience analysis. Use formal methods to gather quantifiable data; informal analysis is appropriate for small or poorly funded projects.

1.3 During formal analysis

- Collect surveys and questionnaire responses,
- Hold structured interviews,
- Conduct usability research, such as focus groups, field studies, or usability tests.

Some organizations perform formal analyses as part of their marketing planning.

1.4 During informal analysis, gather information about the audience indirectly by

- Talking with marketing, development, and other staff who have access to research results and customers;
- Reading notes and reports by product trainers or maintenance personnel who have had contact with the audience;
- Reading periodicals that relate to the product, industry, or audience;
- Talking informally with people who will read the final document.

Identifying Audience Characteristics

1.5 Before you begin writing, identify and consider such important audience characteristics as

- Educational and professional background,
- Knowledge and experience levels,
- English-language ability (see also chapter 2),
- Reading context (the physical and psychological conditions under which the audience reads the document).

Assessing Audience Objectives and Needs

1.6 Use audience objectives and needs to develop an approach to the document:

- Objectives reflect the activity the audience wants to be able to perform after reading the document.
- Needs indicate questions the audience will have that the document should answer.

1.7 Audience objectives may be long-term, short-term, personal, or job-related. They may or may not be directly related to the document.

1.8 Since most readers have job-related objectives for using technical documentation, identify those objectives to determine whether the audience needs the information to perform a task or acquire new knowledge.

1.9 To satisfy the needs of a diverse audience, address both different experience levels and different goals. Follow these general guidelines:

- Rank goals in terms of the questions the document must answer first, second, third, and so on.

- Write for one audience group at a time, and indicate which group you are addressing. Remember that other audiences may need the same information.
- Produce one document for all groups, or divide the information into more than one document.
- Include navigation aids—tables of contents and lists of figures and tables, page headers and footers, headings within the text, appendixes, tab dividers, and the like—to make information easy to find.

Creating an Audience Profile

1.10 Use the audience characteristics, objectives, and needs to develop a profile:

- Group related features in a written sketch of the typical reader.
- For a diverse audience, do a profile for each kind of reader.
- Form mental images of these composite people.
- Get to know the profiles before writing anything.
- Plan the document for typical readers and write to them.
- Provide the kind of information and presentation readers need to achieve their goals.

Analyzing the Information

1.11 To prepare a document that meets your audience's needs, analyze the information to be communicated by

- Identifying information characteristics (1.12),
- Classifying documents by type (1.17),
- Using information characteristics to define documents (1.18),
- Using a typology to guide document development (1.37).

Identifying Information Characteristics

1.12 Documents can be classified into genres, by examining such information characteristics as (see 1.17)

- Purpose and structure,
- Assumptions about audience training,

- Frequency and pattern of use,
- Textual features.

Purpose and Structure

1.13 Identify a document's purpose and structure by asking such questions as

- Does the audience need long-term or short-term knowledge?
- Is the audience familiar with the task's or product's conceptual framework, or should it be presented? A conceptual framework allows readers to generalize from one situation to another.
- Will the audience need peripheral as well as essential information?
- How much breadth and depth of information does the audience require?
- What is the most logical way—chronologically, alphabetically, task-oriented— to organize the document?

Assumptions about Audience Training

1.14 Use the following questions to characterize an audience's training:

- Is the audience familiar with the product, service, or software?
- Is the audience familiar with the task, situation, or problem?
- Can the document assume the audience has a grasp of basic concepts or background issues?

Frequency and Pattern of Use

1.15 Consider the following questions to identify a document's frequency and pattern of use:

- Will the document be referred to often or rarely?
- Will the document be read linearly or in short, disparate sections?

Textual Features

1.16 Textual features can also be used to classify documents. Will the document have

- Overview/summary sections?
- Step-by-step instructions?
- Narrative explanations?

- Conceptual models, analogies, and/or examples?
- Figures, charts, and/or tables?
- Cross-references and/or navigation aids?
- Technical terminology, language conventions, and/or symbolic systems?

Classifying Documents by Type

1.17 Although technical documents can be classified in many ways, the following typology offers a useful starting point for developing one within any organization.

Exhibit 1–1: Document Typology

Genre	Purpose	Characteristics
Marketing	Provides summary and overview information aimed at persuading the audience to initiate an action.	Uses less formal, often nontechnical language to convey information in lists or paragraphs.
Conceptual	Provides background and theoretical information to explain central ideas.	Uses formal and sometimes technical language to convey information in a narrative format.
Procedural	Helps the user accomplish an immediate task, imparting short-term knowledge.	Provides detailed, sequential step-by-step instructions without conceptual background.
Tutorial	Teaches a skill, imparting long-term knowledge necessary to accomplish important tasks.	Presents essential procedures supported by basic concepts, often using a chapter format.
Job aid	Provides quick-reference information for essential tasks.	Provides detailed descriptions of actions in a modular, nonsequential format.
Referential	Provides encyclopedic, in-depth information about a product or service.	Task-supportive but not sequential, providing greater breadth and depth than other genres.

Using Information Characteristics to Define Documents

1.18 Documents can be defined by applying the information characteristics outlined in 1.12 to the document types (exhibit 1–1) found within many organizations. Do not consider either the characteristics or the types to be restrictive. If your organization uses other practices, simply use this section as a methodology for supporting local needs. The following six sections illustrate how to apply this methodology to the major document types.

Defining Marketing Documents

1.19 Marketing materials include documents such as brochures, booklets, public relations newsletters, and websites. While they are similar in some ways to conceptual documents, their purpose is to persuade rather than simply inform.

1.20 The audience for marketing materials has basic conceptual knowledge, although some information may be aimed at those with greater technical expertise. Often, marketing documents are read quickly, randomly, and carelessly.

1.21 The language and tone of marketing documents is generally less formal than conceptual documents, and they usually employ such textual features as

- Paragraph format;
- Semimodular sections;
- Frequent summaries and overviews;
- Extensive graphics, tables, and lists.

Defining Conceptual Documents

1.22 Conceptual information appears in a variety of documents, including letters, memoranda, reports, articles, booklets, and technical manuals.

1.23 Conceptual documents assume the audience has a specific level of expertise that can be used to expand their existing conceptual framework. These documents are intended to be read until the knowledge becomes internalized.

1.24 Conceptual documents usually employ these textual features:

- Narrative, paragraph format;
- Sections focused on a single concept or idea;
- Theoretical models, analogies, and examples;
- Introduction of new terminology;
- Occasional figures, tables, or lists.

Defining Procedural Documents

1.25 Procedural documents explain how to use a product or service. While the level of detail for procedural documents varies widely, they generally try to present the full functionality of a product or service.

1.26 The audience for procedural documents seeks the answer to a problem or question that interrupts the work of its members. They often have some knowledge of the product through interacting with it. Procedural documents support the daily, routine use of a product or service and are referred to repeatedly until the process is committed to long-term memory.

1.27 Procedural documents usually have the following textual features:

- Overviews of product operations;
- Modules presenting sequences that support one or more activities;
- Some conceptual models, analogies, and examples;
- Topic and section links;
- Macro-level retrieval, location, and navigation aids.

Defining Tutorials

1.28 Tutorial documents, including tutorials and training manuals designed for use with or without an instructor, impart long-term core knowledge by presenting essential rather than peripheral information.

1.29 The audience for training materials has little or no experience with the product, although its members may be subject experts. The document should build upon the audience's existing conceptual frameworks. These documents probably will be used only once, whether in one sitting or over an extended period.

1.30 Tutorials and training manuals usually have such textual features as

- Overviews of product operations;
- Conceptual models, analogies, and examples;
- Modules focused on a single activity;
- Topic and section links.

Defining Job Aids

1.31 Job aids include task cards, quick installation guides, and wizards that provide brief, summary concepts and instructions needed to operate, maintain, or repair a product. Unlike procedural documents, they employ modules that are not linked in a specific sequence.

1.32 Quick-reference materials, such as function lists, are referred to frequently, and the audience is assumed to have basic conceptual knowledge through interaction with the product or training materials. In contrast, quick-performance materials, such as installation guides, are read infrequently and the audience is assumed to have no familiarity with the product.

1.33 Job aids generally have the following textual features:

- Short phrases rather than long sentences and paragraphs;
- Frequent lists, examples, and diagrams;
- Some technical terminology and professional conventions/notations.

Defining Referential Documents

1.34 Reference materials, such as user reference manuals, style guides, and policy manuals, provide encyclopedic information, organized alphabetically or numerically, about a product or service by presenting detailed definitions and descriptions of topics, but without linking them into sequences.

1.35 The audience for reference materials has familiarity with basic concepts, gained either through a tutorial or interaction with the product or service. These documents are used sporadically and read in short sections rather than linearly.

1.36 Referential documents generally use specific language conventions and textual features such as

- Reduced vocabulary and technical terminology;
- Diagrams, drawings, and professional notation systems understandable by experienced users or subject-area specialists;
- Topic and section links;
- Retrieval, location, and navigation aids.

Using a Typology to Guide Document Development

1.37 After identifying information characteristics, decide what kind of document to develop. For example, an audience that needs to perform a task with new software will need either procedural or tutorial information. Use the characteristics for these document types (see 1.17–1.36) to present the information appropriately. See exhibit 1–2: "Communication Media Characteristics" for a table associating document types with specific document formats.

1.38 If the audience is diverse with a wide range of needs, consider creating a suite of documents based on several genres. For example, the expected audience might need a training manual, an operator's manual, and a reference manual to comprehend all of the conceptual, procedural, referential, and tutorial information associated with a new software product.

Determining the Appropriate Medium

1.39 Technical and scientific communication employs a variety of media, each with specific characteristics and goals. After analyzing your audience and information, select the medium that best suits your purpose, style, and document type.

Exhibit 1–2: Communication Media Characteristics

Media	Purpose	Style	Document Type
Articles (1.43)	Discuss a single topic in depth, usually for a journal or book	Narrative style—formality varies with audience expertise	Conceptual
Booklets (1.45)	Offer overviews or introductions for one-time or reference use	Less formal, almost conversational in tone	Marketing or referential
Brochures (1.47)	Ask for action—convince readers to do or decide something	Catchy and easy-to-read, with lots of graphics	Marketing
Newsletters (1.49)	Convey information about a group of related topics at regular intervals	Journalistic—level of formality varies with content	Generally marketing, sometimes conceptual
Correspondence (1.51)	Is used for both internal and external communication	Varies in both formality and length	Generally conceptual
Manuals (1.61)	Describe a product or process in detail in a linear format	Action-oriented—formality depends on audience expertise	Often use a mix of document types
Reports (1.89)	Present a detailed response to a proposed question or problem	Formal and objective in tone—extensive use of supporting evidence	Conceptual
Help systems (1.97)	Describe a product or process in an electronic hypertext format	Action-oriented—formality depends on audience expertise	Mostly referential—may include concepts or procedures
Wizards (1.104)	Guide the user through a task in an interactive format	Concise and action-oriented in style, using nontechnical language	Tutorial or job aid
Websites (1.109)	Present information about a topic or group of related topics in a dynamic online format	Catchy and easy-to-use, with many graphics and dynamic elements—formality varies with content and audience	Marketing, conceptual, referential

1.40 When selecting the medium, also consider how you will distribute the document. In addition to print, documents can be delivered electronically in several ways, each with advantages and disadvantages.

Exhibit 1–3: Document Delivery Methods

Delivery Method	Use and Situation
Print	Deliver hard-copy documents within an organization or to external audiences. The distribute-then-print model of digital printing reduces the need to ship physical documents.
Facsimile	Deliver hard-copy documents to remote sites from hard-copy pages or an electronic file. The print quality of facsimiles is generally poor.
Network	Share electronic documents within an organization.
Floppy disk	Share electronic documents among nonnetworked computers. Disks must be hand-delivered to the recipients.
FTP (file transfer protocol) site	Deliver electronic documents within or outside an organization. Compress large documents before moving files. FTP sites are designed specifically for sharing files and are preferred over sending documents as e-mail attachments.
E-mail attachment	Send electronic documents within or outside organization when FTP is not available. Compress large documents before sending them.
Website	Publish documents electronically, making them available to anyone with access to the Internet, intranet, or extranet to which you publish the site.

1.41 Be sure to develop a document in a format that the expected audience will be able to access. Some potential incompatibilities include

- Platform (such as Macintosh™, PC™, or Unix™),
- Operating system type and version number,
- Application type and version number.

Some personal computers and software packages offer cross-platform or cross-version compatibility, while others do not.

1.42 Since some file formats require specific software to access them, consider both the audience's needs and those of your colleagues who may have to interact with your document.

Articles

1.43 Magazine, journal, or book articles, which readers expect to read in a single sitting, discuss a single topic in depth:

Article Type	Use to Report
Professional journal	Formal research results in an academic writing style for subject-area experts.
Trade journal	News about a specific industry or discipline, case histories of practitioners' experiences, and product reviews.
Commercial and popular press	News about products or technical/scientific advances of interest to the general public.

1.44 Only professional, and some trade, journals use specialized vocabulary and symbols (see also chapter 6) as well as complex data displays (see also chapter 11). For commercial and popular press articles, focus on the content's significance for the reader's daily life.

Booklets

1.45 Booklets, which are generally more extensive than brochures, convey introductory or overview, rather than marketing, information about a topic. Depending on its communication goal, a booklet may resemble a brochure or a technical manual. A booklet may range in length from a few pages to 50 pages.

1.46 Booklets often target a specific group. A university, for example, might publish a booklet to introduce new students to campus services and to serve as a reference for continuing students and staff.

Brochures

1.47 Brochures convince readers to take an action or make a decision. To achieve its primary goal, a brochure typically serves as a marketing or promotional tool. In addition to selling products or services, brochures can be used to offer brief descriptive overviews: the features of a company's pension plan, the participants in a professional trade show, and the like.

1.48 Brochures are usually brief—rarely more than 16 pages, regardless of the page size. To attract readers, a brochure often uses photographs, illustrations, clever headings, and color. Because a brochure is generally read quickly, use short sentences and paragraphs and insert frequent headings to mark major topic divisions.

Newsletters

1.49 A newsletter addresses one major topic in a long article or several shorter articles, or a variety of topics in a collection of articles. The four types of newsletters are

- Informational—present valuable information for which readers may be willing to pay. Independent research or publishing firms prepare such subscription newsletters;
- Public relations or promotional—promote an organization's image. For example, a health care facility might produce a newsletter for the general public that describes recent medical advances, health maintenance plan features, and staff qualifications;
- Internal—report on an organization's activities and events. A corporate newsletter, for example, might contain articles on regional sales, carpooling, and employee achievements;
- Hybrids—combine newsletter characteristics in novel ways. For example, a customer newsletter might provide information for using a product as well as promotional previews of new products.

1.50 Newsletters tend to be between 4 and 16 pages in length and are designed for fast reading. Multiple text columns like those used in general-circulation newspapers and magazines are a popular newsletter format.

Correspondence

1.51 Because correspondence often records information exchanges, the opening should contain the date, name and title of the addressee, complete mailing address, and salutation.

1.52 Provide a descriptive title in a subject line to help recipients determine the message's relative importance and to locate it for later review.

1.53 Review all technical correspondence for accuracy and liability issues before signing them. Conclude these documents with appropriate author identification: signature, typed name, and title.

1.54 People who should know about a correspondence but are not directly involved should be placed on a carbon copy (cc) list at the beginning or end of the document.

Letters

1.55 Use a letter for formal correspondence outside an organization, to make and respond to inquiries, and to provide explanations or instructions. Letters that accompany and explain other documents are called cover letters.

1.56 Technical letters use a block format with a line between each paragraph and no indentation. The body is typically one to five pages in length. Letters employ a typical narrative style, but numbered and bulleted lists may be used to itemize points.

Memoranda

1.57 Memoranda (memos) support communication within an organization by providing short pieces of information, reminders, and clarifications. Memos that accompany and explain other documents are called cover memos.

1.58 The body of a memo should be short, ranging from a few sentences to five pages; use a narrative style and/or lists to communicate your ideas.

Electronic Mail (E-Mail)

1.59 Use e-mail for short, informal correspondence or whenever response time is important.

1.60 The speed of e-mail leads to careless spelling, capitalization, and punctuation. This informality may be acceptable for personal e-mails, but all scientific and technical e-mail should follow accepted writing standards.

Manuals (Guides)

1.61 Technical manuals describe in detail a product, system, or process, or a group of related products, systems, or processes. A technical manual frequently contains many internal divisions, as well as numerous illustrations and data displays.

1.62 Few readers need to understand every detail of a system's installation, application, operation, and maintenance. For example, a technician who repairs a heart monitor does not need to know when and why a doctor orders a heart test, or how to interpret one. A variety of manual types fill specific reader needs:

- Tutorial/training guides,
- User guides/operator manuals,
- Reference manuals,
- User reference manuals,
- Job aids/quick-reference guides.

1.63 If an audience varies in levels of experience and responsibility, divide the necessary information among different manual types. If readers' needs may vary based on tasks, take a similar approach.

1.64 When organizing information into books or sections, consider the expected frequency of use:

- Only once, such as installation information;
- Only by first-time or novice users, such as introductions or tutorials;
- Regularly, but by specialized audiences, such as preventive maintenance procedures;
- Only if something goes wrong, such as troubleshooting instructions.

1.65 Several smaller manuals, which are more convenient for readers than one large manual, also offer advantages for the writer:

- Manuals can be produced one at a time, as information becomes available;
- Smaller manuals are easier to update than large manuals;
- Supplementary manuals can be added without reorganizing and reprinting a single, large manual.

1.66 When documenting a modular product, or one with many options, create modular manuals.

Tutorial/Training Guides

1.67 A tutorial uses explanation, repetition, hands-on practice, exploratory learning, and other motivating methods to help the reader attain a desired skill level. Typically, tutorial/training guides are produced either as stand-alone documents or as part of classroom instruction.

1.68 A stand-alone tutorial, which helps readers learn without an instructor, can be used with a variety of products, from personal computers to telescopes. Stand-alone tutorials are generally more extensive than classroom tutorials because the audience must gain the skills without expert assistance.

1.69 A classroom tutorial supports an instructor, who will supply additional information, answer questions, and help solve problems. Materials for classroom tutorials are usually incomplete; that is, the audience cannot gain the desired skill without the instructor's assistance.

1.70 In a tutorial, the sequence of procedures is important and should be designed to aid learning. Simple skills lead to complex skills, and general principles to particular applications.

1.71 Divide tutorials into short sections or training modules. Assume that the audience will read sequentially within a section or module but may complete sections out of order. Include reminders and tips about how to complete tasks learned in previous sections.

1.72 Since users frequently make mistakes when learning new skills, provide recovery information for common errors.

User Guides/Operator Manuals

1.73 User guides and operator manuals present the specific step-by-step procedures and concepts needed to use a product or system. Although more extensive than job aids, these manuals are generally not a complete reference; they include only those tasks that the majority of the audience will need most of the time.

1.74 Use chronology, frequency, or importance to organize the tasks needed to complete daily work.

1.75 Write a procedure for each task or group of related tasks that offers a description of the procedure's purpose, followed by numbered steps that use concise, action-oriented language.

1.76 Depending on the audience profile, assume that experts have some system experience and do not need explanations for basic concepts. For less experienced audiences, discuss the basic concepts and/or theory of operation. Present vital conceptual information before procedures, and secondary or background information after procedures.

1.77 In addition to procedures and concepts, include examples, typical warnings and system responses, and error recovery information.

1.78 Place in appendixes: detailed tables and lists, background concepts, advanced or alternative procedures, and any information not needed by all audience members.

1.79 Include an index to help readers find information quickly (see also chapter 9).

Reference Manuals

1.80 Reference manuals, which provide encyclopedic information about a product, including complete technical details about its operation, are not task-oriented and are organized alphabetically, numerically, or by product feature.

1.81 A reference manual shows how a system is designed and operates. For example, the reference manual for a multiple-line telephone system may contain wiring schematics and switch-setting tables of interest to technicians but not to telephone users.

1.82 Many reference manuals are multipurpose and serve the needs of installation, operations, and maintenance personnel. However, a complicated system may require a separate reference manual for each task or component.

User Reference Manuals

1.83 A user reference manual is a compromise between a user guide and a reference manual; choose it when resources or schedule dictate that a product have only one manual. While most user guides address only primary tasks, a user reference manual describes everything users can do with a system.

1.84 User reference manuals generally include both procedures and concepts, as well as a reference section. Effective headings, cross-references, and a good index help readers, who often have varied technical backgrounds and experience, find specific information.

1.85 Always try to make a user reference manual task-oriented, recognizing that—because it must be a complete information source—this approach is not always possible. For example, if a product has several hundred commands, time and space may not allow them to be organized by audience and application; an alphabetical presentation may be necessary.

Job Aids

1.86 Job aids present task information in a concise manner. Unlike tutorials, a job aid does not teach a set of skills but assists the user in performing a task as quickly as possible. Job aids are often used:

- To present information needed only once or occasionally, such as troubleshooting and installation instructions;
- As a memory aid for tasks learned in a tutorial or a user guide;
- As a quick reference for information used regularly, such as computer command summaries or telephone dialing instructions.

1.87 When designing job aids, use tables, bulleted lists, and step-by-step procedures to guide the reader through the task. Be concise and use an action-oriented approach.

1.88 Although a job aid should fit on a single sheet or task card, task cards may be bundled as a set that can be used independently and in any order.

Reports

1.89 Reports, which answer a question or offer a solution to a problem, generally support four basic activities:

- Scientific research,
- Business research,
- Progress or status,
- Proposals.

Reports generally range from 5 to 20 pages in length. Those written for business or internal audiences are generally shorter and less formal than those written for scientific or external audiences.

Scientific Research Reports

1.90 Scientific research reports communicate the results of formal scientific studies. Since the audiences for such reports are generally experts, use technical terms and concepts without background explanation. Nonexperts may read the abstract and conclusions, however, so write these sections at a more basic level. Typically, these reports include the following sections:

- Title and author attribution—precisely describes the content by using appropriate keywords that can assist in searching for the published report;
- Abstract—summarizes the study's objectives, methods, results, and conclusions;
- Introduction—presents the research objectives and the hypothesis;
- Literature review—provides an overview of the current state of research and the theoretical foundation for the study;
- Procedures—describe the subjects, methods, and materials used in the study;
- Results—summarize the data collected from the study using graphs, charts, and tables, with accompanying narrative explanation. Present this data objectively and without interpretation;
- Conclusions—discuss and interpret the results;
- Appendixes—provide additional information, such as data, that may be inappropriate within the text.

Business Research Reports

1.91 Business research reports provide data and answer questions to support making an important decision or taking a specific action.

1.92 The form of a business research report is similar to that of a scientific research report. It includes some or all of the following sections:

- Title and author attribution—precisely describes the content by using appropriate keywords that can assist in searching for the published report;
- Executive summary—summarizes the entire report in one page or less, emphasizing the recommendations and conclusions. This section is very important because it may be the only section that is read;
- Introduction—explains the report's purpose and the questions or problems it addresses;
- Body—provides a literature review, study results, and/or the author's analysis of the problem or situation;
- Recommendations and conclusions—propose a detailed course of action;
- Appendixes (if needed)—provide additional information, such as data, that may be inappropriate within the text.

Progress/Status Reports

1.93 Progress or status reports are similar to but generally shorter than business research reports. They sometimes take the form of a letter, memo, or oral presentation rather than a formal report. Whatever its form, a progress report

- Identifies the project,
- Describes the project's current status,
- Analyzes and evaluates that status,
- Explains any problems,
- Forecasts future activities.

Proposals

1.94 Use a proposal when offering to furnish goods or services, research a subject, or provide a solution to a problem.

1.95 Proposals usually have a specific format dictated by the person or organization requesting the proposal. In general, they use the following basic format:

- Title and author attribution—precisely describes the content by using appropriate keywords that can assist in searching for the published report;
- Summary—provides, in one page or less, a statement of the problem; describes a potential solution; and characterizes that solution's benefits;
- Introduction—identifies the problem and explains its significance, provides background information, identifies objectives, and summarizes the solution;
- Proposed solution—details the solution and justifies its selection over competing possible solutions;
- Qualifications—present proof of the proposing organization's experience, as well as the names, titles, and qualifications of the personnel who will be involved in the project;
- Evaluation—explains how the proposal will be evaluated;
- Budget—offers a detailed analysis of the project's estimated cost. Solicited proposals, in response to a request for proposals (RFP), generally have specific budgetary requirements;
- Appendixes (if necessary)—include information that is not essential to the proposal. Use appendixes to provide complete information while keeping the main body of the proposal clean and concise.

Feasibility Studies

1.96 A feasibility study compares a number of alternatives and recommends, rejects, or endorses a suggested change or a particular alternative. Feasibility studies usually have the following sections:

- Title and author attribution—precisely describes the content by using appropriate keywords that can assist in searching for the published report;
- Recommendations—summarize the situation and state the alternative the author endorses;
- Original situation—describes the problem or opportunity that prompted the suggested change;
- Background of the investigation—outlines the investigation's method;
- Comparison of alternatives—includes evaluation criteria, assessment of alternatives, and a formal comparison, often in tabular or graphic form;
- Conclusion—reiterates the alternatives, evaluation, and recommendation;
- Appendixes (if needed).

Help Systems

1.97 A help system, an electronic document that supplements technical manuals (see also 12.63), appears on a computer monitor in a series of windows containing text or graphics. These windows contain electronic links that provide random movement among help topics.

1.98 A help system can present any information type; however, due to the limited space available in a help window, it usually presents reference or procedural information. Help windows should generally cover no more than one-third of the screen, so that the user can still view the application beneath the help window.

1.99 Allow users to access help using a standard method such as pressing a function key, clicking a question mark icon, or selecting a help menu. Users should not have to search for help.

1.100 Provide forward and back buttons, a table of contents, and a keyword search index to support help system navigation. Since users commonly report feeling lost in electronic information, develop a way to help users discover their current location in the help system's structure. For example, some authoring tools create help that provides a collapsible table of contents displaying the current topic.

1.101 Make screen- and field-level help *context-sensitive*, directly related to the user's location in the product, so that when users click on a product function, they get immediate help related to that function.

1.102 Layer help system information so that primary information appears in the initial layer (the first window that appears when a user clicks a function). The initial layer should contain links to secondary and peripheral information if necessary. For example, if a user initiates a help request while creating an index, a context-sensitive help system should display:

Creating an Index

    Deleting an Index

Generating an Index

    Building a Custom Index

    Formatting an Index

Marking Index Entries

    Creating Secondary-Level Index Headings

    Working with Index Entries

Troubleshooting

1.103 Help systems are actually collections of many files; be sure to include all of the files when distributing the document. More complex help systems, particularly programming language–based help, may require you to deliver files in addition to the main project file. These additional files allow users to view the help system using various platforms and browsers.

Wizards

1.104 A wizard, an interactive electronic document that supports user tasks, presents a series of electronic forms that require users to make selections. Unlike a help system, which tells users what action to take, a wizard performs an action based on the user's selections.

1.105 A wizard's primary goal is assisting users to accomplish a task quickly and easily. When writing a wizard, make the required responses obvious and define any potentially confusing terminology. If the wizard asks users to supply information, be sure the user knows where to find that information if it is not obvious.

1.106 Wizards, like job aids, guide users through procedures such as program installation. Wizards used for this purpose should minimize conceptual and background information.

1.107 Wizards, like tutorials, can show users how to perform a task for the first time. Wizards used for this purpose generally support the needs of novice audiences and may require some conceptual background.

1.108 A wizard should also provide users with the opportunity to retrace their steps and change their responses if necessary.

Websites

1.109 Use websites on the World Wide Web to make text, graphics, animation, sound, and video available to a large audience. Anyone with Internet access can obtain the information on a website quickly and cheaply. Websites can also be updated easily (see also 12.92).

1.110 Websites consist of a number of webpages organized in a web or tree structure (see 1.138). When designing a site, balance the breadth and depth of this structure. Too much information on one page is distracting, so place general information at the top of the hierarchy and let users drill down to get more detail. However, if information is buried too deep, the user may give up. Once in your site, the user should be no more than three links away from the desired information.

1.111 Writing style for webpages differs from that for printed documents. Because screen space is limited, chunk information into short sections and provide a descriptive heading for each chunk.

1.112 Use the same background, font styles, and placement of navigation buttons for each page in a website. Use consistent wording in headings and links. However, it is also possible to develop a palette of complimentary color options, font variations, and navigational devices that can be used to discriminate among different subsections of a larger website.

1.113 Place navigation links along the top, side, and bottom of the page so that users do not have to scroll to navigate your site. Although scrolling seems undesirable, some websites use a single page because it requires only one download to provide total access. Prominently display a link to the home page on each page.

1.114 Weigh the benefits of dynamic effects, such as video and animations, against the time they require to download. Users who have to wait more than ten seconds for a page to load will probably leave a site.

1.115 Web standards change continually due to the rapid development of the associated technology and techniques. (For detailed information about standards and other resources, see also 13.1).

Site Elements

1.116 The most common elements of websites are

- Home page—provides users a starting point, the top-level page in the site's structure, for a website. A home page usually has a title, which appears in the menu bar; a headline, the organization's name or a welcome banner that appears on the page itself; and an identifying graphic, such as a company logo. The home page sets the site's tone and visual style and creates users' expectations about site content and accessibility;
- Hyperlinks—provide users with the ability to move among and within websites. Links may be either graphics or text;
- Electronic documents—provide a website's content. HTML (HyperText Markup Language), XML (Extensible Markup Language), and other programming languages underlie most text, while other site elements use formats such as *.bmp*, *.gif*, and *.jpg* (images); *.au*, *.mid*, and *.wav* (sound); and *.avi* and *.mpg* (video). These common file formats can be viewed using most browsers. In addition to displayed information, many sites provide archival materials that users can retrieve. Currently, such files use Portable Document File (*.pdf*), Rich Text Format (*.rtf*), or a similar format;

- Forms—allow users to interact with a site by answering survey questions or ordering products. Since long forms can be annoying, restrict them to a reasonable length and indicate both required and optional fields.

Site Types

1.117 The four main types of websites are

- Commercial—provides services or advertises and sells products. Typically, these sites incorporate many graphic and dynamic techniques. Commercial websites have an Internet address that ends with *.com*;
- Educational—represents institutions such as universities. These sites promote the institution and its services and provide access to the institution's research and resources. Educational sites generally use fewer special effects than commercial sites and focus on content. They generally have an Internet address that ends with *.edu*;
- Informational—provides reference information in the form of archives, frequently asked question (FAQ) lists, and live online assistance. They generally reject graphics or dynamic effects and use text to improve the speed of information retrieval. Informational websites have Internet addresses with various endings, such as *.net* for networks, *.mil* for the military, *.gov* for the government, and *.org* for nonprofit organizations;
- Regional—represents political and geographic regions: *.ie* (Ireland), *.de* (Germany), *.uk* (England), and the like.

Site Access

1.118 Access to a website can be limited, depending on whether it is operating as an

- Internet—a network of networks. Publish a website on the Internet to provide unlimited access;
- Intranet—a secure network belonging to an organization. Intranets have firewalls that prevent unauthorized access and control information access and exchange. Publish a website on an organization's intranet when only organization members should have access;
- Extranet—an extension of an intranet used to share selected information with people outside an organization. Publish a website on an organization's extranet to give suppliers, vendors, partners, retirees, and customers access. Extranets also provide employees at remote locations access to a site.

Planning a Document's Content

1.119 After analyzing the audience and choosing the appropriate medium for the document, plan its content. The document's structure—the information order and its level of detail—is important when describing a complicated concept or a technical task.

1.120 Document planning involves at least three activities:

- Collecting information about the subject (1.121),
- Selecting an organizational method (1.131),
- Preparing an outline (1.134).

Collecting Information about the Subject

1.121 Base information collection procedures on the document type being prepared and the source material available. For example, product specifications may be a good information source when writing a technical manual but less useful when writing a brochure. Laboratory notes may supply important source information for a technical journal article but not for a newsletter article about the same research.

1.122 When written source material does not exist, rely on experience, interviews, and product access to collect information:

- If the author is also the researcher or developer of the product to be documented, experience and product knowledge are the major information sources.
- If the author is not the developer, the author usually must interview the developer to obtain information.

1.123 Be aware that someone can know too much about a topic to write about it effectively. If a researcher writes her own material, for instance, she may easily make inappropriate assumptions about audience knowledge, unless she remembers the audience characteristics.

Analyzing Written Source Material

1.124 Find extensive written source material, including marketing materials, earlier versions of documentation, data sheets and specifications, and notes. By analyzing this material, the author can

- Consider where and how to find additional source information,
- Determine whether it contains adequate source information, and
- Formulate questions for interviews and/or subsequent research.

1.125 To prepare for the analysis, inventory the source materials and arrange them in an order that supports the analysis (e.g., order of presentation).

1.126 When assessing source materials, consider these questions:

- What is the purpose of each source document? What does it help users do?
- How is each source document organized? What is the organizational pattern? Does this pattern support the document's purpose?
- What assumptions does each source document make about its readers?
- Does each source document include special terms? Does it define these terms? Should it?
- Does the document contain information that can answer questions the reader might have about a topic in each source document? Imagine a question that a reader might have about a topic in each source document.

During this assessment, keep in mind the audience needs and the document's goals to reduce the risk of missing important information or of spending too much time on unimportant information.

Interviewing Sources

1.127 If a document's author is not the primary researcher or product developer, obtain information from other people.

1.128 To prepare for an interview

- Bring a tape recorder to the session;
- Develop a list of questions;
- Explore with the interviewee whether additional sessions are possible;
- Give the interviewee an advance copy of the questions, if possible, to prompt ideas before the discussion; and
- Schedule the interview for a convenient time, place, and duration.

1.129 Keep in mind that some interviewees will be comfortable and speak easily, while others will have trouble communicating. Participate in and control the interview according to the personality of the interviewee. If the interviewee seems uncomfortable, start the interview with a question based on his or her expertise. Ask for additional explanation as needed, or pose open-ended questions.

Conducting a Hands-on Evaluation

1.130 When evaluating a product before writing about it, pretend to be a target user. Think of tasks the user must perform and then try to do them. Try to use all the basic product functions. Observe whether they operate consistently; note any inconsistencies.

Selecting an Organizational Method

1.131 Document organization depends largely on the media being used (see 1.39). Some organizational methods will be more appropriate than others, depending on the document's audience, type, and purpose. When organizing a document

- Consider the audience's goals by assessing members' expected usage patterns and needs. Make the document's organization rational and consistent, so readers can follow it easily and understand the information it presents;
- Select an organizational approach. Typical organizational approaches can be based on chronology, the time when events occur; spatial relationships, the physical placement of objects; or climatic order, a presentation arranged for emphasis; or can be task-oriented in support of specific activities. Some documents may need an organization that combines more than one of these approaches;
- Develop guideposts to help readers navigate and understand a document's organization. Guideposts can be as simple as headings and cross-references, or as elaborate as graphic icons and color coding incorporated in the page design.

1.132 Task-oriented, a common technique in technical communication, emphasizes the actual activities someone must perform. These activities include

- Physical performance—the movements necessary to cause changes in states. For example, setting switches, moving levers, and entering specific characters all require physical movement;
- Cognitive performance—the ability to apply information to novel situations. For example, selecting switches to configure a machine for local conditions and constructing a unique command string based on a generic example both require readers to understand the idiosyncratic nature of their tasks.

1.133 Four common ways to order tasks are

- Most important to least important—a user's guide for a computer operating system with a graphic interface might first describe how to interpret the interface and interact with the system. Next, it might describe using files;

- Most frequent to least frequent—maintenance schedules often first list the procedures to be performed every 5,000 miles, followed by those performed every 12,000 miles, and so on;
- Simple to complex—a user's guide for a computer graphics program might begin by describing how to copy shapes and then enlarge, reduce, or stretch them. Finally, it might explain how to create original shapes;
- Required to optional—instructions for assembling an exercise bicycle might conclude by explaining how to install the pulsimeter available as an optional feature.

Preparing an Outline

1.134 Outline a document to organize and define its content before writing. The outline provides a baseline for the document; parts of the outline may change during writing. The different types of outlines include

- Formal—a detailed outline for reviewers who will approve the final document;
- Informal—a tool intended only for the author; and
- Nonlinear—based on systems analysis, programming, and structured design analysis.

1.135 Formal outlines show a document's hierarchical structure and are used when reviewers must approve a document plan. Depending on the level of detail, the outline may present chapter titles, section headings, one or two levels of subheadings, and even groups under subheadings. The formal outline's logical structure makes clear the relationships among topics. Most formal outlines

- Catalog the background information needed to write the document,
- Evaluate the scope of the headings, and
- Identify the number of headings in a document.

1.136 There are two types of formal outlines: topic and sentence. Although both types use the same parts and groupings of ideas, a topic outline uses phrases to express the ideas while a sentence outline uses complete sentences. A sentence outline thus conveys the intended style and tone of the final document. The following example shows the beginning of a topic outline for a product tutorial:

1. Data network benefits
    a. Shared software
    b. Efficient data and file exchange
    c. Shared equipment resources

2. Problems of evolving complex networks
    a. Early networks
        (1) Mainframe computers
        (2) Terminals
        (3) Single vendor
    b. Today's networks
        (1) Mainframe computers and terminals
        (2) Microcomputers
        (3) Minicomputers
        (4) Local area networks (LANs)
        (5) Multiple vendors

When reviewers will evaluate the final document's style and tone, the sentence outline includes more detail:

1. Data networks provide modern computer environments with many benefits, including shared resources, efficient file transfers, and electronic mail.
    a. Networks simplify support through shared software.
    b. Exchanging data through networks provides accuracy and reduces costs, because all employees can access the same information.
    c. Network users can also share equipment such as printers, plotters, and storage devices.
2. As complex networks grow, managers face problems.
    a. Early networks consisted of mainframe computers and terminals from one manufacturer.
    b. Now large organizations have also bought microcomputers, minicomputers, and local area networks (LANs), often from many different manufacturers.

1.137 Informal outlines generally function as a writer's tool. An informal outline lists the main, or controlling, idea of each paragraph but does not indicate the hierarchy of ideas. For example, the beginning of an informal paragraph outline about user-interface evaluation might be stated as

1. User-interface design can be evaluated using heuristic evaluation, focus groups, and/or usability tests.
2. Heuristics are a set of accepted usability design standards used to evaluate an interface.

3. Focus groups are planned discussion groups used to obtain perceptions and reactions to an interface.
4. Usability tests are controlled experiments used to see how participants interact with an interface.

Although this sample uses complete sentences, it could have used phrases or simply keywords.

1.138 Some authors may feel uncomfortable with topic, sentence, and paragraph outlines because they are linear approaches to document organization. Another option is to construct a nonlinear outline. Start by writing the primary idea in the middle of a page. As related ideas occur, add them around the primary idea as relational nodes and draw connections among them. These connections extend to new ideas related to the secondary ideas and so on. Work at random within this outline to combine, add, and delete nodes in any order or direction.

1.139 Use the nonlinear outline to write a document by manually arranging index cards based on the nodes on a surface, some separate, some overlapping, and some piled on top of one another; then write the document by adding information to cards and adding cards to accommodate new relationships. An example of a nonlinear outline might look like this:

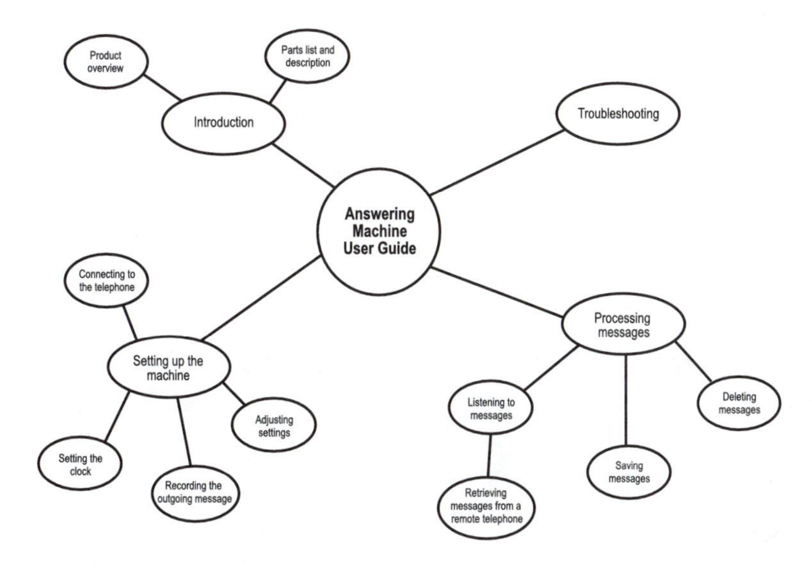

2. Writing for Non-native Audiences

2.1 When you are writing information that will be read in English by non-native speakers or that will be translated into other languages, three major approaches can help you prepare useful texts:

- Minimum Word Strategy (2.3),
- Controlled English (2.4),
- Global English (2.10).

Both Controlled English and Global English impose some restrictions on writing style and terminology. However, Controlled English is much more restrictive than Global English.

2.2 The approach that you use depends on such factors as

- The type of documents you produce;
- The subject matter of the documents;
- Whether you intend to translate the documents or to export them in English to countries in which English is not the native language;
- Whether you intend to use human translators or machine translation or both;
- Whether your translators are familiar with the subject matter;
- Whether you are providing your translators with a list of technical and domain-specific terms, together with their equivalents in the target languages;

- The degree of communication you expect to have with your translators;
- The degree of fluency in English that your audience(s) will have (in cases where you do not intend to translate your documents).

Minimum Word Strategy

2.3 Minimum Word Strategy uses illustrations in place of words as often as possible. This approach is suitable for describing physical mechanisms or simple procedures, but it can be applied to some extent to many types of documentation. In some cases, such as the emergency evacuation placards for aircraft, it can be used as an alternative to translation. In other cases it reduces translation costs by reducing the number of words that must be translated (see also 10.25 and 10.29).

Controlled English

2.4 Several companies, industrial affiliates, and research institutions have developed versions of Controlled English. Controlled English generally imposes too strict limitations on vocabulary, sentence structure, sentence length, verb tenses, and voice. Like the Minimum Word Strategy, it is most suitable for describing physical mechanisms and procedures, and it can be used either to reduce translation costs or as an alternative to translation. For example, the Eastman Kodak™ Company teaches service technicians at its overseas locations enough English so that they can comprehend service manuals that are written in the Kodak version of Controlled English, the Kodak International Service Language. This approach is more cost-effective than translating the service manuals into dozens of languages.

2.5 Other companies find that Controlled English enables them to use machine translation for some language pairs, although human translators must edit this text in order to produce a "production quality" translation. However, this approach can still be more efficient and more cost-effective than using human translators exclusively.

2.6 Developing a Controlled English that is suitable for a particular product line, subject matter, or type of document is not a simple task. A Controlled English typically begins with a basic, nontechnical vocabulary, supplemented by technical terminology to suit the type of products that a company produces. In order to limit vocabulary and simplify the language as much as possible, Controlled English generally does not permit synonyms. Moreover, each word has only one meaning and is used as one part of speech. For example, the verb *allow* would not be used as a synonym for *enable*, and if *default* is used as an adjective, then it would not be used as a noun or as a verb:

- The *default* value of LANG is FRENCH. (adjective)
- The *default* for LANG is FRENCH. (noun)
- The LANG option *defaults* to FRENCH. (verb)

2.7 Controlled English typically limits verb tenses to present, past, and future; permits only active voice; and limits sentence structures to simple statements and questions. The Kodak International Service Language also clearly limits and defines the use of punctuation: the colon is used only to mean *as follows* or *following* and parentheses are used only to indicate conversion of metric units or to enclose circuit-board indicators.

2.8 Another version of Controlled English, called Simplified English, was developed by the European Association of Aerospace Industries (see 13.2).

2.9 Clearly, any Controlled English that follows all of the guidelines mentioned in this section is going to be quite restrictive. However, there is nothing inherent in the definition of Controlled English that requires it to have so many restrictions. It is possible, at least in theory, to develop a Controlled English in which most sentence structures are allowed, some words can be used as more than one part of speech or with more than one meaning, passive voice is permitted (when there is a good reason for using it), and verb tenses are unrestricted (though kept as simple as possible). However, when more options are allowed than are not allowed, you are venturing into what—in this chapter, at least—is called Global English.

Global English

2.10 In an ideal world, *all* technical and scientific documents that are not written in Controlled English would be written in Global English, because Global English is English that is precise, consistent, unambiguous, and eminently readable—qualities that benefit all readers. Unlike Controlled English, Global English is suitable for all types of documents, regardless of whether they will be translated or not. It does not limit the expressive capacity of the English language, and its style and rhythm do not differ noticeably from a conventional technical writing style except that it is clearer and more readable.

2.11 Global English encompasses three types of guidelines:

- General sentence structure guidelines (2.12),
- Syntactic cues procedure guidelines (2.56),
- Terminology guidelines (2.88).

General Sentence Structure Guidelines

2.12 Some of these guidelines are familiar to many writers as guidelines that they already follow to improve the readability of their documents. However, those that are particularly important to translators and non-native speakers are listed here.

Use Short, Simple Sentences

2.13 Keep sentences as short and simple as is possible and appropriate for your subject matter and audience. A long and complex sentence can be difficult even for native speakers of English to comprehend:

> The log shows the "uninitialized variable" warning for variables whose values come from ISPF services when a variable is neither retained with an initial value nor appears on the left side of the equal sign in an assignment statement.

2.14 However, a long sentence can be clear and reasonably translatable if

- It consists of more than one clause,
- The structure and meaning of each clause are obvious,
- The logical relationship between the clauses is clear:

> If a variable is neither retained with an initial value nor appears on the left side of the equal sign in an assignment statement, then the ISPF service assigns a value to the variable. However, because the value was not assigned by a statement, the log shows the "uninitialized variable" warning.

In the revision, the first sentence has been divided into two clauses. The second sentence establishes its logical relationship to the preceding sentence and adds another two clauses that provide additional detail. Thus, a long and complex sentence has been divided into four clauses, each of which is clear, readable, and more readily translatable.

Say What You Mean

2.15 Do not expect readers or translators to understand what you "meant" instead of what you "wrote."

> This report compares the salaries of different departments for employees who have the same education level.

This sentence is illogical because departments do not earn salaries—employees do. What the author meant was:

> This report compares the salaries of employees who have the same education level, grouped by department.

Simplify Verb Phrases and Tenses

2.16 Keep verb phrases and tenses as simple as possible:

> Scrolling to the right *should not be being performed* by any CLIST or REXX exec.

Should be revised as

> No CLIST or REXX exec should scroll to the right.

Untangle Long Noun Phrases

2.17 Avoid noun phrases that contain more than three words. Even native readers often find a long noun phrase difficult to interpret, because it may not be obvious which words in the phrase are closely related. Moreover, in some languages, English noun phrases have to be translated as a series of prepositional phrases (one prepositional phrase for each noun or adjective in the noun phrase), which rapidly becomes unwieldy.

2.18 It is usually possible to divide a long noun phrase into a shorter noun phrase plus a prepositional phrase or relative clause, as in the following sentence:

> Specify a file to be dynamically allocated for the *host sort utility message print file.*

This noun phrase combination should be shortened as

> Specify a file to be dynamically allocated for the *message print file of the host sort utility.*

2.19 Alternatively (or in addition), use hyphens to indicate which words in the noun string are closely related (see 2.81 and 4.83):

> The SYSID option tells the *cross memory services communication facility* to use the subsystem ID that was chosen during the installation process.

This noun phrase combination should be hyphenated as

> The SYSID option tells the *cross-memory-services communication facility* to use the subsystem ID that was chosen during the installation process.

Make Sure It Is Clear What Each Pronoun Is Referring To

2.20 Two pronouns, *it* and *them*, often have obscure antecedents. In addition to confusing readers, ambiguous pronouns cause problems for translators of languages such as French in which nouns and pronouns have gender. The pronoun must have the same gender and number (singular or plural) as its antecedent. If the antecedent is not clear, the translator does not know which pronoun to use. In the second sentence of the following example, the reader's natural tendency is to regard *initial value* as the antecedent of *it*:

> If a variable name is specified only in the KEEP statement, and if you do not specify an *initial value*, then the *variable* is not written to the table. If you specify an initial value, then *it* is written to the table.

But in fact, *it* refers to *variable* (from the preceding sentence). In order to eliminate the ambiguous reference, use that noun instead of *it*:

> ... If you specify an initial value, then *the variable* is written to the table.

Place a One-Word Modifier in front of the Word That It Modifies

2.21 In general, place a one-word modifier such as *only*, *not*, and *all* in "front" of the word that it modifies. Consider the following sentence:

> Artificial Neural Network forecasting *only* works with MainState Axcel version 6.0.

The author does not mean that this type of forecasting only works, as opposed to only doing something else. He means that it works only with MainState Axcel version 6.0, not with some other application. In other words, *only* is modifying the prepositional phrase *with MainState Axcel version 6.0* and should therefore be placed immediately in front of it:

> Artificial Neural Network forecasting works *only* with MS Axcel version 6.0.

2.22 *Not* is another one-word modifier that often occurs in the wrong position. In the following sentence, *not* has gone very far astray!

> *All* information requests have *not* crossed my desk, but quite a few have.

Saying that *all information requests have not crossed my desk* is equivalent to saying that none of them has crossed your desk. What the author meant is:

> *Not* **all** information requests have crossed my desk, but quite a few have.

Make Sure It Is Obvious What Each Prepositional Phrase Is Modifying

2.23 Most (if not all) prepositions can be used in phrases that modify either nouns or verbs. For this reason, it may be difficult to determine whether a prepositional phrase that follows a noun modifies that noun or whether it actually modifies a verb that occurs earlier in the sentence.

2.24 In scientific and technical writing, ambiguous modification is likely to be subtle and might not even be noticed by most readers. However, translators, machine translation software, and readers who are more sensitive to sentence construction will notice the ambiguity in such sentences as

> When you access a function *in an external DLL*, you transfer control to the external function.

The prepositional phrase could be modifying either *access* or *function.* That is, either the function is being accessed in an external DLL, or else the function is located (or stored) in an external DLL. A subject-matter expert would know which interpretation is correct, but a translator or nonexpert reader might not. If the latter meaning is correct, then we could simply write

> When you access a function *that is stored in an external DLL*, you transfer control to the external function.

If the former meaning was intended, there is no way to make that meaning obvious without considering the surrounding context and making substantial revisions.

Make Sure It Is Clear What Each Relative Clause Is Modifying

2.25 When two or more nouns are joined by *and* and are followed by a relative clause, it might not be obvious whether the relative clause modifies all of the nouns, or only the last one. For example, consider this sentence:

> The audience consisted of employees, quality partners, and customers *who are participating in beta testing.*

Does the relative clause *who are participating in beta testing* modify only *customers*, or does it also modify *employees* and *quality partners*? That is, are all three groups participating in beta testing? If only customers are participating, then the ambiguity can be eliminated by separating the items in the series with semicolons:

> The audience consisted of employees; quality partners; and customers *who are participating in beta testing.*

If all three groups are participating, then the sentence must be restructured in order to make it unambiguous:

> The audience consisted of three groups of people who are participating in beta testing: employees, quality partners, and customers.

Make Sure It Is Obvious What Each Infinitive Phrase Is Modifying

2.26 In the following sentence, the infinitive phrases that begin with *to allocate* could be modifying either the preceding gerund, *writing*, or the preceding verb, *can use*:

> There are three service routines that programmers *can use* when *writing* user exits *to allocate memory, free memory, or access variables.*

Readers and translators are likely to interpret the infinitives as modifying *writing* because they are closer to *writing* than they are to *can use.* But if that interpretation is correct, then it would be better to eliminate the potential ambiguity by revising the sentence as

> When writing user exits to allocate memory, free memory, or access variables, programmers can use three service routines.

If *to allocate*... modifies *can use*, then it would be better to place the infinitives closer to *can use* by "fronting" (moving it to the beginning of the sentence) the adverb phrase:

> When writing user exits, programmers can use three service routines to allocate memory, free memory, or access variables.

Consider *Fronting* Prepositional Phrases and Other Sentence Constituents

2.27 When a prepositional phrase modifies a verb or verb phrase, consider "fronting" the prepositional phrase (moving it to the beginning of the sentence) to simplify or clarify the structure of the main part of the sentence. In the following example, *on a standard tape label* could modify either the noun phrase *table name* or the verb phrase *are available*:

> Only 17 characters are available for the table name *on a standard tape label.*

To make it obvious that the prepositional phrase modifies *are available,* we can "front" it as follows:

> *On a standard tape label,* only 17 characters are available for the table name.

2.28 You can sometimes "front" other sentence constituents as well:

> A server must have significant computing resources *in order to process large tables rapidly.*

Can be "fronted" as

> *In order to process large tables rapidly,* a server must have significant computing resources.

Make Each Sentence Syntactically and Semantically Complete

2.29 Writers occasionally leave out words or phrases that they believe are "understood" in a particular context:

> Space is tracked and reused in the compressed file according to the REUSE= value when the file was created, not when you add and delete records.

However, when writing for non-native readers or translators, you should be syntactically and semantically explicit:

> Space is tracked and reused in the compressed file according to the REUSE= value *that was in effect* when the file was created, not *according to the REUSE= value that is in effect* when you add and delete records.

Avoid Unusual Grammatical Constructions

2.30 Avoid using grammatical constructions that are unfamiliar to many non-native speakers of English or that might be problematic for machine translation systems (see 2.31–2.37).

The "Get" Passive

2.31 Do not use *get* as an auxiliary verb to form the passive voice:

> When you press F6, your program *gets* submitted for execution.

Use a form of the verb *to be* instead:

> When you press F6, your program *is submitted* for execution.

Causative "Have" and "Get"

2.32 Do not use *have* or *get* as auxiliary verbs meaning *to cause someone to do something* or *to cause something to happen*:

> *Have* the sponsor send a copy of the document to each member of the Project Team.

Both *have* and *had* are combined with past participles to form the perfect tenses. To avoid translation confusion, simply select a more descriptive term for the activity:

> *Ask* the sponsor to send a copy of the document to each member of the Project Team.

"Given" and "Given That"

2.33 In some contexts or subject areas (mathematics, for example), *given* and *given that* are perfectly appropriate and unavoidable. However, avoid using them in other contexts:

> *Given* some of the complications that are involved in *having* the central server receive the request, you might consider the following approaches: (Note that this sentence also uses *having* in the causative sense.)

Often, *given* can simply be deleted:

> Directing requests to the central server can cause complications; therefore, consider using one of the following approaches instead:

The basic test for editing *given that* is the content of the sentence:

> *Given that* the application is in the early phases of testing, you should not regard any of your updates as permanent.

If the sentence expresses causes or reasons as part of a rational argument, then use *because*; if the sentence expresses time, then use *since*:

> *Because* the application is undergoing testing, you should not regard any of your updates as permanent.

"In That"

2.34 Whenever possible, use *because* instead of *in that*. You might also be able to find other ways to avoid using *in that*:

> The order in which the libraries are specified is important, *in that* several libraries define the same symbol.

Edit this sentence to read:

> The order in which the libraries are specified is important *because* several libraries define the same symbol.

"Should" as a Conditional Tense Modal Verb

2.35 Do not use *should* as a conditional tense modal verb:

> The next section describes how to submit statements for compiling and executing, *should* you prefer to use that method.

In most (if not all) cases, you can easily substitute *if*:

> The next section describes how to submit statements for compiling and executing, *if* you prefer to use that method.

"Need Not"

2.36 Consider the following sentence:

> Remote procedure calls *need not* return results.

A non-native speaker might interpret the sentence as

> Remote procedure calls *need to not* return results.

Or it could be rephrased as

> Ensure that remote procedure calls do not return results.

Clearly that is an incorrect interpretation, and the *need not* construction can easily be avoided:

> Remote procedure calls *do not need to* return results.

Inverted Sentences

2.37 Except in poetry and questions, it is unusual for the main or an auxiliary verb to precede the subject of a clause in English. Although inverting a sentence in this manner can produce a desirable change in emphasis, do not use this type of construction if it is likely to be unfamiliar to some of your readers or if you intend to use machine translation:

> Only when being stored in an integer variable *is* the value truncated.

This sentence can be revised as

> The value *is* truncated only when it is being stored in an integer variable.

Beware of Inherently Ambiguous Constructions

2.38 A few English words are often the source of subtle ambiguities. Whenever you use these words, you should ask yourself whether multiple interpretations are possible (see 2.39–2.41).

"Require" + an Infinitive or "Need" + an Infinitive

2.39 Any form of the verbs *require* and *need* followed by an infinitive can result in an ambiguity that is difficult to explain but easy to illustrate. Consider the following sentence:

> Two technicians are *required to perform* maintenance on fuser assemblies.

This sentence could be interpreted in either of the following ways:

> Of the entire staff of technicians, two must perform maintenance on fuser assemblies. (The other technicians do not have this responsibility.)
>
> In order to perform maintenance on fuser assemblies, two technicians are required. (One technician cannot do the job alone.)

This type of ambiguity often confuses readers even when the meaning seems obvious to the writer. Human translators will not necessarily know which interpretation is correct (if indeed they even notice the ambiguity). And machine translation software cannot be expected to "know" which interpretation the author intended.

2.40 Changing *to* to *in order to* is usually an acceptable remedy if such an edit supports the intended meaning, although you might also want to restructure the sentence. If some other interpretation is intended, then you generally need to revise the sentence substantially in order to find an unambiguous but stylistically acceptable alternative.

"Must Have Been"

2.41 The modal verb *must* can imply either obligation (I must be going) or logical necessity (You must enjoy doing things the hard way). The latter interpretation is unusual in scientific and technical writing, but the potential for ambiguity is perhaps worth noting. Consider the following sentence:

> If you are accessing a view that is owned by another user, you *must have been* granted access privileges on the view.

If *must* indicates a logical necessity, then we could make that connotation more explicit by using the word *presumably* instead:

> If you are accessing a view that is owned by another user, then *presumably* you must have been granted access privileges on the view.

If (as is undoubtedly true) *must* is intended to imply an obligation, then we could rephrase the sentence as

> In order for you to access a view that someone else owns, the owner *must* first grant access privileges to you.

Avoid Adverbial Interrupters

2.42 In general, avoid interrupting a clause with a conjunctive adverb such as *however*, *therefore*, or *nevertheless*, or with other adverbial interrupters that can interfere with the reader's processing of a clause; instead, place such terms at the beginning of the clause.

2.43 Keep the structure of each clause as simple as possible by "fronting" modifiers that add to clause length and complexity (see 2.26–2.27). Moreover, many conjunctive adverbs express logical relationships:

> You can, *however*, use a RETAIN statement to assign an initial value to any of the previous items.

If a clause is logically related to the preceding clause, make the logical connection obvious at the beginning of the clause:

> *However*, you can use a RETAIN statement to assign an initial value to any of the previous items.

2.44 If you cannot put the conjunctive adverb or other adverbial interrupter at the beginning of the clause, then try to find another way to revise the sentence:

> As each step executes, notes and, *if required*, error messages or warning messages are written to the message area.

Can be edited as

> As each step executes, notes are written to the message area along with any applicable error messages or warning messages.

Use Standard, Conventional English

2.45 The flexibility of the English language makes it relatively easy for writers to invent new words or to use existing words in new or unusual ways. That is one reason English has become so widely used outside the countries in which it is an official language. However, the use of nonstandard or unconventional English could be a hindrance to translators and to non-native speakers who are learning English. In addition, it can cause a machine translation system to translate a clause or sentence incorrectly, because the lexicons and linguistic rules of machine translation systems are based on standard usage.

However, do not be excessively conservative. If a nonstandard usage has been used in a particular subject area so frequently that subject-matter experts regard it as standard, it is probably best to accept that usage.

Use Standard Verb Complements (see also 13.2)

2.46 Many writers do not realize that in English, as in other languages, each verb is linguistically associated with one or more types of complements. Here are some examples of different types of complements that verbs can take:

- Suddenly, the aliens started to speak. (infinitive)
- Suddenly, the aliens started speaking. (gerund)
- Lola became smug. (predicate adjective)
- Lola became an engineer. (predicate noun)
- Give me a break! (indirect object + direct object)

2.47 Many verbs can take more than one type of complement, but in some cases the meaning of the verb differs when different complements are used:

Ali ran to the store. (prepositional phrase)

Ali ran a high-speed printing press. (direct object)

Pressing the red button initiates the "destruct" sequence. (direct object)

The professor initiated Jason into the mysteries of psycholinguistics. (direct object + prepositional phrase)

2.48 In contrast, here are some examples of unconventional verb complements that you should avoid:

- If a client *requests to delete* a file, you first need to authenticate the client.

  If a client *asks to delete* a file, . . .

  If a client *submits a request to delete* a file, . . .

- If performance is an issue, we recommend *to use* the web version of the application.

  If performance is an issue, we recommend *using* the web . . .

Often the best remedy is to use a different verb or to revise the sentence more substantially:

You can *select to display* a command dialog box instead of a command bar.

You can *choose to display* a command dialog box instead of a command bar.

You can *display* a command dialog box instead of a command bar by selecting `Command bar` from the Preferences menu.

Do Not Use Transitive Verbs Intransitively, or Vice Versa

2.49 Most, if not all, dictionaries specify whether a verb is transitive, intransitive, or both (see 13.2). Often a verb has different meanings depending on whether it is used transitively or intransitively. For example, when the verb *display* is used intransitively, its only conventional meanings are "to show off" or "to make a breeding display." Therefore, the following sentence could be considered nonstandard:

> If the X libraries cannot locate the required bitmaps, your program might *display* strangely.

Because the performer of the action is unimportant, the use of passive voice is acceptable:

> If the X libraries cannot locate the required bitmaps, your program might *be displayed* strangely.

Alternatively, you could use a different verb:

> If the X libraries cannot locate the required bitmaps, your program's image might be *distorted* on the display monitor.

2.50 When a machine translation system encounters a verb that is listed in its lexicon as transitive, it expects a direct object to occur in the sentence; when it encounters an intransitive verb, it will not expect a direct object. Therefore, a sentence like the following is likely to be mistranslated:

> If you do not specify the X11R5_PREFIX, then the X libraries will have difficulty locating bitmaps and application-defaults files. This can cause your program to *display* strangely.

Can be revised as

> . . . This can cause your program to *be displayed strangely.*

Do Not Violate Semantic Restrictions

2.51 When a verb is used to convey a particular meaning, the subject or object of the verb might be restricted to a particular semantic category in conventional English. For example, when the verb *prints* is used intransitively, it typically requires an animate subject such as a human being (someone who works as a printer) or a machine (a printing press). It is logically and semantically impossible for an inanimate object to perform the action of printing. Therefore, the following use of *print* is unconventional:

> Notice that the page number *prints* on the first line of each page.

The performer of the action is unimportant, so the use of passive voice is appropriate and is probably the easiest remedy:

> Notice that the page number *is printed* on the first line of each page.

Use Singulars and Plurals Appropriately

2.52 When two semantically or pragmatically related nouns are used in a sentence, pay close attention to the subtle distortion in meaning that can arise when you use a singular noun instead of a plural noun or vice versa. For example, in the following sentence, it is illogical and incorrect to say that all data items (plural) have a (single) numeric value, unless they all have the *same* numeric value:

> All the data *items* shown in figure 7 have *a numeric value.*

If the data items all have the same value, then you should state that explicitly:

> All the data *items* shown in figure 7 have the *same numeric value.*

If the data items all have multiple numeric values, then the sentence should read

> All the data *items* shown in figure 7 have *numeric values.*

If each data item has a single numeric value, then the sentence should read

> Each data *item* shown in figure 7 has *a numeric value.*

If each data item has a unique numeric value, then the sentence should read

> Each data *item* shown in figure 7 has a *unique numeric value.*

Use the Article "the" Only When It Is Clear Which Instance of the Following Noun You Are Referring To

2.53 When you use the definite article *the* plus a noun, the implication is that your reader can identify specifically which instance of the noun you are referring to. If you have not mentioned the noun before, and if there is no other way for readers to understand which particular instance of the noun you are referring to, your readers will be confused.

For example, if someone says to you, "I saw *the* dog on my way to work," it clearly implies that you and the speaker have talked about this dog before and that you know specifically which dog the speaker is referring to. If you and the speaker had not discussed the dog before, then the speaker's use of *the* would seem strange and bewildering.

Incorrect or inconsistent use of articles can make it more difficult for some non-native speakers to learn to use articles appropriately in English. Speakers of languages that do not use articles to indicate definiteness or indefiniteness (Chinese, Japanese, and Korean, for example) have particular difficulty learning the English article system.

2.54 In the following example, the use of *the process's* in the second sentence will confuse readers if there has been no previous mention of a process:

> The DBE option specifies how many disk blocks RMS adds when it automatically increases the size of a table during a 'write' operation. If you specify DBE=0, then RMS uses *the process's* default value.

In order to make it clear which process is being referred to, you could add a relative clause that provides additional modification:

> ... If you specify DBE=0, then RMS uses the default value for *the* process *that is performing the 'write' operation.*

2.55 In the next example, the use of *The alias* in the second sentence implies that all options have aliases:

> The PAGELENGTH option specifies the page length of the output file. *The alias* for this option is PL.

If that is not true, then it would be better to write:

> The PAGELENGTH option specifies the page length of the output file. PL is *an* alias for this option.

Syntactic Cues Procedure Guidelines

2.56 Syntactic cues are function words—articles, prepositions, relative pronouns, and the like—and other elements or aspects of language that

- Enable readers (especially non-native speakers of English), translators, and machine translation systems to analyze sentence structures quickly and accurately;
- Make it easier for all readers to predict the structure of subsequent parts of a sentence;
- Eliminate certain ambiguities that cause translation problems.

For more information about research and rationales that support the use of syntactic cues, see 13.2.

2.57 You can develop your ability to use syntactic cues effectively by following this procedure or by using it as a reference. Each step helps you identify places where syntactic cues might be appropriate. With practice, you will soon be able to decide where to insert syntactic cues without referring to this procedure.

2.58 Use syntactic cues with discretion. Be careful not to change meaning or emphasis, and do not use so many syntactic cues that your text sounds unnatural to native speakers of English.

Do Not Write Telegraphically

2.59 Make sure you have not used a telegraphic writing style in which the articles *a, an,* and *the* and other function words are omitted. This writing style is often used in lists, in descriptions of functions or parameters, and in definitions. In the following example, which comes from a set of instructions, the article *a* was omitted for no particular reason:

> Provide *a* base class for all stream buffers.

Consider Expanding Past Participles into Relative Clauses

2.60 Consider expanding past participles—verb forms that usually end in *-ed,* such as *described, provided,* and *specified*—into relative clauses by inserting either *that* or *that* plus some form of the verb *to be* (*that is, that are, that was, that has been,* etc.). However, do not feel that you have to expand *all* past participles, and do not expand them if doing so would make the sentence sound unnatural to native speakers or would change the emphasis in the sentence:

> Timeout values [*that are? that have been? that were?*] *set* by specifying CSSET_TIMEOUT override timeout values [*that are? that have been? that were?*] *set* by specifying CLNT_CALL.

2.61 If a past participle precedes the main verb in a sentence or clause, make a special effort either to expand the participle or to revise the sentence in some other way:

> A socket [*that is*] *opened* by one task is not available to another. (*Socket* is the subject; *is* is the main verb.)

2.62 If a sentence or clause contains more than one past participle, then make a special effort to expand one or more of the participles or to revise the sentence in some other way:

> If a FREE LIBRARY command is executed for a library [*that has been*] *defined* by a LIBRARY statement in the server's session, then a CLEAR LIBRARY statement [*that is*] *executed* after the SERVER statement produces an error.

Consider Editing Present Participles

2.63 Look for present participles (verb forms that end in *-ing*) such as *corresponding, describing,* and *using.* If the participle follows a verb such as *begin, start,* or *continue* that can take an infinitive complement, then consider changing it to an infinitive:

> DATAPRO/2000 continues *processing* program statements after it repairs the data set.

Changing the participle to an infinitive results in

> DATAPRO/2000 continues *to process* program statements after it repairs the data set.

2.64 If the participle follows and modifies a noun, then always either expand the participle into a relative clause or find some other way of eliminating it:

> Conclude the terminal description with comment lines to explain the capabilities [*that are*] *being* defined.

2.65 You can also eliminate any confusion by deleting the participle:

> Conclude the terminal description with comment lines to explain the capabilities *being* defined.

Can be revised as

> Conclude the terminal description with comment lines to explain its defined capabilities.

2.66 If the *-ing* word follows and modifies a noun, then always either expand the *-ing* word into a relative clause or find some other way of eliminating it:

> Conclude the terminal description with comment lines to explain the capabilities [*that are*] *being* defined.

2.67 If the *-ing* word follows a temporal conjunction such as *when, while, before,* or *after,* then make sure the subject of the main clause or superordinate clause is also the implied subject of the *-ing* word:

> When *registering* a control-window class, *you* must specify how many window words will be associated with that class. (Correct, because the subject of the main clause, *you,* is also the implied subject of *registering.*)

> If an *error* occurs while *updating* a VSAM file, the operating system may be able to recover the file and repair some of the damage. (Incorrect, because the subject of the superordinate clause, *error,* is not the implied subject of *updating.*)

This problem can be corrected by revising the sentence as:

If an error occurs while *you are updating* a VSAM file, . . .

2.68 If the *-ing* word *using* introduces a participial phrase but is NOT at the beginning of a sentence, then either eliminate it or clarify it:

These same *options* can also be defined *using* the Options dialog window. (Incorrect, because the subject of the main clause, *options*, is not the implied subject of *using*.)

This sentence can be revised as

You can define these same options *by using* the Options dialog window.

2.69 If an *-ing* word functions as an adjective (occurring before a noun), then consider whether it could be mistaken for a gerund (acting as a noun), or vice versa. If so, revise the sentence or phrase. For example, in the following sentence, the word *switches* could be interpreted as a noun—that is, ". . . when [the interface is] processing switches . . ."—but it is actually a verb, with *processing* as its subject:

The interface must rebuild its buffer pool and restart I/O when *processing* switches from one file to the next.

One way to prevent misreading would be to revise the sentence drastically:

After one file has been processed, the interface must rebuild its buffer pool and restart I/O. It can then process the next file.

2.70 Not all words that end in *-ing* are participles, and in some contexts, *-ing* words are syntactically unambiguous. Therefore, don't change any *-ing* word that

- Is not derived by adding the suffix *-ing* to a verb: *during, spring, wing.*
- Is preceded by a preposition:

  For more information *about printing* Help files, see chapter 3.
- Ends in *-s*:

  The correct *settings* for the latitudes and longitudes of major U.S. cities are listed in appendix 1.
- Is preceded by any form of the verb *to be* (e.g., *is, are, was, were, has been, have been, had been*):

  If you *are using* browse mode, an error message appears at the top of the window.
- Is the subject of a clause or sentence or is otherwise acting as a noun (gerund):

Specifying the system password gives you full administrative access. (When it is the first word of a simple sentence, an *–ing* word can only be a gerund.)

- Introduces a participial clause:

The second argument has been set to NULL, *indicating* that the future amount is to be calculated. (Note that, unlike a participial phrase, a participial clause is—or should be—preceded by a comma.)

Search for All Occurrences of the Word *"And"*

2.71 Whenever *and* appears, there could be a "scope of conjunction" problem. That is, it might be difficult or impossible for a reader or translator to determine which parts of the sentence are being conjoined. Therefore, consider whether you could insert an infinitive marker *to*, a preposition, a modal verb, an auxiliary verb, or another syntactic cue to make the structure of the sentence unambiguous and predictable:

The application can use the window to establish a dialog with the user and [*can? to?*] *format* text responses. (*Can* is a modal verb; *to* is an infinitive marker. This sentence is ambiguous.)

2.72 If the *and* joins two noun phrases, and if an adjective precedes the first noun, then consider whether that adjective modifies either both nouns or only the first one. If it modifies only the first noun, then try to make that obvious by

- Inserting an article and/or an article plus an adjective in front of the second noun,
- Reversing the order of the noun phrases, or
- Using a bulleted list.

This type of ambiguity is known as "ambiguous scope of modification." With some other types of ambiguity, translators can find equally ambiguous parallel structures in their own languages. But translators in languages like French and Spanish cannot do that, because adjectives in those languages must agree in number and case with the noun(s) they modify. So we have to make it obvious which noun(s) an adjective is modifying. In the following example, if *logical* modifies only member name (and not member type), then you can make that relationship obvious by inserting *the* in front of member type:

FILEMEMN and FILEMEMT contain the logical member name *and* [*the*] member type, respectively.

If *logical* modifies both nouns, then you must repeat it, as well as the article *the*:

> FILEMEMN and FILEMEMT contain the logical member name *and* [*the logical*] member type, respectively.

2.73 If *and* is joining two noun phrases, then consider whether a nonexpert reader or translator could determine whether the word that precedes *and* is a noun or an adjective. If not, then either insert the appropriate noun, reverse the order of the noun phrases, or find some other way of resolving the ambiguity. For example, in the following sentence, it is not clear whether *text* is a noun or whether it is an adjective that modifies *areas:*

> The remaining portion of the window contains text *and* line-number areas that are used to view source code.

If *text* is an adjective, then we could revise the sentence as

> The remaining portion of the window contains text [*areas*] and line-number areas. . . .

If *text* is a noun, then we could make that obvious by reversing the order of the noun phrases:

> The remaining portion of the window contains line-number areas and text. . . .

2.74 Consider whether you could replace *and* with *as well as* or *plus*, either of which almost always introduces noun phrases. By contrast, *and* can introduce many different constructions (a new clause, a prepositional phrase, a noun phrase, etc.). Thus, *as well as* and *plus* make it easier to predict and analyze the structure of the rest of the sentence:

> LABEL= enables you to specify the type and contents of the label of either a tape data set or a disk data set[,] **and** [*as well as*] other information such as the retention period or expiration date for the data set.

2.75 Consider whether you could you use *both ... and* to help readers analyze the sentence:

> This information must specify [*both*] the syntactic category of the word **and** the categories of the phrases that it can take as complements.

2.76 If *and* is joining two verb phrases, then consider inserting a pronoun or noun subject so that you will have a compound sentence (two independent clauses) or two separate sentences instead. Commas, periods, and other punctuation marks also act as syntactic cues by marking clause or sentence boundaries:

Push buttons are rectangular or oval, include only text labels, and may appear pushed in after being selected.

Can be edited as

Push buttons are [*either*] rectangular or oval[.] [*They*] include only text labels[,] and [*they*] may appear to be pushed in after being selected.

Search for All Occurrences of "*Or*"

2.77 Search for all occurrences of the word *or* to determine whether it joins equivalent terms (synonyms). Use synonyms joined by *or* only if you are not sure which of two terms your readers will recognize. Insert an article (*a, an,* or *the*) or use *either . . . or* (or do both) if the terms that are joined by *or* are not synonyms:

The system displays a message and prompts for user input when [*either*] a hard error **or an** exception occurs.

Emphasizing alternatives makes it obvious that *hard error* and *exception* are two different concepts, if indeed that is the case.

2.78 Use parentheses to indicate that two terms or expressions are equivalent, if your company is willing to adopt this as a standard (see also 4.127 and 7.18):

a hard error (or exception)

3 in. (7.62 cm)

2.79 If *or* is joining two noun phrases, ask yourself whether a nonexpert reader or a machine translation system could readily determine whether the word that precedes *or* is a noun or an adjective. If not, then either insert the appropriate noun or find some other way of resolving the ambiguity. For example, in the following sentence, it is not obvious to someone who is unfamiliar with the subject that the author means both *fact tables* and *dimension tables*:

Table Name is the name of each fact *or* dimension table.

To eliminate any doubt, edit the sentence as

Table Name is the name of each fact [*table*] or dimension table.

2.80 If *or* is joining two verb phrases, consider inserting a pronoun or noun subject so that you will have a compound sentence (two independent clauses) or two separate sentences instead.

Look for Long Noun Phrases

2.81 Long noun phrases, sometimes called "stacked nouns," are often difficult for nonexpert and non-native readers to understand. In addition, translators need to know which words in a long noun phrase are related. For example, a *terminal interrupt handler* is a compound adjective + noun construction (a handler of terminal interrupts), not a compound noun + adjective (an interrupt handler that is terminal). It would be hyphenated as *terminal-interrupt handler.* At a minimum you should provide translators with a list of noun phrases that are used in your document, along with explanations of those phrases.

2.82 Hyphenate compound adjectives that are a part of a noun phrase unless you feel that expert readers, those who have encountered the noun phrase before and who will immediately understand it, will find the hyphens intrusive.

Look for Specific Verb Forms

2.83 Look for the verbs *assume, be sure, ensure, indicate, mean, require, specify,* and *verify.* Ask yourself whether you could insert the word *that* after these verbs to make the sentence structure clearer. In this context, the word *that* tells the reader that a noun clause is coming next in the sentence:

> It is important to *ensure* [*that*] the LDRTBLS value is large enough to contain your X client.

Look for *"Give"* and *"Assign"* as Verbs

2.84 Look for the verbs *give* and *assign,* and determine whether there is an indirect object that could be made grammatically explicit by using the word *to*:

> Before printing a date, you usually *assign* it a format.

Can be edited as

> Before printing a date, you usually *assign* a format *to* it.

Look for the Word *"If"*

2.85 If the subordinating conjunction *if* introduces a conditional clause, then consider beginning the following clause with *then* to reinforce the idea of a condition being followed by a result or consequence:

> *If* you have not assigned a logical name to the data file, [*then*] specify the physical filename in the statement that refers to the file.

2.86 If the first *if* clause is followed by a second conditional clause, then the second clause should generally also start with *if*:

> If any single lookup might take several seconds[,] and [*if*] many clients might request service simultaneously, [*then*] the server must be able to handle multiple clients concurrently.

Consider Expanding Adjectives into Relative Clauses

2.87 When an adjective follows a noun, consider expanding the adjective into a relative clause. Make a list of each such adjective that you find in your company's documentation, and then search for other occurrences of those words:

> This organizational structure is an aid to users [*who are*] familiar with X implementations on other systems.

Terminology Guidelines

2.88 When you are writing for an international audience, consistency and precision should be your primary concerns. These two characteristics can lend their own kind of elegance to your writing style. As with the other guidelines in this chapter, these terminology guidelines are goals to strive for, not absolute rules that must be followed in every document.

Develop a Multilingual Terminology Database

2.89 A multilingual terminology database is an extremely useful tool for controlling terminology, and the more detailed and comprehensive you make the database, the better. The database saves translators the trouble of finding appropriate translations for technical terms that have been used and translated previously. In addition, if more than one translator is translating a document into a particular target language, or if different translators translate documents that deal with the same subject matter, then the database helps ensure that all translators translate technical terms in the same way. For additional information about multilingual terminology databases and translation software, see 13.2.

2.90 If your company or institution uses a translation agency, then consider asking the agency to develop the terminology database for you or to collaborate with you on its development. In your contract with the agency, specify who will own and maintain the database after it is created.

2.91 Ideally, the terminology database should contain the following elements, along with definitions in English and foreign-language equivalents in the languages that you typically translate your documents into. If possible, also include examples of sentences in which the terms are used:

- All technical words that are used in your documents:

  You can use *wildcard characters* in file specifications to operate on a group of files instead of on one particular file.

- All nontechnical words or phrases that are typically used with specialized, technical meanings:

  Superblocking options work by setting aside *pools* of *memory* for different *classes* of use.

- All nontechnical words that have multiple meanings but are consistently defined in your documentation and used as only one part of speech:

  The DD option does not *honor* the SELECT and EXCLUDE statements.

- All noun phrases that are likely to be unfamiliar to translators or whose meanings might not be readily apparent:

  cross-memory-services communication facility

- Any other words or phrases that are likely to be unfamiliar to translators:

  Consider using a *flipchart* to facilitate the *brainstorming* process.

- All acronyms, initialisms, and abbreviations:

  MAGVAR (magnetic variation)

- All names (or other terms) that should not be translated: companies, organizations, products, and personal names. For these words, list the same word as the foreign-language equivalent for all foreign languages that your database includes:

  Schottke diodes have very little junction capacitance.

  An *ORACLE* database is the repository for some of the warehouse data.

Each Word Should Have Only One Meaning

2.92 Controlled English typically limits you to using each word as only one part of speech. However, in Global English you can relax that restriction somewhat by permitting a word to be used as more than one part of speech as long as the term's "core semantic meaning" remains the same. For example, you could permit the use of the word *sort* as either a verb or a noun as long it refers to *the process of putting data into a particular order*:

Specify ORDER=A to *sort* the data alphabetically.

A large *sort* may require a significant amount of memory.

2.93 By contrast, the following use of the word *sort* would not be permitted, because it has an entirely different meaning:

This *sort* of product often reduces the number of I/O requests.

Standardize Your Terminology

2.94 In addition to using each word to convey only one meaning, you should avoid using more than one word or phrase to mean the same thing.

Do Not Use Synonyms

2.95 For example, if you refer to something as a *concept* in one place, do not refer to it as an *idea* elsewhere in the document. Similarly, if you use *enable* to mean *to give the ability*, then do not use *allow* to convey the same meaning:

To *allow* DMA transfers to use the DMA interrupt, the pDmaMode-Flags variable is created.

Should be revised as

To *enable* DMA transfers to use the DMA interrupt, the pDmaMode-Flags variable is created.

Use Conventional Word Combinations and Phrases

2.96 Use conventional word combinations and phrases consistently. For example, the noun *procedure* is most often used with the verbs *follow*, *implement*, and *develop*. It would seem unusual to read about someone *designing* a procedure or *devising* a procedure. If you are not sure whether a particular combination of words is conventional or not, consult the dictionary sources listed in 13.2.

Standardize Your Use of Prepositions

2.97 Many prepositions have a wide variety of meanings. It is entirely possible, although not particularly easy, to catalog how prepositions are used in your documents and to standardize them. For example, you could prohibit the use of *with* when *that have* or *that has* could be used instead:

A center-tapped transformer is a transformer *with* a connection at the electrical center of a winding.

Using this guideline to replace *with* results in

A center-tapped transformer is a transformer **that** *has* a connection at the electrical center of a winding.

Avoid Using Words and Phrases That Have Multiple Common Meanings

2.98 If a particular word has multiple common meanings, whereas a synonymous word has fewer meanings (or, ideally, only the single meaning that you are trying to convey), then use the latter. For example, the word *another* is often used to mean *an additional.* However, in the following sentence, it is being used to mean *a different*:

You can store temporary files in directories other than TEMP. Here are two methods for directing temporary files to *another* directory:

Because *a different* has only one possible meaning, it would be better to use that term instead:

... Here are two methods for directing temporary files to *a different* directory:

2.99 This guideline applies to common prepositions and adverbs, as well as to nouns, verbs, and adjectives:

For more information *on* data modeling, see "An Overview of Data Modeling" on page 46.

For more information *about* data modeling, see "An Overview of Data Modeling" on page 46.

Avoid Using Idiomatic Two- or Three-Word Verbs

2.100 Scientific and technical documents often contain idiomatic two- and three-word verbs as in

You could *go about supporting* a DB2 data repository in two ways.

You can usually substitute a single, nonidiomatic verb:

You could *support* a DB2 data repository in two ways.

Avoid Using a Phrase When There Is a Single Word That You Could Use Instead

2.101 Single words are preferable because they are syntactically simpler than phrases. In particular, avoid phrases that are idiomatic:

Quite a few different reports can be generated using applets that are included with the product.

Can be revised as

> *Several* different reports can be generated using applets that are included with the product.

Avoid Slang and Colloquialisms

2.102 Most scientific and technical documents do not contain slang or colloquialisms, but other types of documents do. Avoid such colloquialisms as *in a nutshell* and *the whole nine yards.*

2.103 In addition, avoid using language that is excessively informal:

> A multicast is *kind of like* the packet types that were discussed in chapter 2.

Can be revised as

> A multicast is *similar to* the packet types that were discussed in chapter 2.

Avoid Using Figurative Language

2.104 Words that have figurative meanings have too much "semantic breadth" (see 2.92):

> The topic was *tabled* until more input could be obtained from Marketing.

Can be revised as

> The topic was *postponed* until more input could be obtained from Marketing.

Avoid Humor, Satire, and Irony

2.105 Humor, satire, and irony are often difficult, if not impossible, to translate. Something that is funny in one language or culture is not necessarily funny to readers from other cultures. And in a particular type of document, satire and irony might be acceptable in one culture but unacceptable in another.

Avoid Referring to Culture-Specific Ideas

2.106 Do not assume that your international readers will know as much as you know about your native country and culture. Most sports analogies are inappropriate, as are references to political institutions and historical events. The concepts of eating disorders, dieting, and wellness clinics are also foreign to many cultures.

Use Nouns as Nouns, Verbs as Verbs, and So On

2.107 In English, many words can be used as more than one part of speech. For example, *light* can be a verb (to light a candle), a noun (the speed of light), or an adjective (a light touch). Other words are commonly used only as a single part of speech. A dictionary or terminology database is almost certain not to list unconventional parts of speech for those words, and a machine translation system is less likely to analyze a sentence correctly if the sentence contains a word that is used in an unconventional way. Therefore, follow the specific guidelines in 2.102–2.104.

Do Not Use Verbs as Nouns

2.108 Because most, if not all, dictionaries are clear on the acceptable functions words can serve as parts of speech, be sure to use words appropriately (see 13.2):

> The *quiesce* of the system is an abnormal shutdown that causes any executing job to be requeued.

The term *quiescence* refers to an abnormal shutdown of the system:

> *Quiescense* causes any executing job to be requeued.

Do Not Add Verb Suffixes to Nouns, Conjunctions, and Adjectives

2.109 In other languages it is not possible to convert a noun, conjunction, or adjective into a verb by adding a verb ending to it. Therefore, this type of word is untranslatable, and it might be very difficult for a translator to rephrase it. In the following sentence, the writer has used the noun *VDEFINE* (which is the name of a software routine or function) as a verb by adding the past-tense verb suffix *-d*:

> If a variable is *VDEFINEd* more than once in any step, then the next reference to that variable will result in a storage overlay.

> If a variable is *defined by more than one VDEFINE statement* in any step, then the next reference to that variable will result in a storage overlay.

Do Not Use Adjectives as Nouns

2.110 By omitting the noun in a noun phrase, a writer can force an adjective into the role of a noun. However, a machine translation system probably won't analyze the word correctly:

> This section explains how to change the text of items on the *pop-up*.

Editing the text to make the word function correctly would result in

> This section explains how to change the text of items on the *pop-up menu.*

Beware of Words That May Be Grammatically or Semantically Incomplete

2.111 For example, if appropriate, use *in order to* instead of *to*; *in order for* instead of *for*; and *so that* instead of *so.* Also, consider moving a phrase that begins with *in order to* or *in order for* to the beginning of the sentence (see 2.27–2.28):

> You need to check each line as it is read *to know* which statement to use.

> *In order to know* which statement to use, you need to check each line as it is read.

3.

Grammar, Usage, and Revising for Publication

3.1 Scientific and technical writing must be clear, consistent, and concise. Various communication efforts beginning with the initial draft and continuing through proofreading, editing, and revising help authors achieve these goals. This chapter considers three major issues that underlie these techniques:

- Grammar basics (3.3);
- Common problems in grammar, style, and usage (3.7);
- Revision strategies (3.38).

3.2 In addition to this style guide, technical and scientific writers should be familiar with such general style guides as the *Chicago Manual of Style*, as well as those appropriate for their professions (see 13.3).

Grammar Basics

3.3 In order to revise a document successfully, a writer needs to know how best to evaluate the grammatical function of words, phrases, clauses, sentences, and paragraphs, as well as entire documents. This section briefly defines basic grammatical terms used in the remainder of this section.

Phrase Types

3.4 Phrases are groups of related words without a subject or finite verb or both. Phrases cannot make a complete statement as a clause can but instead act as nouns, verbs, or modifiers in a sentence.

Exhibit 3–1: Grammatical Functions of Parts of Speech

Part of Speech	Definition and Role in Sentences
Nouns	A word or phrase used to name a person, place, thing, idea, quality, or action. Nouns function as either a subject or an object and can be singular or plural.
Pronouns	A word used in place of a noun. Pronouns usually function as either a subject or object and change case depending on their grammatical role. Pronouns can be personal, demonstrative, relative, reflexive, indefinite, interrogative, intensive, or reciprocal. *Possessive* pronouns, like possessive nouns, always function as adjectives.
Verbs	A word that expresses a sentence's action. Verbs must agree with number of subject, tense, voice, mood, and person. A *helping* verb works with another verb to indicate tense and other meanings.
Modifiers: adjectives and adverbs	Words that describe aspects of subjects, objects, or actions. Their usage depends on their relationship to the modified word or phrase. Adjectives modify nouns and pronouns only; adverbs modify verbs, adjectives, and other adverbs. Modifiers can be individual words, phrases, or clauses. *Special case*: Conjunctive adverbs link independent clauses only.
Prepositions	Words used with nouns and pronouns to form modifying phrases. Use prepositions to link a noun or pronoun to some other word in a sentence. All prepositions must have a noun or pronoun, which functions as the object of the preposition within a prepositional phrase. The assertion that a preposition never ends a sentence is untrue. Occasionally, prepositions follow instead of precede their objects, and they can be placed at the end of a sentence: What is this program *for*?
Conjunctions	Structures that connect words, phrases, and clauses within a sentence. Conjunctions may be coordinating, correlative, or subordinating. • Coordinating conjunctions connect similar grammatical units: nouns with nouns, prepositional phrases with prepositional phrases, and so on. These conjunctions include *and, but, for, so, yet, or, nor.* • Correlative conjunctions connect parallel structures within a sentence. They consist of the following pairs: *either . . . or, not only . . . but also, neither . . . nor, both . . . and, whether . . . or.* • Subordinating conjunctions signal a dependent clause and connect it to an independent clause within a sentence. Some subordinating conjunctions are *because, if, although, when, as, while.*
Interjections	These words express emotions or surprise. For the most part, they are inappropriate in technical and scientific writing.

Exhibit 3–2: Phrase Types

Phrase	Characteristics
Noun	Noun phrases include nouns and their adjectival modifiers: The *system program manual* contains *program code.* This sentence has two noun phrases: the *system program manual* and *program code.*
Appositive	An appositive or appositive phrase, another noun or pronoun that renames or defines its referent, directly follows the noun or pronoun to which it refers: The SAVE command writes files in ASCII format to a disk file, *DSK in Version 3.1,* in the system directory. *DSK in Version 3.1,* an appositive phrase, renames or elaborates on *disk file.*
Verbal phrases	A verbal phrase contains a verb that is not expressed in a simple tense (in which the verb is only one word). For example: The corporation *may have had* trouble at that point. (conditional) The line *is connected* by the adapter. (passive) Laboratory personnel *will monitor* the test. (future) This office *is consuming* too much coffee. (progressive)
Verbals	Verbals include participial, gerund, and infinitive phrases. Verbals are created from nonfinite verb forms that look like verbs but do not act like them. *Providing directional indicators for data exchange,* the DMAACX signal can request data. (participial phrase) The crew enjoyed *flying experimental aircraft.* (gerund phrase) *To service the machine accurately,* use this access code. (infinitive phrase)
Prepositional	Prepositional phrases act as adjectives and adverbs. They consist of a preposition as a headword (the phrase's first word) and a noun or pronoun acting as the object of the preposition: The preliminary report should go *to the Marketing Department.* The prepositional phrase *to the Marketing Department* acts as an adverb modifying the verb phrase *should go.*
Absolute	Absolute phrases modify the entire sentence and therefore act as adverbs. They can be a confusing structure in the midst of complex information because they seem like elliptical constructions: *The error messages being correct,* I can finish executing. Turn the absolute phrase *the error messages being correct* into a dependent clause to create a more obvious connection: Because the error messages are correct, I can finish executing.

Clause Types

3.5 A clause contains both a subject and a predicate. There are two types of clauses:

Exhibit 3–3: Clause Types

Clause	Characteristics
Independent	An independent clause can stand alone as a sentence: Mary sat on her hat.
Dependent	A dependent clause cannot stand alone. A subordinating word introduces a dependent clause and defines the clause's function. For example: • An adverbial dependent clause *Because she was distracted,* Mary sat on her hat. • An adjectival dependent clause Writers *who don't revise their work* are seldom published. • A nominal (noun) dependent clause (as object of a preposition) Jones knew nothing *about what the team had decided.*

Most of the problems that arise with dependent clauses involve punctuation (see also 4.19).

Sentence Types

3.6 English has four sentence types: simple, compound, complex, and compound/complex. The number and kind of clauses in a sentence determines its type.

Exhibit 3–4: Sentence Types

Sentence Type	Characteristics and Examples
Simple	Contains one independent clause: Jane wrote a letter. Jane wrote and sent a letter. Jane and Sue carefully wrote and edited the report and letter. Struggling to meet their deadline, Jane and Sue carefully wrote and edited the report and cover letter.
Compound	Contains two or more independent clauses and can be joined by coordinating conjunctions, semicolons, and colons: This library provides maximum performance; however, it can also be used with the APC Math Library.
Complex	Contains one independent clause and one or more dependent clauses: When a transposition occurs, two numbers are switched.
Compound/Complex	Contains two or more independent clauses and one or more dependent clauses: When the handler routine detects an error, the designator routine assigns an error code, and this becomes the error record.

Common Problems in Grammar, Style, and Usage

3.7 Grammatical rules give structure to a piece of writing so that the writer's intention is effectively conveyed. Correct grammar alone is not sufficient for good writing. Matters of style and usage must also be considered. This section reviews four common grammar problems:

- Problems with nouns (3.8),
- Problems with pronouns (3.17),
- Problems with verbs (3.25),
- Problems with modifiers (3.30).

Problems with Nouns

Abstract versus Concrete Nouns

3.8 When possible, limit the use of abstract nouns, or, when this is impossible, further define the abstract noun with concrete nouns. Abstract nouns refer to things that cannot be discerned by senses, while concrete nouns refer to things that can be discerned by the five senses.

Plural versus Possessive Nouns (see also chapter 5)

3.9 Add *s* (*es*) to nouns to create the plural form: chemical-chemicals, potato-potatoes.

3.10 Add *s* with a preceding apostrophe to create the possessive form:

> This *chemical's* structure is unstable.

Sometimes the writer does not make a distinction between the plural and the possessive, especially in instances in which the apostrophe is used with acronyms in plural forms:

> The *DMAs* must be registered.

3.11 Use an apostrophe after plural nouns ending in *s* (*girls'*) to show possession. For compound nouns, use the apostrophe only with the final noun:

> John, Terry, and *Joe's* board . . .

Such possessive nouns always function as adjectives.

Appositives and Parenthetical Definitions

3.12 An appositive—a noun, pronoun, or noun phrase—renames a noun, pronoun, or noun phrase that it directly follows. Appositives are useful for defining an acronym, mnemonic, abbreviation, or new term:

> The host control status register, *HCSR*, assigns memory locations to incoming data.

Gerunds and Infinitives

3.13 Gerunds, nouns created from a verb plus *-ing*, function as nouns:

> Loading a program is tiresome.

Loading a program, the gerund phrase, acts as the sentence's subject.

3.14 Infinitive constructions can function as nouns, adjectives, or adverbs. Usage in the sentence determines the function of any part of speech, particularly infinitives:

> To write a program in OCCAM can be difficult.

Here, the infinitive phrase *To write a program in OCCAM* acts as the sentence's subject.

Noun Stacks

3.15 Noun stacks, an all-too-common occurrence, make it difficult for readers to understand which nouns act as modifiers for other nouns:

> The request acknowledgment exchange data bit is always HIMEM.

To make the sentence understandable, unstack these nouns by using some of them in prepositional phrases (Also, notice that changing the verb helps clarify meaning):

> The data bit used for exchanging request acknowledgments resides in HIMEM.

3.16 Preparing the same sentence for translation might require additional revisions (see also chapter 2):

> The data bit, which exchanges request acknowledgments, resides in HIMEM.

Problems with Pronouns

The Problem with Pronoun References

3.17 Every pronoun has an antecedent, the noun it replaces. The sentence with the pronoun must be arranged to make the antecedent obvious. Two particularly troublesome pronouns are *this* and *it.* Avoid using the word *this* by itself as a pronoun. *This* often stands for a situation or event and rarely does so obviously. Instead, follow the word *this* with a clarifying noun (making *this* an adjective):

> This sequence repeats until all errors are detected and a message is sent announcing that all subroutines have been completed. *This* allows the user to run the computer most efficiently.

In the second sentence, *this* can refer to the sequencing, to the message, or to both. Adding a noun to refer to the original subject (with an additional adjective for extra clarity) solves the problem:

> *This automatic sequencing* allows the user to run the computer most efficiently.

In like manner, keep the pronoun *it* specific, unambiguous, and close to its antecedent.

Compound Pronoun Antecedents

3.18 When the pronoun has a compound antecedent in which *and* joins the nouns, use a plural pronoun:

> Ms. Robbins and her secretary are at a meeting *they* had to attend.

3.19 If *or* or *nor* joins a compound antecedent, use a singular pronoun if both nouns are singular, and plural if both nouns are plural. When one noun is singular and one plural, the pronoun agrees with the noun closest to it:

> A slide or a printout will be available; it should be sufficient. (singular antecedents)

> The slides or the printouts will be available; they should be sufficient. (plural antecedents)

> A slide camera or printouts will be available; they should be sufficient. (plural direct antecedent)

Collective Nouns as Pronoun Antecedents

3.20 Collective nouns can be either singular or plural, depending on the sense of the noun:

The project group decided to delay *its* plans for ship date. (singular)

The project group met and returned to *their* cubicles. (plural)

Problems with Second-Person Pronouns

3.21 Imperative sentences seem to be missing a subject; but, in fact, an implied *you*, the second person pronoun, acts as the subject. Use the imperative in procedures, instructions, and tutorials:

Mark the end of a block with ^C and press the ENTER key.

3.22 Be careful with the imperative; used inappropriately, it can make readers feel uncomfortable, as if they were being continually ordered to perform. In addition, imperatives may be culturally unacceptable for some international texts (see also chapter 2).

Agreement of Pronouns

3.23 Pronouns must agree with their antecedents in number and gender:

Writers are always sure *they* are right. (number)

Ms. Jackson offered *her* legal opinion. (gender)

Sexist Language and Pronouns

3.24 The judicious use of pronouns can help avoid sexist language. Usually antecedents such as *each, either, neither, one, anyone, anybody, everyone, everybody, a person* take singular pronouns. Traditionally, the singular male pronoun has been used:

Anyone should be able to find *his* way out of the building.

Writers have tried to remedy this sexist approach in several ways:

Anyone should be able to find *his or her* way out of the building.

Anyone should be able to find *their* way out of the building.

Anyone should be able to find *her* way out of the building.

Although some editors will reject it, the first solution is acceptable if the pattern is not repeated too often. The second solution, although common, is unacceptable because it does not agree in number with the antecedent. The third solution is acceptable. However, some writers have taken to varying the pronoun, using *his* in one sentence, *her* in the next, and so on. An easier solution to the problem is to make the antecedent plural whenever possible:

> People should be able to find *their* way out of the building.

Depending on context, a variety of gender-neutral plural antecedents could be used: patrons, clients, personnel, students, researchers, and the like.

Problems with Verbs

Verb Agreement with Collective Noun Subjects

3.25 Collective nouns, although plural in form, take singular verbs when thought of as one unit or entity:

> The *number* of calculations is extraordinary.

However, if the noun is intended as a plural entity, the verb is plural:

> A *number* of calculations were made.

3.26 Sometimes writers get confused when a collective noun follows a singular subject, but the verb is singular because the subject is:

> *One* of the oscilloscopes is missing.

Verb Agreement with Subject Complements

3.27 In sentences with a subject complement, the verb agrees with the subject, not the subject complement. For example, in the following sentence, *object* is the subject and *molecules* is the subject complement:

> The *object* of his studies was *molecules.*

Verb Agreement with Compound Subjects

3.28 If the words *each* or *every* modify the compound subject, the verb is singular:

> Each circuit and board *has* been tested.
>
> Every circuit and board *has* been tested.

3.29 A compound subject joined by *or* or *nor* may contain a singular and plural noun or pronoun. In that case, the verb agrees with the noun or pronoun closest to it:

Jerry or the other managers *are* able to sign for it.

Problems with Modifiers

Restricting, Intensifying, and Absolute Modifiers

3.30 These types of modifiers are often misused in casual speech.

Exhibit 3–5: Common Modifier Problems

Modifier	Typical Examples	Usage Problems
Restricting	hardly, scarcely, simply, exactly, just, most, mostly, almost, nearly, merely	Placement of these terms is critical to meaning. Place them as close as possible to the word or phrase modified.
Intensifying	very, really, rather, quite, so	These words give only a vague sense of intensification. In editing, check to see if these words contribute to the sentence's sense.
Absolute	unique, complete, pure, perfect, wholly, final, ultimate	Despite the absolute denotations of these words, the temptation to intensify them seems hard to resist: *very unique, completely perfect.* Writers often abuse *unique* by treating it as synonymous with the less limiting adjective *unusual.*

Adjective versus Adverb Forms

3.31 Occasionally, writers confuse adjective and adverb forms. In the case of complements, this can be particularly confusing:

She smells good.

This sentence says that her smell is good. But the following sentence, which uses the adverb form, produces

She smells well.

The sentence now suggests that her smelling ability is highly developed. *Well* is an adverb modifying the verb *smell.*

3.32 Again, in an object complement position, the same problem can occur:

We considered the project hurried.

The sentence says that the project was done in haste. But the adverb form produces a different meaning:

We considered the project hurriedly.

Now the sentence says that we were in a hurry when we considered the project. *Hurriedly* is an adverb modifying the verb *considered.*

Misplaced Modifiers

3.33 A complex sentence contains at least one dependent clause and only one independent clause, which should contain the sentence's main idea. One of the most common writing errors is putting the sentence's main idea in the dependent rather than the independent clause:

When a transposition error occurs, two numbers are switched.

The sentence's intent is more apparent and actually altered when revised as

A transposition error occurs when two numbers are switched.

3.34 Modifiers must be placed in relation to the term(s) they modify. In technical or scientific writing of any complexity, unclear modifiers can seriously undermine writing clarity:

The technician reported that the sample was ready with alacrity.

With the modifier placed closer to the item it modifies, the sentence becomes

The technician reported with alacrity that the sample was ready.

Dangling Participles

3.35 A dangling participial phrase has no noun or pronoun to modify. A common problem in technical writing is an introductory participial phrase followed by a clause in passive voice, often one in which the subject has been dropped:

Being high true, the bus is granted access by the modem signal.

This sentence says that the bus is high true when, in fact, it is the signal that is high and true. After the passive voice is rewritten, the participle has the correct noun to modify:

Being high true, the modem signal grants the bus access.

Dangling Modifiers

3.36 Dangling modifiers have nothing to modify. This problem often occurs when the writer uses passive voice and drops the subject:

Executing a program, code was first checked.

In this example the participial phrase *executing a program* modifies the noun *code,* but that word is not the subject of this sentence. Actually, a missing subject—the user—executes the program:

Executing a program, the user checked the code.

Squinting Modifiers

3.37 Modifiers are said to "squint" when they can modify either the word preceding or following them:

Performing *often* tires him.

Often can function as an adverb modifying *tires* (a verb) or as an adjective modifying *performing* (a gerund). The solution is to move the modifier:

Often, performing tires him.

Or:

Performing tires him often.

Revision Strategies

3.38 Since many organizations use electronic methods for creating documents, the use of paper-based editing techniques has declined. Organizations that rely on these traditional methods should consult *The Chicago Manual of Style,* or similar style guides (see 13.3). Writers need to develop revision strategies tailored for the situation in which they work. This section suggests methods that can make the revision process efficient, effective, and more professional based on four approaches:

- Examining components (3.39),
- Examining basic grammar ("triage") (3.40),
- Eliminating sentence faults (3.41),
- Establishing consistency (3.48).

Revision Strategy 1: Examining Components

3.39 Professional editors review a document during multiple revision or editing sessions—variously called levels, sweeps, or passes. For example, there might be passes for organization; layout; legal issues; and for grammar, spelling, and punctuation. Such focused editing sessions are more effective than one or two extended general editing sessions that attempt to identify and account for all the diverse aspects of a document.

Exhibit 3–6: Component Edits

Edit Focus	Typical Editing Question	Actions to Solve Editing Problems
Audience (see also chapters 1, 2)	Who will be reading this document? Will they be experts, technicians, laypersons, or executives?	Ensure that word choice and writing style is appropriate for the audience.
Purpose	What goals should the document accomplish? For example, should it teach, explain, persuade, or entertain?	Test audiences can help measure how well a document accomplishes its purpose. Select readers who fit the expected audience profile.
Use	What should readers know or be able to do during or after reading the document? Make an executive decision? Use a computer application? Change their opinion on a topic?	Study the document to see that it is organized and formatted to facilitate the primary use. For example, a software application manual should have easy-to-find-and-follow instructions; an executive summary should have easy-to-find key points.
Organization	Can readers understand both global organizers that unite the entire text and local organizers that occur at chapter or lower levels? Can readers follow the logical arrangement of individual sections?	Create an after-the-fact outline to ensure that all the necessary points have been discussed in the proper order. Write down the topic of each paragraph. Evaluate the resulting list to see that the sequence of ideas makes sense given the document's audience, purpose, and use.
Layout	Are pages formatted appropriately for the material? For example, a reference document should have access aids: index, section tabs, table of contents.	Determine layout based on the conventions of the target publication. Base layout decisions on the purpose, use, and expected audience.

Exhibit 3–6: Component Edits *(continued)*

Edit Focus	Typical Editing Question	Actions to Solve Editing Problems
Graphics, illustrations (see also chapters 10, 11, 12)	What kinds of graphic conventions, for both data displays and illustrations, will support the text?	Consider each graphic and illustration as a minidocument. Just as with any full document, a graphic must be appropriate for a given audience, purpose, and use.
Legal issues	Does the document deal both fairly and legally with the issues of copyright, trademarks, and safety notice?	Copyrighted material is the property of the copyright owner. For material copied from other sources, ensure either that permissions have been obtained or that your use is legitimately based on the "fair use" principle. By law, copyright can extend to online materials. The first use of a trademark should carry either the trademark symbol ™ or the registered trademark symbol ® as appropriate. If safety is an issue, state appropriate warnings and attach any necessary labels. Include a review that focuses specifically on safety. For further information about organizations that have established standards for indicating hazards, see 13.3.
Sexist language	Does the document use sexist language in a blatant manner? Do any graphics present information in a sexist or culturally biased manner?	See 3.24 and chapters 10, 11, and 12 for suggestions on handling these issues in graphic materials.
Grammar and style (including "house" style)	Does the text reflect good basic grammar conventions appropriate to the topic and the intended audience?	See 3.7 and following. Some word processing applications have grammar checkers; however, these can be unreliable (see also 5.8).
Spelling and punctuation	Does the text reflect current spelling and punctuation practices, as well as any idiosyncratic conventions—legal, technical, or scientific terminology—within a corporation?	Use available electronic and hard-copy dictionaries to check a document. Corporate writers and editors should record special terms in local dictionaries.

Revision Strategy 2: Examining Basic Grammar ("Triage")

3.40 Some grammar and usage authorities recommend a type of edit referred to as "first aid" or "triage" to check for basic grammar errors.

Exhibit 3–7: Typical Error Types Found in Triage Edits

Part of Speech	Editing Action
Nouns (see 3.8)	Mark abstract nouns and check for noun stacks.
Pronouns (see 3.17)	Circle all pronouns and check to see that their antecedents are close enough to be obvious to the reader and that the pronouns are of the same number and gender as their antecedents.
Verbs (see 3.25)	Mark all forms of the verb *to be* when combined with a past participle; also circle *like, seem, become, have,* and *exist.* Substitute active verb forms when possible.
Modifiers (see 3.30)	Mark all restricting, intensifying and absolute modifiers. Also ensure that opening adjectival or adverbial phrases do not have modifier faults.
Prepositions (see 3.3–3.4)	Assess all prepositions to ensure that they are the most logical preposition possible. Limit chains of prepositional phrases to two or, at most, three in a row.
Connectives (see 3.3, 3.5–3.6)	Assess all coordinating and correlative conjunctions to see that the joined elements are parallel in structure and in form.

Revision Strategy 3: Eliminating Sentence Faults

3.41 Sentence faults, basic errors in sentence construction, occur in two ways:

- Fragments (incomplete sentences),
- Run-on sentences (improperly joined clauses in a sentence).

Dependent Clauses as Fragments

3.42 Dependent clauses cannot stand alone as complete sentences. Here is an example of a dependent clause as a sentence fragment:

> The I/O bus is an assembly of 150 signal lines. *That* connect the I/O Device Adapters to the I/O port.

The faulty period and capitalization breaks the close relationship between the two units. In this case, the closeness of the relationship is evident in the double usage of *that* as a pronoun (subject) standing for *lines* and as a subordinating conjunction. The sentence should be corrected as

> The I/O bus is an assembly of 150 signal lines that connect the I/O Device Adapters to the I/O port.

Using an appositive can often solve the same problem:

> The I/O bus, an assembly of 150 signal wires, connects the I/O Device Adapters to the I/O port.

Phrases as Fragments

3.43 Unless a fragment occurs in a list or other appropriate situation, do not punctuate a phrase as if it were a sentence.

Run-on Sentences

3.44 A run-on (or fused) sentence consists of two complete sentences combined without proper connectives:

> Five slides were prepared with solution *REB six* slides were prepared with solution TRP.

Use a connective between the words *REB* and *six*:

> Five slides were prepared with solution REB, and six slides were prepared with solution TRP.

The sentence could also be divided into two separate sentences or joined with a semicolon:

> Five slides were prepared with solution REB. Six slides were prepared with solution TRP.

> Five slides were prepared with solution REB; six slides were prepared with solution TRP.

The choice would depend on the intended relationship between the two sentences.

Fragments as Used in Lists

3.45 A list is an indented series of words or phrases that present a group of similar items. The list can make complex material easier to read by separating information so that the reader can select important points at a glance. The sentence fragment leading into a list can often act as a subject with the list providing a series of predicates:

This test
1. Sets the counter,
2. Increments the pointer to 1,
3. Checks the register contents.

3.46 A list can also be a series of noun phrases preceded by a complete sentence:

Shared processor systems have three features:
- A number of keying systems,
- A controlling processor,
- Disk-stored data.

3.47 Sometimes a list includes a mixture of fragments and complete sentences. In the following two-column list the fragment on the right describes the text on the left. Since the words *is the* are implied by the space between the columns, the fragment is punctuated as a complete sentence (see also 4.68).

LSTMSG	Address of the last message received. The supervisor uses this address when answering a message from an unspecified address.
RCLOCK	Right pointer for this task in the clock queue. If this task is waiting at the clock queue, RCLOCK points to the next task in the queue. If this task is not waiting at the clock queue, RCLOCK points to itself.

Revision Strategy 4: Establishing Consistency

3.48 Ensuring consistency can be achieved by
- Establishing logical transitions from sentence to sentence and from paragraph to paragraph;
- Deleting unexpected topical shifts that may distract or confuse readers;
- Creating parallelism within sentences and paragraphs to ensure readability and cohesion.

Establishing Logical Transitions between Grammatical Structures

3.49 Since readers often search for and read an isolated passage or section to help perform a specific task, transitions between sentences tend to be abrupt in technical writing. Nonetheless, the use of transitional words and phrases is important for creating continuity among grammatical units. Common transitional words and phrases include conjunctive adverbs, coordinating conjunctions, and phrases (see 3.3).

Exhibit 3–8: Common Transitional Words and Phrases

Purpose	Word or Phrase
addition	moreover, further, furthermore, besides, and, and then, likewise, also, nor, too, again, in addition, equally important, next, first, second, third, in the first place, finally, last
comparison	in like manner, similarly, likewise
contrast	but, yet, and yet, however, still, nevertheless, on the other hand, on the contrary, even so, notwithstanding, for all that, in contrast to this, at the same time, otherwise, nonetheless
location	here, beyond, nearby, opposite to, adjacent to, on the opposite side
purpose	to this end, for this purpose, with this object
result	hence, therefore, accordingly, consequently, thus, thereupon, then
summary	in short, that is, for instance, in fact, indeed
time	meanwhile, at length, soon, after, later, now

Deleting Topical Shifts

3.50 Consistency is one of the hallmarks of good technical and scientific writing. Not only should organization and use of key terms be consistent, but paragraphs and sentences should not have shifts in verb tense, mood, voice, person and number, and tone and point of view.

Verb tense

Avoid changing verb tense unnecessarily:

> When the technicians *removed* the air intake valves from the combustion chamber, two testing methods *are repeatedly attempted.*

The two verb phrases should be rewritten to match:

> After removing the air intake valves from the combustion chamber, the technicians used two methods to test them.

Mood

Sentence moods include the indicative (statement of fact or opinion); the imperative (a command or direction—omits the pronoun you); and subjunctive (states a condition, suggestion, etc.):

> Press the Command key, and then you should press Enter.

This sentence mixes the imperative and the subjunctive moods. A correct version would be

> Press the Command key, and then the Enter key.

Voice

Sentences can be in active voice (Mary hit the ball) or passive voice (The ball was hit by Mary). Active voice sentences are generally preferred, particularly in instructional documents. Passive voice can, however, play a useful role in scientific writing because it places more emphasis on the objects than on the researchers.

Generally avoid voice shifts. For example:

> From the File menu, *choose* the Save command and your work *is saved* to disk.

Should be revised as

> To *save* your work, *choose* the Save command from the File menu.

Person and number; pronouns (see 3.17)

Use person and number consistently. A change in person and number can create problems in subject and verb agreement and in pronoun reference.

Tone and point of view

Tone reveals the writer's attitude toward the reader and the subject. Tone is a combination of syntax, diction, and subject matter. The following example mixes formal with informal style:

> During the detailed system investigation, the *analyst* should gain a thorough understanding of the concepts that *make up* the proposed new system architecture. So *you* might want to *get going* on a prototype of the new system after *you* finish.

An informal version of this same text would be

> After *you* investigate the proposed system in detail, *you* should thoroughly understand its architecture. So *you* might want to get going on a prototype of the new system *asap.*

Creating Parallelism

3.51 Parallel structure presents items of equal importance in the same grammatical form and remains consistent in form throughout a series. For example, the following sentence lacks parallelism because it is unclear if the tire, which lacks a verb, was repaired, replaced, or cleaned:

He repaired a valve, replaced a hose, cleaned the spark plugs, and a tire.

3.52 Parallel structure creates certain expectations, giving the reader a familiar structure from which to extract information and the writer a boilerplate (standard) form in which to insert similar pieces of information.

3.53 Use parallelism to create balance between a set of words, phrases, or clauses within a given sentence. Faulty parallelism hampers the readability of a sentence. The faulty structure interferes with the presentation of the information as in

> The engine is overheated, corroded, and it is damaged.

This sentence can be rewritten in parallel as

> The engine is overheated, corroded, and damaged.

3.54 In a long sentence, repeat a key word that introduces parallel elements to help the reader understand the relationship between the elements:

> The processor *cannot* work automatically and *cannot* achieve good performance without some programmer involvement.

3.55 Use parallel structures to describe processes. For example, when writing installation instructions, begin each step with an imperative form of the verb. This technique emphasizes the step-by-step nature of the process:

1. *Place* a copy of the system disk in the disk drive,
2. *Observe* the flashing prompt line on the screen,
3. *Type* FINSTAL, and
4. *Press* any key.

A parallel structure, such as this one, helps the reader locate and use information. Similarly, consistency in punctuation, sentence structure, and item names will make it easier for the reader to follow the instructions.

3.56 Although parallelism is most often used at the sentence level, in technical writing it can also be used in a larger context. Use boilerplate (standard) paragraphs for situations in which information changes rapidly or remains similar in many contexts. For instance, in a diagnostic manual describing similar testing procedures, a boilerplate form might be used:

> Test 34BX performs the following sequence of steps:

1. Resets the BIRP,
2. Clears the MDP pointer,
3. Increments the AP pointer, and
4. Calculates the pointer differences.

In this example, the author has used active, imperative verbs and parallel sentence structure. Since this example represents a common user action, it can be repeated throughout the text in exactly the same form. In addition, similar listings can be used to describe other activities that have the same structure. Readers will soon learn the pattern and can use it to improve their performance.

3.57 The use of boilerplate and parallelism depends upon how readers use a document. If they use it to fix an engine, write a program, or troubleshoot a circuit board, they will appreciate easy-to-locate information. This ease in locating depends upon predictability, upon users knowing where to find what they need for their specific task.

3.58 Parallelism can also help readers locate information in sections, chapters, manuals, or even manual sets. When this information structure remains consistent, readers will be able to use the same retrieval methods each time they need to find additional information.

3.59 Begin planning parallel structure in the outlining stage. Usually in a larger document the tables of contents—both for the document as a whole and for individual chapters—develop from the outline. Good parallel structure in the outline puts related ideas together or effectively contrasts ideas. The outline hierarchy reflects these relationships (see also 1.131).

4.

Punctuating Scientific and Technical Prose

4.1 Writers have considerable flexibility in choosing the appropriate punctuation mark (or lack thereof). Most of the punctuation rules have one or two qualifications, and few can be applied inflexibly in all situations. Several might seem arbitrary and even illogical. However, most are sensible, and all of them foster ease in reading and help clarify the text.

4.2 At present, grammar checkers in word processing software provide little help with punctuation. Current word processing software, however, does offer typeset-quality punctuation marks not previously available, including *printer* quotation marks (“ ”), em (—) and en (–) dashes, inverted question mark (¿) for Spanish text, and so on. But beware when converting a word-processed document into another format such as hypertext markup language (HTML); some of these special characters may not convert.

Punctuation of Common Sentence Structures

4.3 The following guide summarizes how to punctuate the typical grammatical structures of scientific and technical prose:

Comma (4.4–4.42)

Subject + verb [,] {*and, but, etc.*} subject + verb.

Subject + verb [no comma] {*and, but, etc.*} + verb.

Introductory expression [,] subject + verb.

Subject [,] transitional expression [,] + verb.

Subject [,] nonrestrictive clause [,] + verb.

Subject [no comma] restrictive clause [no comma] + verb.

Adjective [,] + adjective + noun

Semicolon (4.43–4.49)

Subject + verb [;] subject + verb.

Subject + verb [;] {*however, consequently, etc.*}, subject + verb.

Subject + verb [;] subject + verb[;] *and* subject + verb.

Subject + verb + series of three items separated by commas [;] *and* subject + verb.

Colon, Dash, Parentheses (4.50–4.66, 4.127–4.137)

Subject + verb (complete thought)[:] illustration or amplification.

Subject + verb (incomplete thought)[no colon] illustration or amplification.

Subject [—]intervening word, phrase, or clause[—] + verb.

Subject + verb [(] secondary information to main thought [)].

Period, Question Mark, Exclamation Point (4.67–4.82)

Subject + verb [.] [? for a question or ! for special emphasis]

Hyphen (4.83–4.105)

Adjective or noun [-] + adjective or noun + head noun

Adjective or noun [-] + participle + head noun

Adverb ending in *-ly* [no hyphen]+ adjective or participle + head noun

Adverb not ending in *-ly* [-]+ adjective or participle + head noun

Apostrophe (4.108–4.116)

Possessive singular noun ['s] + head noun

Possessive plural noun ending in *s* ['] + head noun

Possessive plural noun not ending in *s* ['s] + head noun

Quotation Marks (4.117–4.126)

Subject + verb: ["] quoted matter.["]

Subject, ["] quoted matter,["] + verb.

Subject + verb + "quoted matter": illustration or amplification.

Subject + verb + "quoted matter"; subject + verb.

Comma

4.4 The comma, semicolon, colon, dash, and parenthesis prevent parts of a sentence from running together in such a way as to obscure the author's meaning. The comma—the most frequently used of these marks—provides the smallest interruption in sentence structure. It also has, by far, the most rules.

Compound Sentence and Comma

4.5 Insert a comma before the coordinating conjunction (*and, but, nor, or, for, so,* and *yet*) in a compound sentence containing two independent clauses:

> The rules control the use of nonflammable liquids, *but* provisions exist for handling flammable liquids as well.

This comma is optional for a sentence with short clauses:

> The samples have arrived [no comma needed here] and testing will begin shortly.

Compound Predicate and Comma

4.6 Do not separate compound predicates (two verbs for a single subject) by punctuation:

> The samples *were suspended* by a wire [no comma here] and *held* in the center of the study chamber.

4.7 Compound predicates also appear within dependent clauses, and these do not require a comma:

> We began our work with methane, which *is also known* as marsh gas [no comma here] and *is* a primary part of the natural gases.

4.8 Occasionally, inserting a comma before the conjunction separating a compound predicate is needed for clarity:

> This version of the programming language provides extensive tracing and debugging capabilities, [comma helpful here for clarity] and has windows that can be used simultaneously for editing code.

Introductory Phrase or Clause Punctuated with a Comma

4.9 Use a comma after a phrase or clause that begins a sentence:

> For sorting more than one database, use the sort program in the batch mode.

4.10 Omit the comma after very short phrases if no ambiguity results:

> In 2002 [no comma needed here] we will investigate the effect of water flow on bridge pilings.

Introductory Expression after Conjunction and Comma

4.11 Separate by commas an introductory word, phrase, or clause that immediately follows the coordinating conjunction of a compound predicate:

> In the late 1960s, the National Radio Astronomy Observatory implemented FORTH and, *even today*, employs this programming language.

4.12 An introductory word, phrase, or clause following a coordinating conjunction in a compound sentence leaves the writer with several options:

> The curves in figure 2 are less familiar geometrical objects than those in figure 1, but, *other than the top curve*, they have a reasonable shape.
>
> ... in figure 1, but *other than the top curve*, they have ...
>
> ... in figure 1, but *other than the top curve* they have ...

The first option includes all the punctuation that the sentence structure suggests. However, at least one comma could be eliminated (second option) without sacrificing clarity. Use the third option when the introductory expression is short and its function in the sentence is apparent without the final comma.

Restrictive Phrase or Clause and Comma

4.13 Set off a nonrestrictive phrase or clause (one that is not essential to the sentence's meaning) with a comma:

> Sodium sulfate changes to sodium phosphate by addition of potassium oxide, *a typical reagent chemical.*

4.14 Do not separate a restrictive phrase or clause (one that is essential to the meaning of the sentence) by commas unless it precedes the main clause:

> You will be more productive *if you invest time in learning how to use these tools.*

But:

> *If you invest time in learning how to use these tools,* then you will be more productive.

Participial Phrase as Subject and Comma

4.15 A participial phrase does the work normally reserved for a noun alone. Do not set off a participial phrase used as the subject when it immediately precedes the verb:

> *Storing batch jobs on card decks* [no comma here] has gone the way of the dinosaur.

Items in Series and Comma

4.16 In a series of three or more items with a single conjunction, use a comma after each item:

> The Computer Services Group provides assistance in graphics applications, equipment procurement, data analysis, and personal computer networking.

4.17 Some authorities recommend omitting the comma before the conjunction except when misreading might result. They argue that the conjunction signals the final break in the series and, therefore, the final comma is unnecessary. However, including the final comma leaves no doubt about the author's meaning.

4.18 Do not use a comma if conjunctions ("... one *and* two *and* three") join all the items in a series.

Phrases and Clauses in Series and Comma

4.19 With three or more parallel phrases or dependent clauses in a series, insert a comma after each one:

> This program will automatically read the file, apply the user's instructions, and create an updated file.

Transitional Word or Phrase and Comma

4.20 Set off transitional words (*however, thus, accordingly, hence, indeed, nevertheless,* and similar words) or transitional phrases (*for example, in other words, that is*) when there is a distinct break in thought:

> This reaction, *in turn*, leads to the development of other opinions.
>
> *Finally*, we finished repairing the computer code.

But omit the commas when the sentence seems sensible without them:

> They *thus* believe that some other kind of strong reaction acts between the subatomic particles.

Transposed Terms and Comma

4.21 Set off any words, phrases, or clauses placed out of their normal position for emphasis or clarity:

> If, *after a period of time,* metal crystallization appears, then metal failure will happen.

Apposition and Comma

4.22 An apposition is a grammatical construction in which two adjacent nouns (or noun phrases) refer to the same person, thing, or concept. Separate the second part of the apposition by commas (or parentheses) if it is not restrictive:

> A peritectic reaction, *the reverse of a eutectic reaction,* occurs in many metals.

4.23 Follow this same rule with mathematical symbols and expressions:

> Substitute the initial temperature, T_i, by . . .

4.24 Do not separate the second part of an appositive if it is restrictive:

> The British Association for the Advancement of Science debated Darwin's treatise *On the Origin of Species* at an 1860 meeting.

Contrasting Phrase and Comma

4.25 Separate by commas any phrase inserted to contrast with a preceding word or phrase:

> This unlikely, *though at the same time logical,* conclusion started the debate.

Deliberate Omissions and Comma

4.26 A comma can indicate the deliberate omission of a word or words for the sake of brevity:

> The first two digits indicate flying time in hours; the next four, the fuel usage in pounds.

4.27 A comma indicating omission is unnecessary when the sentence reads smoothly without it:

> The first sample came from France, the second from Japan, and the third from Canada.

Similar or Identical Words and Comma

4.28 To avoid confusion, separate two similar or identical words with a comma, even when the grammatical construction does not dictate such a break:

> In 1988, 377 coal samples were analyzed for sulfur content.

Two or More Adjectives in Series and Comma

4.29 Separate by a comma or commas two or more adjectives modifying the same noun:

> The material consisted of dense, coarse particles.

This comma may be deleted when the expression can be readily understood without it:

> For these tests, we designed and built a high-temperature [no comma needed here] high-pressure apparatus.

4.30 Do not use a comma if the first adjective modifies the idea expressed by the second adjective and noun combined:

> The material consisted of coarse dust particles.

Here, *coarse* refers to *dust particles*, not just *particles*.

Direct Quotations and Comma

4.31 Use a comma to introduce direct quotations that are a complete sentence or two in length:

> In his opening address, the physicist Henri Poincaré asked, "Does the ether really exist?"

However, in several similar situations do not use a comma:

- If the quotation is the subject of the sentence or a predicate nominative (that is, it completes the meaning of a linking verb such as *is* or *was*):

  > Marshall McLuhan's message *was* "The medium is the message."

- If the quotation is a restrictive appositive (that is, it follows and further describes a noun or noun substitute and is essential to the sentence's meaning):

  > Louis Sullivan followed the dictum "Form follows function."

- Or if the conjunction *that* introduces the quotation:

  > Although I am not sure what it means, the rule states *that* "Nonstandard replacements violate company standards."

Direct Questions and Comma

4.32 Direct questions are usually set off by a comma (see 4.78):

The question is, how do we get there from here?

Dates and Comma

4.33 Punctuation for dates depends on the date style. If the date is given as month, day, year, then place commas after the day and year:

A paper printed in *Nature* on April 22, 1990, asserts that sea turtles produce fewer than two hundred eggs per gestation.

4.34 Do not use commas when the day does not appear in the date:

Plutonium was discovered in March 1940 at the Lawrence Radiation Laboratory.

4.35 These punctuation rules for dates assume that they will be reported in a typical American style. The date style used in many countries outside the United States is day, month, and year with no punctuation: 25 September 2001 (abbreviated as 25 Sept 01 or 25.ix.01). Gaining popularity is a style used in some computer programs: year, month, and day separated by hyphens, as in 2001-09-25 (see 7.8–7.10).

Specifying Names with Commas

4.36 Set off a term that makes the preceding reference more specific:

The thermal data are shown in table 4, column 3, and in figure 2, top curve.

4.37 Use a comma to separate the elements in a geographical location:

The meeting in Chicago, Illinois, was postponed because of a blizzard.

Numbers and Commas (see also 7.13)

4.38 Counting from right to left, insert commas between groups of three digits in numbers greater than or equal to one thousand:

1,731 345,657 3,972,999,029,388

4.39 Note that this rule applies only to digits to the left of a decimal point. Digits to the right of a decimal should not be separated by commas:

11,496.01 0.0040

4.40 The style for some publishers of technical documents is not to insert commas for numbers with four digits or less:

1000 9999

4.41 These rules reflect common American and British practice. In other countries, the decimal marker is a comma instead of a period. The recommended style for international publications, therefore, is to insert blank spaces in setting off groups of three digits (to the right and left of decimal markers):

4 586 782,02 or 4 586 782.02

Reference Numbers and Commas

4.42 Set off reference numbers by commas unless they signify a range of numbers, which requires an en dash:

Recent research[6, 9, 11–15] has revealed that . . .

Semicolon

4.43 The semicolon divides parallel elements in a sentence when the text's meaning calls for a separation stronger than a comma but weaker than a period.

Semicolon and Clauses Joined without Conjunction

4.44 Insert a semicolon between the clauses of a compound sentence when they are not connected by a conjunction:

With this software, you need not write a program to create a graph; after a little training, nearly anyone can produce graphs.

Semicolon and Clauses Joined by Conjunction

4.45 Use a semicolon to separate the clauses of a compound sentence joined by a coordinating conjunction when they are long or are themselves internally punctuated:

A vapor analyzer measured atmospheric moisture, humidity, and dew point; and a sniffer instrument sampled heavy metals, pollen, and gases.

4.46 Use a comma in place of the semicolon if the sentence will still be understandable:

These hardness tests are accurate and reproducible, and if sound judgment is used, they give a good estimate of the strength for most metals.

Transitional Adverbs and Semicolon

4.47 Use a semicolon (often followed by a comma) to precede transitional adverbs (*accordingly, consequently, furthermore, hence, however, indeed, instead, moreover, nevertheless, so, then, therefore, thus,* and similar words) when these words link clauses of a compound sentence:

> The values from the two studies disagree; *therefore,* definitive results cannot be given at this time.

Transitional Terms and Semicolon

4.48 Use a semicolon before expressions such as *for example, in other words,* or *that is* and follow these words with a comma when they introduce an independent clause:

> No refinement is necessary; *that is,* one merely has to derive a single set of rules to describe the wing's shape.

Elements in a Series and Semicolon

4.49 Use a semicolon to separate parallel elements in a series if they are long or internally punctuated:

> The zinc ores include sphalerite, ZnS; smithsonite, $ZnCO_3$; willemite, Zn_2SiO_4; calamite, $Zn_2(OH)SiO_3$; and zincite, ZnO.

Colon

4.50 The colon usually introduces an illustration or elaboration of something already stated.

Elements in a Series and Colon

4.51 Use a colon to introduce words, phrases, or clauses in a series if the clause preceding the series is grammatically complete:

> The metals tested were treated with three salts: LiCl, KCl, and NaCl.

4.52 Do not use a colon when the elements in a series are needed to complete the clause introducing them:

> The metals tested were [no colon here] Mo, Hg, and Mn.

Amplification and Colon

4.53 Use a colon to separate two statements, the second of which is introduced by the first and amplifies or illustrates it:

> Two simple observations underlie relativity theory: a body's mass depends on its total energy, and a body's energy is proportional to its mass.

4.54 An introductory phrase like *the following* or *as follows* signals that an amplifying or illustrating statement is about to be made and a colon is needed:

> One of the simple modifications to the ideal gas law is *the following* equation: $pV = RT + \alpha P$.

If *the following equation* is omitted from the above example, then the colon is also dropped:

> . . . modifications to the ideal gas law is $pV = RT + \alpha P$.

4.55 If the material introduced by a colon contains more than one sentence, capitalize the first letter in the first complete sentence after the colon:

> An outline of our experimental procedure follows: First, the samples were heated at high temperature for one hour. Second, a new reactant was used. Third, . . .

Long Quotation and Colon

4.56 Use a colon to introduce long quotations:

> In the spring of 1915, Einstein wrote to a friend: "I begin to feel comfortable amid the present insane tumult, in conscious detachment from all things which preoccupy this crazy community." [trans. by Abraham Pais]

Ratios and Colon

4.57 Use a colon between the parts of a ratio:

> The procedure calls for an atmosphere with an $H_2O:H_2$ ratio of 1 : 115.

Time of Day and Colon

4.58 Use a colon between the hours and minutes in expressing time on a 12-hour scale (see also 7.8):

> Our experiment ran from 7:05 A.M. to 9:43 P.M.

Note that the British use a period instead of a colon between the hour and minutes, and use lowercase type for a.m. and p.m.

4.59 The convention is not to use a colon for separating the hours and minutes on a 24-hour scale:

> Our experiment continued from 0705 hours to 1343 hours.

4.60 Use colons to separate hours, minutes, and seconds on a 24-hour scale:

> 07:05:55.1 = 5 minutes and 55.1 seconds after 7 A.M.

Dash

4.61 The two kinds of dash are the em dash and the en dash. The characters (—) and (–) signify the em and en dash, respectively. If these characters are not available on your typewriter or word processing software, then use a double hyphen (—) for the em (—) dash and a single hyphen (-) for the en (–) dash.

Amplification or Illustration and Dash

4.62 Em dashes can introduce and conclude a short enumeration or amplification. They can also enclose asides and brief interruptions in thought. Unlike commas (or parentheses), em dashes direct the reader's attention to what has been set off:

> Users have available to them several interactive systems—Orville, Wilbur, and Kitty Hawk—twenty-four hours a day.

Em Dash and Other Punctuation

4.63 Normally, an em dash both precedes and follows an enumeration or amplification. However, the second dash should not be immediately followed by a comma, semicolon, colon, or period. If the grammatical structure indicates a comma, use the dash alone:

> Because this alloy bears a misleading name—beryllium copper—[no comma here] the uninformed reader might assume that it is a beryllium-rich alloy.

4.64 If the grammatical structure calls for a semicolon, colon, or period, use that punctuation mark and leave out the dash.

Range of Numbers or Time and En Dash

4.65 Use the en dash for a range of numbers or a time span unless it is preceded by the prepositions *from* or *between*:

> in the years 2001–20
>
> for the period December–June 2008

But:

from 2001 to 2020

between December and June 2008

Two-Noun Combinations and En Dash

4.66 Use the en dash to link a compound term serving as an adjective when the elements of the compound are parallel in form (that is, the first does not modify the second):

heat capacity–temperature curve

metal–nonmetal transition

A slash can also be used for this purpose.

Period

4.67 The period marks the end of a sentence or sentence fragment. For example, the following footnotes to a table should all end with periods:

[a]From studies by Wheeler and coworkers.[9]

[b]Results calculated by Leonard.

The period also has several specialized uses.

Vertical Lists and the Period

4.68 Use a period after numbers or letters that enumerate the items in a vertical list.

These lectures on technical communication offer instruction in three areas:

1. Constructing readable and attractive tables and figures,
2. Reducing technical material to its simplest terms without compromising the author's meaning, and
3. Eliminating technical jargon and wordy expressions.

Instead of periods, you can also use parentheses around the numbers or replace the numbers with bullets. In identifying items within a vertical list, consistently use either bullets or numbers.

4.69 Punctuation of items in a vertical list has several options. If the items are not complete sentences, the options are to

1. Punctuate as though the items were running text.
2. Eliminate the commas or semicolons separating the items in series and drop the final period.
3. End each item with a period, as in this list.

The first letter of each item is usually uppercased.

4.70 If one or more items contain a complete sentence, then all sentences and sentence fragments should end with periods:

Users have access to three types of data storage:

- HAL 3350 disk drives. There are 35 such drives.
- HAL 3330 disk drives. There are 80 such drives.
- HAL 2314 disk drives. There are two such drives, which must be set up as needed for a job.

Abbreviations and Periods (see also chapter 6)

4.71 Many abbreviated words end with periods (see also 6.2). These include

Academic degrees (Ph.D., A.B., B.S.)

Parts of firm names (Corp., Co., Inc., Ltd.)

Names of countries (U.S., U.K.)

Scholarly Latin abbreviations (e.g., i.e., ibid.)

Journal names used in references (J. Electrochem. Soc., Phys. Rev., J. Chem. Phys.)

4.72 Abbreviated Latin names for microorganisms, plants, and animals end with periods: *E. coli* for *Escherichia coli.*

4.73 In general, periods are not included with *initialisms,* abbreviations formed by combining the first letters of each word in a technical expression:

NMR nuclear magnetic resonance

DNA deoxyribonucleic acid

4.74 If an abbreviating period ends a sentence, do not add another period to mark the end of the sentence:

The battery was manufactured by Harmon Inc.

Raised Periods

4.75 The raised period, *center dot*, serves several purposes in scientific prose. It indicates multiplication ($a = b \cdot c$) in mathematics, an addition compound in a chemical formula ($Ni_2O_4 \cdot 2H_2O$), the unshared electron in the formula of a free radical ($HO\cdot$), and associating base pairs of nucleotides (T·A). You should become familiar with any specialized usage for this unusual punctuation in your own specialty.

Ellipsis Points

4.76 Use three ellipsis dots for any omission within a quoted passage:

> One of the most famous papers in all of science begins, "We wish to suggest a structure for . . . deoxyribose nucleic acid."

Most word processing software will automatically generate a set of three periods after you strike the assigned keys.

4.77 The ellipsis is also used in mathematical expressions:

$$x_1, x_2, \ldots, x_n$$

$$x_1 + x_2 + \ldots + x_n$$

Note how the commas and mathematical symbols enclose the ellipsis points.

Question Mark

4.78 The question mark indicates a direct question. Research problems are sometimes expressed in the form of a question:

> The question we examined is, what is the effect of these market conditions on corporate profits?

4.79 Spanish introduces a direct question with an inverted question mark:

> ¿Fue el estrés la causa de los síndromes de la Guerra del Golfo?

4.80 An indirect question within a statement does not end with a question mark:

> We had to ask ourselves why their calculated results coincided so closely with the measured ones.

Exclamation Point

4.81 The exclamation point signifies that a vehement or ironical statement has been made:

> Mixing these chemicals together at higher temperatures could result in an explosion!

4.82 Science and technical writing—especially that in specialized technical periodicals—assumes a restrained tone. Emotional statements followed by exclamation points are usually out of place. However, the exclamation point can be used effectively when the author assumes a folksy tone, as is often done in science and technical writing aimed at a wide audience:

> TEX's new approach to this problem (based on "sophisticated computer science techniques"—whew!) requires only a little more computation than the traditional methods, and leads to significantly fewer cases in which words need to be hyphenated. —Donald E. Knuth

Hyphen

4.83 The hyphen links words in compound terms and unit modifiers and separates certain prefixes and suffixes. A "compound term" expresses a single idea with two or more words (*cross-contamination, face-lift, time-sharing*). Two or more words modifying a noun or noun equivalent constitute a "unit modifier" (*one-dimensional computer model*). In general, science and technical writing relies more heavily on unit modifiers than do other kinds of writing.

4.84 The hyphen helps readers identify the relationship between any words modifying a noun.

Unit Modifier Containing Adjectives and Hyphen

4.85 Hyphenate unit modifiers consisting of

- An adjective plus a noun—*high-temperature* process,
- A noun plus an adjective—*radionuclide-free* sample,
- An adjective or noun plus a participle—*far-reaching* consequences, and
- A noun plus a noun—*random-access* disk file.

4.86 When the meaning is obvious, use of the hyphen is unnecessary:

> light water reactor
>
> molecular orbital calculation

4.87 Achieving consistent hyphenation in technical documents can be difficult. If an author decides to hyphenate *long-term leaching* in a report, then she should hyphenate not only all instances of this expression but also any similar expressions like *long-term durability*. Use your computer's "find" function to search for compound terms and unit modifiers and check for consistency in hyphenation. Be aware that a text littered with hyphens probably indicates excessive use of unit modifiers.

Unit Modifier Containing Number and Unit of Measure (see also 7.20)

4.88 Insert a hyphen between a number and its unit of measure when they are used as a unit modifier:

15-g sample

5-mm thickness

4.89 If a number precedes a number with its unit of measure, write out the first number:

three 10-g samples

two 20-mL aliquots

Unit Modifier Containing Adverb and Hyphen

4.90 Do not use a hyphen in unit modifiers containing an adverb ending in *-ly*:

previously known conditions

recently given procedure

4.91 Hyphenate unit modifiers when the adverb does not end in *-ly:*

still-new battery

well-kept laboratory

4.92 If another adverb is added to a unit modifier beginning with an adverb, then no hyphens at all need appear: very well defined structure.

Unit Modifier Containing a Proper Name

4.93 Do not insert a hyphen in a unit modifier if one of the elements is a proper name:

Fourier transform infrared analysis

Raman scattering experiment

Unit Modifier Longer than Two Words

4.94 Use a hyphen to join multiple-word unit modifiers:

up-to-date files

easy-to-learn software

root-mean-square value

4.95 Avoid unit modifiers longer than three words: use *well-defined structure at high temperature* instead of *well-defined high-temperature structure.*

Single Letter and Hyphen

4.96 Hyphenate single letters attached to nouns, adjectives, or participles:

Student's t-test

L-shaped

O-ring

Foreign Phrase Used as Adjective

4.97 Do not hyphenate foreign phrases used as adjectives unless they are hyphenated in the original language:

a priori argument

in situ spectroscopy

Predicate Adjective

4.98 Hyphenate unit modifiers that serve as predicate adjectives:

This step in the procedure is *time-consuming.*

The package must be *vacuum-sealed* before shipment.

4.99 Do not hyphenate compound verbs formed with a preposition acting as an adverb:

We *burned off* the binder after the tape-casting procedure.

Self- Compound and Hyphen

4.100 Hyphenate all *self-* compounds:

self-consistent

self-discharge

Suspended Hyphen

4.101 A suspended compound is formed when a noun has two or more unit modifiers, all linked by the same basic element. Use a hyphen after each such modifier:

10-, 20-, and 30-mm diameter

long- and short-term tests

Numbers and Hyphens (see also 7.20)

4.102 Use a hyphen in fractions and in numbers between twenty-one and ninety-nine:

one-third (fraction)

thirty-three hundred (whole number)

three thousand thirty-three

Technical Terms and Hyphens

4.103 If a compound technical term (names of chemicals, diseases, plants, and the like) is not hyphenated when used as a noun, it should remain unhyphenated when used as a modifier unless misreading might result:

potassium chloride vapor

lithium oxide solid

But:

current-collector material

Prefixes and Suffixes and Hyphens

4.104 In general, do not use a hyphen if a prefix or suffix is added to a word:

stepwise

nonuniform

multifold

4.105 Consult a dictionary to determine whether or not to use a hyphen with a word containing a prefix or suffix. If that word is not in the dictionary, spell it without a hyphen except for the following situations:

1. Use hyphens after prefixes added to proper names or numbers:

    non-Newtonian, un-American, post-World War II, mid-2010

2. Use hyphens after prefixes added to unit modifiers:

   non-radiation-induced effect, non-temperature-dependent curve, non-tumor-bearing tissue, non-load-bearing material

3. Use hyphens when the suffix *-fold* is added to a number:

   40-fold, 25-fold (*but:* tenfold)

4. Use hyphens with prefixes or suffixes to avoid a confusing repetition of letters or similar terms:

   non-negative, bell-like, sub-micromolar, anti-inflation

5. Use hyphens if their omission would alter the meaning:

   multi-ply, re-solve, re-pose, re-cover, re-fuse

6. Use hyphens with prefixes that stand alone:

   pre- and post-operative examination

7. Hyphenate prepositions combined with a word to form a noun or adjective:

   built-in, trade-off

   Recognize, however, that some organizational practices allow:

   buildup, cleanup, setup

Slash

4.106 Use the slash to link two terms when an en dash or hyphen might be confusing:

IBM™ 370/195

electrode/electrolyte interface

go/no-go decision

4.107 The slash also serves an important function in URLs (universal resource locations):

http://www.tech.com.nasa.gov/Chapt2.html

This URL has three basic components, separated by slashes. First, *http* signifies that the URL resides on a web server. Second, *www.tech.com.nasa.gov* specifies the location of that server (in this case, a NASA computer) and the first page of the website. Third, *Chapt2.html* specifies a subsidiary link to that first page.

Apostrophe

4.108 The apostrophe indicates the possessive case or a contraction. It is also used in forming certain plurals.

Possessives and Apostrophes

4.109 Form the possessive of a singular noun by adding apostrophe *s*:

What could be lovelier than a calm sea on a *summer's* day?

4.110 Form the possessive of a plural noun that ends in *s* by adding the apostrophe alone:

The government failed to heed the *scientists'* warnings.

4.111 If the plural ends with a letter other than *s*, add apostrophe *s* to form the plural possessive:

spectra's variance

4.112 Form the possessive of closely linked proper names by changing the last name only:

Einstein and *Planck's* law

Lewis and *Randall's* classic textbook

4.113 Time expressions are formed in the same way as possessives:

three *weeks'* time

Contractions and Apostrophes

4.114 An apostrophe can also indicate the omission of a letter or letters:

it's	it is
cont'd	continued

Plurals and Apostrophes

4.115 Use apostrophe *s* to form the plural of abbreviations with periods, lowercase letters, numerals, and abbreviations where confusion would result without the apostrophe:

Btu's

Ph.D.'s

x's and y's

4.116 The *s* alone is usually sufficient to form the plural of all-capital abbreviations and years, although some style guides recommend apostrophe *s* here too:

CPUs

2010s

Quotation Marks

4.117 Quotation marks enclose material spoken or written by someone other than the author, or they give a word or expression special emphasis. They are also used around the titles of certain types of publications, such as journal articles and book chapters (see also 8.29, 8.73).

Single versus Double Quotes

4.118 Use double quotation marks to enclose primary quotations, and single quotation marks to enclose quoted or emphasized matter within a quotation:

> The senator's question was simple and direct, "What do you think about when you hear the words 'nuclear energy'?"

Placement of Quotes with Other Punctuation

4.119 Common practice in the United States is to place closing quotation marks after commas and periods, but before colons and semicolons:

> Science involves much more than the application of "organized common sense," according to Stephen Jay Gould.

4.120 Great Britain and other countries follow the more logical approach of inserting commas and periods outside the closing quotation marks unless they are part of the quoted matter:

> The title to Einstein's first relativity paper is translated as "On the Electrodynamics of Moving Bodies".

But:

> Einstein wrote that "The whole of science is nothing more than a refinement of everyday thinking."

Even within the United States, the American Chemical Society and a few other publishers favor the British style.

4.121 Question marks or exclamation points always appear within a closing quote when they are part of the quoted matter; otherwise, they appear outside the quotes:

> Who said that "Editing is like bathing in someone else's water"?

4.122 France and other European countries use a pair of *caret* marks to indicate a quotation and follow the British style for placement of other punctuation:

> On peut distinguer l'eau «libre» de l'eau «liée».

Specialized Usage for Quotation Marks

4.123 Put quotation marks around words or phrases that constitute new or special usage. These quotes need not be repeated for a given word or phrase after its first appearance in the text:

> We determined the effect of the income loss on the "corporate profile" for various sales campaigns. The required corporate profile for acceptable losses was determined to be . . .

4.124 Quotes are also used in naming characters the reader should enter to execute some computer commands:

> Type "HELP ME" for online assistance in changing your password.

4.125 Make certain that the meaning of all new or special terms is obvious from the context in which they appear; if not, define them:

> X-rays passing through crystals undergo "diffraction"; that is, the rays bend from a straight-line motion.

So-Called Expressions

4.126 The meaning of words or phrases following *so-called* is usually apparent without quotation marks:

> The data illustrate the *so-called* principle of insufficient reason.

Parenthesis

4.127 Parentheses—like commas, colons, and dashes—separate material that clarifies, elaborates, or comments upon what preceded them. Parentheses usually enclose material that is of secondary importance:

> We recovered the remaining gases (SO_2, CO_2, and O_2) by changing our procedures.

Abbreviations and Parentheses (see also 6.2)

4.128 Place acronyms and other abbreviations in parentheses when they are preceded by the spelled-out term:

> The Source Term Experiments Program (STEP) consists of four accident simulations run in the Transient Reactor Test (TREAT) facility.

References and Parentheses (or Brackets)

4.129 One method for citing reference numbers in text is to enclose them in parentheses (or brackets):

> Further details on their experimental technique can be found in the literature (2, 5–8).
>
> . . . in the literature [Tillis 1982a].

Numbered Lists and Parentheses

4.130 Use parentheses in pairs, ( ), when numbering items in a list within running text:

> After an error message, you have several options: (1) type "help!"; (2) type "x" or "X"; or (3) type "e" or "E".

4.131 If the house style is to cite references by numbers enclosed in parentheses, use of a single parenthesis, instead of a pair, helps avoid confusion in numbering items:

> Burris (3) attributes this discrepancy to one of three factors: 1) the absorption influence identified by Johnson (4), 2) the wettability factor, or 3) the effect of high temperature on the system.

Punctuating Parenthetical Comments

4.132 Do not place a period inside the closing parenthesis of a sentence in parentheses inserted into another sentence:

The reaction $2SO_2 + O_2 = 2SO_3$ (sulfuric acid is obtained by dissolving SO_3 in water) occurs in the desired direction only at low temperatures.

4.133 A sentence enclosed by parentheses but not placed in another sentence should end with a period inside the closing parenthesis; it should also begin with a capital letter:

Hadrons are subatomic particles called quarks. (The word *quark* was adapted from a line in a novel written by James Joyce, *Finnegans Wake.*) Murray Gell–Mann was . . .

Brackets

4.134 Brackets set off an editorial interjection or parenthetical material within parentheses.

Editorial Interjections and Brackets

4.135 Use brackets to enclose editorial interjections within quoted matter:

Robert Boyle told young scientists that they would likely ". . . meet with several observations and experiments which, though communicated for true by candid authors or undistrusted eye-witnesses, or perhaps *recommended by your own experience* [italics added], may, upon further trial, disappoint your expectation . . ."

Parentheses within Parentheses

4.136 Use brackets to enclose a parenthetical remark within parentheses:

(We must still examine the mixture's effects on several species [particularly, Ba, Cs, and Sr].)

4.137 Use brackets to enclose a parenthetical remark containing a term requiring parentheses:

Initial experiments were performed with nonradioactive solutions [e.g., solutions containing Ho(III)].

Iconic Characters

4.138 In recent years, messages posted in electronic mail, news groups, webpages, listservs, and chat rooms have been including iconic characters named *emoticons* or *smilies*—playful combinations of punctuation meant to express emotion:

:-)	happiness or pleasure
:-(	sadness
;-)	winking
:-))	extreme happiness
=)	surprise
<:-)	dumbness

While suitable for e-mail and the like, such punctuation is rarely appropriate in formal technical documents.

Punctuating Equations

4.139 See 7.82–7.84.

5. Using Acceptable Spelling

Good Spelling Practice

5.1 Everyone who writes should own a suitable dictionary. *Merriam-Webster's Collegiate® Dictionary*, preferably an unabridged version, is widely accepted among usage authorities. However, creating scientific and technical documents requires more than a good dictionary. Topics to consider include

- Plurals (5.9);
- Proper names, possessives, and abbreviated terms (5.23);
- Compound words and hyphenation (5.24);
- Word division (5.25);
- Suffixes (5.28);
- Capitalization (5.39);
- Troublesome words (5.58).

Preferred Spellings

5.2 If a dictionary provides more than one spelling, the first spelling is the preferred. American rather than British spelling should be used for material published in the United States. To check words that have recently entered the language, consult the current edition of a regularly updated dictionary.

Spelling Checkers

5.3 Many word processing applications have spell-checking functions. Some of these functions spell-check a text based on a certain command—others work during typing by flagging suspect entries. Spelling checkers play a useful role in the editing and revising process; however, the careful writer must remain aware of their inherent limitations.

5.4 Be aware that a document may have no misspellings and still communicate only nonsense. In the words of James Joyce, "No birdie aviary soar any wing to eagle it."

5.5 An automated spelling checker may flag words it does not recognize, but it will not flag a homonym. This failing is particularly vexing in the case of common words such as *there, they're,* and *their*, as well as *two, to*, and *too*. Further, if you write the word *casual*, but you intended to write *causal*, the spelling checker will not flag this error.

5.6 Spelling checkers usually have limited source dictionaries—that is, few of them carry the full content of even a standard dictionary such as *Merriam-Webster's Collegiate Dictionary*. This shortcoming can be mitigated to some extent by adding words to a spelling checker's dictionary as you work—although not all spelling checkers have this capability. You might need to add such terms as:

- Proper names;
- Discipline-specific terms and symbols (see also chapters 6 and 7);
- Acceptable acronyms, abbreviations, and initialisms (see also chapter 6);
- Words identified as problematic through litigation;
- Words identified as problematic for translation (see also chapter 2);
- New, and acceptable, words created by basic changes in terminology.

5.7 Even worse, spelling checkers give the writer a sense of security vis-à-vis spelling, while giving no help in the matter of appropriate word choice. There is no substitute for the vocabulary and diction building exercise of compulsive dictionary checking.

5.8 Online spelling checkers, such as Research-It! and One-Look, while useful, are heirs to the same pitfalls as the spelling checkers built into word processing applications. In addition to these general dictionaries, discipline-specific dictionaries, such as Inductel's MicroLibrary, in a variety of professional fields are available. Some of these electronic dictionaries function as stand-alone spell checkers; others can be integrated directly into a word processor.

Plurals

5.9 Form plurals by adding *-s* or *-es* to the singular. Use *-es* for words ending in *-sh, -x, -z: rushes, appendixes, buzzes.*

5.10 Some words that end in *-f, -ff*, or *-fe* may change the plural to a *-ve* form; some even have two forms: *half-halves, leaf-leaves, wife-wives.*

5.11 Add *-s* to words ending in *o*, if a vowel precedes the *o*, as in *ratios*; however, if a consonant precedes the *o*, add *-es: mosquitoes.* Exceptions do occur, however*: zeros.*

5.12 Words ending in *-y* generally change the *-y* to an *-i* and add *-es: counties, harpies, puppies.*

5.13 Some words change form altogether:

child	children
foot	feet
louse	lice
tooth	teeth
woman	women

Plurals of Foreign-Language Words

5.14 Many technical words come from foreign languages and form the plural irregularly.

5.15 Retain the original language plural for these foreign terms:

crisis	crises
erratum	errata
hypothesis	hypotheses
ovum	ova
stratum	strata

5.16 In formal writing—especially in scientific writing—*datum* is considered the correct singular and *data* the correct plural. However, in less formal writing, common usage has accepted *data* as a mass noun taking a plural verb. For example, the following sentence might be appropriate for a news story:

Polling *data* reveals that Smith lost support in every precinct.

However, in a formal report, the sentence should read

Polling *data* reveal that Smith lost support in every precinct.

5.17 Some non-English words have, in most cases, one preferred spelling in the plural but have another acceptable but less frequently used plural form:

Singular	Preferred Plural	Permitted Plural
appendix	appendices	appendixes
criterion	criteria	criterions
formula	formulas	formulae
phenomenon	phenomena	phenomenons
matrix	matrices	matrixes
vortex	vortices	vortexes

5.18 Some words change meaning based on their endings:

indexes (tabular content)

indices (algebraic signs)

Singular/Plural Dilemma

5.19 Some spelling rules must be memorized. For example, some singular words can also be used as plurals:

deer, fish, sheep, moose

5.20 Some words that can be made plural in the usual way can also be plural without any change:

couples	couple
dozens	dozen
heathens	heathen
peoples	people

5.21 Some words can be either singular or plural with no change:

aircraft, chassis, series

5.22 The word *species* exists only in the plural. The singular noun *specie* refers not to a single species but to "money in coin."

Proper Names, Possessives, and Abbreviated Terms

5.23 See 4.108–4.117 for proper names and possesives; 6.2–6.39 for abbreviated terms.

Compound Words and Hyphenation

5.24 See 4.83–4.105.

Word Division

5.25 Divide words according to the American system of pronunciation.

5.26 Many contemporary word processors provide sophisticated hyphenation tools with many options. However, all these tools can make mistakes; there is no substitute for careful proofreading.

General Word Division Guidelines

5.27 Do not divide at single letters (*i-dea*), and never divide the suffix *-ed* from the parent word. Neither gains a space advantage on the line, and both influence reading patterns. In dividing words

- Avoid breaking any word in such a way that only two letters carry to the following line.
- Never divide a word between one page and the next. Try to avoid breaking a word from one column to the next.
- Try not to break proper names, and although it is both permissible and possible, do not even break between title, initials, and the last name.
- Hyphenated compounds should be broken only at the hyphen:

  pre-Columbian.

- Once compounded, but now solid, words ordinarily should be broken at the natural compound:

  over-all, data-base

- If prefixes exist, break at that point rather than elsewhere in the word:

  quasi-technical, anti-matter, super-conducting, re-examine, pre-exponential

- For using hyphens with large numbers, see 7.20.
- For numbered lists in text, carry the number to the following line to appear with the material directly following the listing:

  The components included (1) hydrogen, (2) oxygen, and (3) various unidentified solids.

  Not:

  The components included (1) hydrogen, (2) oxygen, and (3) various unidentified solids.

Suffixes

5.28 Words ending with a silent *-e* usually drop the *-e* before a suffix beginning with a vowel, with some exceptions:

create, creating

force, forcible

But:

mile, mileage

enforce, enforceable

And, to avoid misreading:

dyeing, not dying

singeing, not singing

5.29 If the silent *-e* precedes a consonant, the *-e* is usually retained:

movement, wholesome

Of course, there are exceptions:

wholly, truly

5.30 Most confusing, perhaps, are words for which either dropping or retaining the *e* is correct. Note that, as in dictionary entries, the first word represents the preferred (American) spelling; the second, British spelling:

acknowledgment, acknowledgement

judgment, judgement

Soft Endings ("*-ce*," "*-ge*")

5.31 Words with soft endings—*-ce* or *-ge*—retain the *-e* before suffixes beginning with *a*, *o*, or *u*, to retain the soft sound:

advantageous, changeable

The "*-y*" Ending

5.32 Most words ending in a vowel followed by *-y*, keep the *-y*:

buy, buyer

enjoy, enjoyable

5.33 When a consonant precedes *-y*, an *-i* usually substitutes for the *-y*:

creepy, creepier

fatty, fattiness

fly, flier

Consonant Endings

5.34 Many words that end in a single consonant preceded by a single vowel double the final consonant:

control, controlled

occur, occurrence

prefer, preferred

But:

cancel, canceled, cancelable

transfer, transferable

travel, traveled

5.35 For words ending in a hard *-c*, a *-k* is often added to ensure proper pronunciation:

mimicking, picnicking, shellacked

5.36 The *-ise* ending is the commonly accepted American ending for many words (and the preferred British ending for words that often end in *-ize* in America—such as *characterize* or *criticize*):

advertise, comprise, exercise, improvise

5.37 Contemporary practice creates verbs from nouns by adding *-ize*: finalize, prioritize. But these words tend to be jargon for which meaningful words already exist: complete, rank.

The "*-Ceed*," "*-Sede*," "*-Cede*" Rule

5.38 The rule for these endings is relatively easy to remember: Three English words end in *-ceed* (*exceed, proceed, succeed*), only one ends in *-sede* (*supersede*), others with the same pronunciation end in *-cede* (e.g., *precede*).

Capitalization

Beginning Sentence

5.39 The first word in a sentence is normally capitalized (capped). Capitalize all proper nouns. Caps are also sometimes used for emphasis (as in a *Good* and *Loyal* subject).

5.40 When a complete sentence follows a colon, the first word is normally capped:

> The mainspring was rusted: Excessive forces created the possibility of an equipment breakdown.

Listings

5.41 Depending on the format, listings can be uppercase or lowercase. Ordinarily, listings within a sentence are not capitalized:

> . . . consists of (1) methods and materials, (2) procedures, and (3) test results.

Items in a bulleted or numbered list, however, whether they consist of complete sentences or not, are usually capitalized. Incomplete sentence listings do not need end punctuation; sentence listings do (although they may contain internal punctuation; see also 4.68–4.70):

1. Methods and materials
2. Procedures
3. Test results

But:

1. Methods and materials will be listed on the basis of MILSPEC 15-1802.
2. Procedures follow MILSPEC 28-1717.
3. Test results must be reported to STATSOFF within 72 hours.

Parentheses

5.42 A sentence enclosed in parentheses within another sentence need not use an initial capital letter. However, long or complex parenthetical sentences would be more understandable if rewritten as independent sentences (see also 4.127–4.133).

Quotations

5.43 When quoting a passage, retain the source's capitalization practice. Begin a complete quotation with a capital letter. If the quotation begins in the middle of a sentence, you should so indicate by using ellipses (see also 4.31, 4.56, 4.117–4.125):

> "... using results found in earlier tests. However ..."

Salutations and Closings

5.44 In the salutation and closing of letters, usually the first word only is capitalized:

> My dear friend,
>
> Yours sincerely,

5.45 However, capitalize the noun used in place of a name, or a formal title:

> Dear Aunt Margaret,
>
> Dear Mr. President:
>
> Dear Sir:

Book and Article Titles

5.46 See chapter 8.

Proper Nouns

5.47 One of the more confusing areas of capitalization lies in the use of proper nouns and adjectives. Some cases are uncontroversial:

> American, Farouk, Fourier, Vermont

However, when proper nouns are used to produce a common classification, the tendency is toward lowercase, and sometimes the distinction is not so easy to make:

> afghan robe, dutch oven, italic (or roman) type, turkish bath

5.48 Measurements derived from proper names are usually lowercase—but note that the abbreviated version retains the capital (see also chapter 7):

ampere, A	pascal, Pa	kelvin, K	watt, W
hertz, Hz	tesla, T	newton, N	

5.49 Both elements (see also 6.112) and vitamins (see also 6.80) use initial capital letters in their abbreviations, but their full names are lowercased.

5.50 The typical practice capitalizes

- Formal titles or names, particularly if they refer to specific offices or people

  Note that *naval* is always set in lowercase type.

  Court of Claims, King James, the Senate, the President . . . Army. . . Navy (of the United States)

- Honorary titles and academic degrees, if they follow a proper name

  Frank Smith, Fellow of STC, Paul Nash, Ph.D.

- Organizations names

  Boston Pops℠, Giants, Girl Scouts

- College colors and national flags

  the Black and Orange, Old Glory, Union Jack

- Holidays or other special days

  Christmas, July Fourth, Passover, Thanksgiving, Kwanza

- Names of the months and days of the week

  October, January; Monday, Wednesday

  But the seasons (*spring, summer, fall, winter*) are lowercase.

- Names of races and tribes

  Asian, Caucasian, Indian, Kiowa, Masai

5.51 For information on capitalization practices in specific disciplines, see also chapter 6.

Trade Names, Trademarks, Registered, and Copyright

5.52 Trade names are protected by law, but many fall into common use in casual writing. Xerox™ is an example of a trademark that is casually treated as a generic term even though it is still an active trademark. While people often refer to "a *xerox* copy," the correct term would be "*Xerox* copy." To avoid such difficulties, use a generic term instead:

photocopy or xerographic copy

Over time, companies generally cease protecting trademarks and allow them to pass into the vernacular:

aspirin, yo-yo, linoleum

5.53 Products or corporate names include the trademark symbol ™ if they are not registered with the U.S. Patent and Trademark Office; once they satisfy registration requirements, the registered trademark symbol ® is used. Use the appropriate symbol for the first reference in a text; subsequent references in the same document need no mark. Set these symbols in a smaller type size than body text; set trademark symbols about 50 percent smaller, registered trademark symbols about 60 percent. Note that the registered trademark symbol can be set either superscripted or on the baseline:

Acme Rocket Skateboard™

Acme Rocket Skateboard®

Acme Rocket Skateboard®

5.54 In reporting test results based on specific equipment, use the equipment's trade name in order to make replication of the test completely accurate. Some companies require a disclaimer—often as a footnote—to indicate that mention of a product does not imply endorsement:

Benzedrine™ (amphetamine sulfate)

Demerol™ (meperidine)

Fiberglas™ (fiber glass)

Pyrex™ (heat-resistant glass)

5.55 The copyright symbol ©, an identifier of text ownership, appears on the "verso" of the full title page of a published document. Set this symbol, on the baseline, in approximately 70 percent of the type size for body text:

©2000. All rights reserved.

Computer Documentation

5.56 In computer documentation, capitalized terms are likely to proliferate, if for no other reason than new terms are invented daily. The careful technical editor needs to regulate this invention process and discern precisely which new terms need capitalization.

5.57 In general, limit capitalization to as few categories of terms as possible. For example, you might capitalize only those words that describe commands found in a software application' s menus.

Troublesome Words

Sound (Homonyms)- and Look-Alike Words

5.58 English has a number of words that sound alike, but are not spelled alike or are spelled alike, but do not sound alike. Here are a few common homonyms (words that sound alike but are spelled differently):

beet, beat

cite, site

feet, feat

5.59 English has a number of endings that are the same but with different sounds:

cough, enough, plough, through

5.60 Because the English language is replete with single words that have vast, complex, and occasionally contradictory sets of denotations, the careful writer takes note of diction, using the correct word for a given context. While the primary tool for considering appropriate terms is *Roget's Thesaurus*, a number of specialized thesauri supports specific disciplines.

5.61 Diction is an issue not only for words with multiple connotations, such as *game*, *space*, and *well*, but in the choice of the proper preposition. Problems generally arise because many prepositions can have multiple, overlapping meanings.

For example, the prepositions *as* and *since* can have the same meaning as *because*. However, *as* can refer to duration (*as* we sailed along), and *since* can refer to from a time in the past (*since* the dawn of time). Consider the sentence:

Since the experiment failed, research has been discontinued.

Does this mean that research was discontinued *from the time that* the experiment failed or *because* the experiment failed? In technical and scientific writing, the use of *because* in such sentences substantially increases clarity.

Spelling Variations

5.62 The trend toward simpler spellings can sometimes be confusing (*ax* or *axe*). Most usage authorities recommend using the first (preferred and American) spelling given in a dictionary listing; for example, *Merriam-Webster's Collegiate Dictionary* lists *ax* as the first spelling.

5.63 Avoid some simplified spellings that have entered common parlance via advertising or highway signage:

light not lite

night not nite

through not thru

slow not slo

5.64 Because of an interest in health, everyone knows what *lite* means—whether applied to beer or mayonnaise. Similarly, commuters understand what *thruway* means on a road sign. However, these spellings should not appear in formal writing.

American versus British Spelling

5.65 American and British spelling differences are often found in the endings of words (see also chapter 2):

American	British
center	centre
check	cheque
connection	connexion
defense	defence
humor	humour
judgment	judgement
labor	labour
program	programme
theater	theatre
organize	organise

Easy Misspellings

5.66 Some words, because of letter doubling, pronunciation variations, or assumptions about vowel use, are almost predictable misspellings:

absorbent	commercial	lightning	preventive
accommodate	desiccate	liquefy	privilege
adsorption	embarrass	maintenance	prominent
all right	fluoride	nevertheless	recur
analogous	forty	nickel	seize
arctic	fulfill	occurred	separate
breathe	grammar	parallel	siege
canister	hangar	personnel	specimen
chromatography	irrelevant	preceding	supersede

Foreign Words

5.67 Many foreign words and phrases are in common use. Convention dictates setting uncommon foreign words and phrases in italic type. Words that have been accepted into English (that is, that appear in English-language dictionaries) are set in roman type. The following list provides examples of already accepted words or phrases (see also 6.71–6.76, 6.117, 6.123–6.124):

ad hoc	caveat	modus operandi	prima facie
ad libitum	de facto	par excellence	pro forma
a priori	dossier	passé	quasi
alma mater	en route	per annum	résumé
bona fide	ex officio	per capita	versus
briquette	in situ	per diem	via
carte blanche	in toto	per se	vice versa

6. Incorporating Specialized Terminology

6.1 Scientific and technical communication relies on writers and readers sharing a common language. Use terminology familiar to readers in an accepted and consistent manner. This chapter discusses creating useful abbreviations and provides guidance on using specialized terminology for specific subjects, the meaning conveyed by those terms, and standard conventions used to print those terms:

- Abbreviations as professional shorthand (6.2),
- Organization names (6.40),
- Military terminology (6.58),
- Health and medical terms (6.69),
- Scientific terms and symbols (6.81), and
- Technology terms (6.126).

Abbreviations as Professional Shorthand

6.2 Abbreviated terms—whether abbreviations, initialisms, acronyms, or symbols—save time, space, and needless repetition. However, decisions about acceptable and appropriate abbreviations remain ambiguous in modern publishing. While a few general rules apply, in most cases writers need to be familiar with the rules and conventions recognized by specific usage authorities.

6.3 Initialisms and acronyms are both created from letters found in a compound phrase. Initialisms are read letter by letter (CIA, FDR); acronyms are generally read as if they formed a single word (COBOL, common business oriented language):

Abbreviations	Initialisms	Acronyms
Dr., tsp., mm.	YMCA™,	UNICEF,
	DDE	MOMA™

6.4 Writing in specialized fields or academic disciplines requires familiarity with their specific abbreviation conventions. For example, in writing about business, *ROI* (return on investment) and *COGS* (cost of goods shipped) may be everyday terms; in the study of biology they are not.

6.5 While observing small stylistic differences might seem unnecessarily pedantic, writers and scholars should be aware that reviewers of their work can interpret neglect of these standards as evidence of general unfamiliarity with a profession. The following abbreviations indicate how four highly regarded authorities can differ:

Chicago Manual	*MLA Handbook*	*Words into Type*	*CBE Manual*
Ph.D.	PhD	Ph.D.	PhD
John Jacobs Jr.	John Jacobs, Jr.	John Jacobs, Jr.	John Jacobs Jr
10:30 A.M.	10:30 a.m.	10:30 A.M.	10:30 AM

Despite these differences, most style guides acknowledge the general trend toward eliminating periods in abbreviations of more than one letter, especially in technical and scientific fields. The guidelines presented in this chapter are appropriate for most technical and scientific writing, including some relevant specialized academic disciplines.

Style Guides for Abbreviated Terms

6.6 Organizations that create or use abbreviations often in their work should develop an abbreviation database or glossary to ensure consistency. This database will also serve as the basis for decisions about abbreviations among different disciplines and organizations. Typical disagreements such as the following can be resolved with a large database:

Term	American Institute of Physics	National Aeronautics and Space Administration
National	Natl.	Nat.
Ultrahigh frequency	uhf	UHF (IEEE)
Ultraviolet	UV (ACS, IEEE)	uv
Electromagnetic radiation	x ray	x-ray (ACS™)
ampere hour	A • h or A h	Ah (IEEEsm)

(ACS—American Chemical Society; IEEE—Institute of Electrical and Electronics Engineers)

6.7 Many word processing applications have a function (referred to as *local dictionaries*) that imports special words and abbreviations. If it is not available, specialized electronic dictionaries can be added to most word processors.

6.8 Follow either the abbreviation style of a particular professional organization or use consistent local preferences.

Abbreviated Terms in Technical and Scientific Texts

6.9 Abbreviated terms, initialisms, acronyms, and symbols are common to all technical and scientific disciplines. The following section illustrates the principles underlying the formation of these terms and the stylistic rationale of various organizations.

Initialisms

6.10 Initialisms are created from letters found in a compound phrase and are read letter-by-letter: CIA, FDR. The following list provides examples of initialisms used in scientific fields:

Initialism	Meaning
AAO™	American Academy of Ophthalmology, Inc.
	American Academy of Otolaryngology Head and Neck Surgery, Inc.
CERN	European Center of Nuclear Research
CFC	Chloroflurocarbon
CIC	Combat Information Center
DJIA	Dow Jones Industrial Average

Initialism	Meaning
DoD (DOD)	Department of Defense
fob (f.o.b.)	Free on board
GPO	Government Printing Office
NMFS	National Marine Fisheries Service
PNG	Papua New Guinea
PVC	Poly vinyl chloride

Acronyms

6.11 Acronyms are created from letters found in a compound phrase and are read as though they represent a single word:

UNICEF MOMA

6.12 Nested acronyms contain one or more other acronyms:

flir	forward-looking infrared radar
radar	radio detection and ranging

Unless the intended audience is highly trained, nested acronyms may be confusing and should be avoided.

6.13 Some acronyms have become words in their own right and may be used routinely in technical and scientific text. Some more common acronyms include:

laser	light amplification by stimulated emission of radiation
maser	microwave amplification by stimulated emission of radiation
quasar	quasi-stellar radio source

Indirect Articles and Acronyms

6.14 To decide whether to use *a* or *an* before an acronym, the author must decide on a pronunciation rationale for initialisms:

Initialism	Pronunciation	Indirect Article
CAM	CAM	a
	cee-a-em	an
FET	FET	a
	eff-ee-tee	an

Symbols (see also chapter 7)

6.15 Symbols are a special type of abbreviation. A symbol has often been adopted or created by international organizations such as the *Système Internationale* (SI) or the modern metric system (see also 7.41). Some organizations still have not resolved the difference between symbols and other abbreviated terms.

6.16 Usually a symbol is composed of a single letter or specialized sign selected from the symbol set available for all abbreviated terms, and is not punctuated. Symbols also do not form the plural by adding s; for example, the symbol for gram—g—also stands for grams.

6.17 While symbols composed of letters seem to be classified as abbreviated terms, mathematic symbols do not share the same confusion. They are always classified as symbols.

Arbitrary Changes in Symbols

6.18 Some organizations change symbols depending on where the symbols occur. For instance, the American Society of Mechanical Engineers uses lowercase symbols in text and uppercase in illustrations. This practice leads to the use of notations such as ML on drawings, which in general use indicates megaliter rather than ml—milliliter. The Institute of Electronic and Electrical Engineers has similar guidelines.

These guidelines seem to be aimed at authors who do not have access to special symbol sets. However, these shorthand practices should

- Be limited to internal audiences,
- Be fully explained in a glossary, and
- Not appear in public documents.

Creating Suitable Abbreviations

6.19 Consider the audience's professional background when selecting abbreviations. Also, find out if this audience usually reads documents that conform to an existing style guide. The most often abbreviated types of terms include

Titles	Names of National and Corporate Entities and Famous Individuals	Measurements: Numbers, Amounts, Time	Scholarly and Scientific Terms
Ms., Mr., Dr., Sen., Pres.	CBS℠, NAACP℠, UPI℠, FDR, JFK, GOP™	A.M., P.M., B.C., A.D., tsp., gal., cm., kb	Q.E.D., etc., et al., i.e., e.g.

6.20 The most common abbreviation forms are single-letter, two-letter, and three-letter abbreviations. The customary letters used are first, first two, or first three. Somewhat less frequently, the initial letter and the final letter are used. Occasionally and especially with longer words, additional internal consonants and sometimes vowels are included:

Abbreviation Type

Single-Letter		Two-Letter		Three-Letter	
t.	temperature	Fr.	French	abs.	absolute
T.	absolute temperature	sl.	slight	pos.	positive
		Ct.	Court	vol.	volume

Abbreviating Single Words

6.21 Abbreviations consist of one or more letters of the abbreviated word followed by a period. Including a terminating period depends on an organization's style. For instance, the American Institute of Physics uses *Assoc.* for Association, while a contemporary commercial publisher uses *Assn.*

Abbreviating Two Consecutive Words

6.22 When abbreviating two words that form one concept, leave a blank space between the abbreviated terms (see also 7.48):

at. wt.	atomic weight
sp. gr.	specific gravity

Frequency of Use for Abbreviated Terms

6.23 Spell out words to avoid too many abbreviated terms. In planning abbreviated terms, establish an optimal and maximum ratio for these terms. Excessive abbreviations can turn otherwise readable prose into gibberish. A rule of thumb for writing directed to a lay audience is one acronym per page.

Explaining Abbreviated Terms

6.24 Circumstances demanding presentation of the full word/phrase and its abbreviation include

- First appearance in a text,
- Subsequent appearances in the same text but at some distance (i.e., in a different major section) from the first appearance, or
- In a glossary.

6.25 Identify all specialized abbreviated forms a reader is unlikely to recognize. Accepted units of measure need not be included, but unusual signs or symbols should be defined:

> The World Health Organization (WHO) has established a committee to investigate the recent outbreak.

6.26 Do not create an abbreviation that contradicts an accepted standard. For instance, use micrograms (μg)—based on *Système Internationale* (SI) units (see also 7.41)—rather than mcg, an abbreviation suggested to avoid misreading μg for milligram (mg).

Formatting Abbreviated Terms

6.27 In addition to spelling and capitalization conventions, abbreviations, acronyms, and initialisms use specialized typographic and punctuation conventions.

6.28 Science, engineering, and mathematics also use the Greek alphabet and graphic symbols such as mathematical operation signs, biological indicators, and the like.

Beginning a Sentence with Abbreviated Terms

6.29 Although a sentence does not usually begin with an abbreviated term, it can be started with an initialism:

> AAA™ members can get service for their cars on the road.

6.30 Convert numbers to text:

> 29 g of taconite were found in the water sample.

Becomes:

> Twenty-nine grams of taconite were found in the water sample.

Punctuating Abbreviated Forms

6.31 Place a blank space between a number and the unit of measure or symbol that follows it:

> 3 g
> 5 m

Some organizations, like the American Institute of Physics, remove the space when reporting temperature: 67°C. The American Chemical Society retains the space: 67 °C. The symbols for degree (of angles) and percent share similar confusions.

6.32 The spacing and period usage convention have many exceptions. In the following abbreviations, no space occurs between terms and they always use periods:

e.g.	for example (exempli gratia)
i.e.	that is (id est)
q.v.	which see (quod vide)
v.i.	verb intransitive
v.t.	verb transitive
etc.	and others (et cetera)

Periods in Abbreviated Forms

6.33 Omit periods after unit symbols (symbols representing units of measure, such as metric and SI):

m	meter (meters)
g	gram (grams)

6.34 Some abbreviations, including symbols, are punctuated with periods to avoid confusing them with similar words:

at.	atomic
fig.	figure
in.	inch(es)
no. (NO.)	number (Number)

6.35 This practice creates its own confusions. Weight, for instance, is usually abbreviated as wt, without a period. Thus, atomic weight would be abbreviated as at. wt—without a terminating period. To avoid confusion, both the National Aeronautic and Space Administration (NASA) and the American Society of Mechanical Engineers use at. wt. for atomic weight; however, NASA still abbreviates weight, when used alone, without the terminating period.

6.36 Professional organizations add their own nonpunctuated abbreviations to these lists. Thus, the American Chemical Society uses

sp ht	specific heat
sp vol	specific volume

But also uses

i.d.	inner diameter
o.d.	outer diameter
e.g.	for example (tables and figures)

i.e.	that is (tables and figures)
(e.g.)	for example (in text)
(i.e.)	that is (in text)

The last two terms must be spelled out if they are not placed in parentheses.

Slash in Abbreviated Forms

6.37 In reporting units associated with sound, use the slash (/) to replace the letter p (per):

c/s	cycles per second
c/m	cycles per minute
r/m	revolutions per minute

Apostrophe Plurals of Initialisms

6.38 A plural is often indicated by adding a lowercase *s* to an initialism:

CFCs	chloroflurocarbons
TAs	teaching assistants

Some organizations, however, still prefer an apostrophe before the lowercase s:

CFC's	chloroflurocarbons
TA's	teaching assistants

6.39 Unless required to follow a specific style, reserve the apostrophe for the possessive case:

The AAA's rules require a driver to obey the law.

Organization Names

6.40 Since they are proper names, capitalize all organization names. When present, prepositions take lowercase.

Academic Organizations

6.41 Capitalize the names of universities, colleges, and schools:

City University, King's College, London University, Merchant Tailors' School, Northeastern University, Thomas Aleyn's Grammar School, Trinity College, University of London

6.42 Certain universities are often well known by their abbreviated or mnemonic form:

MIT	Massachusetts Institute of Technology
SMU	Southern Methodist University

6.43 Such abbreviated forms should not be used without an initial definition no matter how well known the institution may be. It is also acceptable practice to refer to extremely well known universities in their simple form:

Harvard, Yale, Princeton

6.44 When a university has campuses in different cities, indicate this as

University of California at Los Angeles (UCLA)

University of California at San Diego (UCSD)

University of Texas at Arlington (UTA)

6.45 Typical terms for different departments or functions within a college or university are

the Faculty of Engineering, the Engineering Faculty, the School of Journalism, the Sloane School of Business Management

Commercial Organizations

6.46 All company names, from large corporations to small independent consultants, are registered with some level of government agency.

6.47 The official form of company names may include the company's legal status, such as *Incorporated (Inc.)* or *Limited (Ltd.)*. Generally, company names use initial capital letters. However, many companies may use lowercase and specialized typography, punctuation, and other devices for legal or marketing reasons.

6.48 Many companies also have a registered abbreviated form of their company name:

GM™	General Motors Corporation
IBM™	International Business Machines Corporation

6.49 Beginning a sentence with an acronym or initialism is permissible:

IBM introduced its newest mainframe computer today in New York City.

However, some style guides, notably the American Medical Association, do not allow such usage. The following sentence is unacceptable:

TSS has been identified in four commercial products.

6.50 Many companies have separate operating companies or divisions. These will be apparent in the name:

AT&T Wireless Services℠, Bell South Mobility™

6.51 Indicate legal status after the company name:

PPG Industries™, Inc.

Note particularly the placement of the comma with the designation *Inc.*

6.52 Common legal designations for companies in European countries include the following, most of which have the meaning of Limited or Incorporated:

Country	**Incorporated Designation**
Belgium, Netherlands	NV and BV
Belgium, France, and Switzerland	SA
Britain	plc (in lowercase)
Germany	GmbH (in mixed case)
Scandinavia	AB

For example:

EC Erdölchemie GmbH, Reed International plc

6.53 The definitive reference source for company names is literature issued by the companies. Caution may be necessary, because some companies have an inconsistent corporate image. Standard trade directories are also a good reference source.

6.54 Stock exchange listings should not be used as reference since they identify companies by marketing codes that have no legal value. The EDGAR Database, maintained by the United States Securities and Exchange Commission (SEC), provides information about companies required to file forms with the SEC.

Government and Legislative Names

6.55 Capitalize common names for departments and offices when they form part of the name of a government or legislative agency. Where acronyms or abbreviations exist, omit periods. Examples of government and legislative names include

- Congress, Department of Defense (DOD), Department of External Affairs, the Foreign Office (British), the House of Commons, the Commons (colloquial), the House, this House (in direct speech), House of Representatives, Parliament, the Senate

- Canadian Security and Intelligence Service (CSIS), Central Intelligence Agency (CIA), Federal Bureau of Investigation (FBI), MI5 (Military Intelligence 5, British)
- United Nations (UN), United Nations Educational and Scientific Council (UNESCO), World Health Organization (WHO)

6.56 Normally, include the full name of a department, office, or corporation within the text and add the abbreviation or acronym immediately after that citation in parentheses:

> The National Aeronautics and Space Administration (NASA) has primary responsibility for spacecraft exploration.

After this reference, the abbreviated term can be used routinely unless a considerable amount of text occurs between references—typically, more than a chapter. In that case, repeat the full citation of the abbreviated term as a reminder for the reader.

Societies and Associations

6.57 Names for societies and associations take the following forms. Many of these organizations also have abbreviated forms of their name:

> Boy Scouts of America (BSA™), the British Medical Association (BMA), the Institute of Electrical and Electronic Engineers (IEEE℠) (often shown incorrectly as IE^3), the National Geographic Society (NGS), the Society for Technical Communication (STC)

Military Terminology

6.58 Clear communication is of the utmost importance in the military. Terminology is therefore carefully controlled and frequently revised.

Military Standards Related to Terminology

6.59 A number of United States military standards govern Technical Manuals Specification and Standards (see also 13.6), for example:

MIL-STD-2301	Computer Graphics Metafile (CGM) Implementation
MIL-STD-40051	Preparation of Digital Technical Information for Multi-Output Presentation of Technical Manuals

6.60 Use the adjectival form for generic descriptions of the armed forces of other countries:

> the French Navy, the Indian Air Force, the Israeli Army

Military with a United Nations mandate are referred to as UN Forces or United Nations Forces. It is not meaningful to refer to the UN Air Force or similar unit.

6.61 The main services have specific organizational units: army, battalion, corps, detachment, division, fleet, regiment, and squadron. When used as part of a proper name of a military unit, capitalize these terms; numeric identifiers are also commonly used:

Army of the Rhine

8th Army or Eighth Army

84th Foot Regiment, Royal Highland Emigrants

501st Parachute Battalion

U.S. Third Fleet

6.62 Mnemonic forms identify some units:

4CMBG 4th Canadian Mechanized Brigade Group

Names for such units may be further abbreviated by using the cardinal form of number instead of the ordinal:

4 Canadian Mechanized Brigade Group

Aircraft Terminology

6.63 Aircraft designators include

F-117A, FB-11A, CF-18 (C indicates Canadian), B-52D, TU20, Vulcan, MiG21 (note lowercase *i*), Mirage F2, Apache helicopter

Note plural form with lowercase *s* and no apostrophe:

F-111s, CF-18s

Armament

6.64 Types of armament include

Armament	**Initialism**
air-to-air missile	AAM
air-to-surface missile	ASM
anti-ballistic missile	ABM
intermediate-range ballistic missile	IRBM
medium-range ballistic missile	MRBM
multiple independent re-entry vehicle	MIRV

(continued)

Armament	Initialism
multiple re-entry vehicle	MRV
submarine-launched ballistic missile	SLBM
surface-to-air missile	SAM

Naval Vessels

6.65 Warships, the only military equipment with individual names, italicize those names, but their weapon designations, such as USS, are set in roman type:

USS	United States Ship	USS *Cuttlefish*
HMS	His/Her Majesty's Ship	HMS *Ark Royal*
HMCS	His/Her Majesty's Canadian Ship	HMCS *Endeavour*

6.66 Use lowercase for the common types of vessels:

aircraft carrier, battleship, cruiser, destroyer, frigate

6.67 Indicate vessel classes as

Ticonderoga-class cruiser

Tribal-class destroyer

Yankee-class submarine

Surveillance Terms

6.68 Some of the terminology for surveillance equipment is well established, even in lay language. For example, the term *radar* is a prime example of an acronym term that has become a common noun. Generic terms for surveillance equipment include:

radar	radio detection and ranging
sonar	sound navigation ranging system

The term radar may be qualified as

- Airborne radar,
- Land-based radar,
- Over-the-horizon back scatter radar,
- Ship borne radar, and
- Space-based radar.

Examples of complete radar systems and installations include

AWACS	airborne warning and control system aircraft

Health and Medical Terms

6.69 The medical profession uses Latin and Latin-derived terms extensively. Because the terminology is standard for the discipline, the terms are generally set in lowercase roman. Examples of typical health and medical terms include

tibia, thorax, appendectomy, tracheotomy, pulmonary vein, gluteus maximus

Plurals are formed according to the rules of Latin grammar:

vertebra vertebrae

6.70 Some medical vocabulary—such as heart, eye, ear, lung, kidney, liver, and brain—are everyday terminology.

6.71 Some Latin terms should be set in italics to avoid confusion with the word *in*; however, base any italic decision on the term's dictionary treatment. Italicize *In articulo mortis*, but not in vitro, in vivo, and in utero.

Anatomical Terms

6.72 Set anatomy terms in lowercase roman unless they include a proper name:

thyroid, pancreas, islets of Langerhans, aorta, ventricle, Achilles tendon

Diseases and Conditions

6.73 Set terms for diseases and medical conditions in lowercase roman unless they include a proper name:

axanthopsia, anemia, apoplexy, dyslexia, rheumatic fever, Alzheimer's disease

Drug Terminology

6.74 Use generic names for medical drugs and set them in lowercase roman. Examples of generic names include

aspirin, digitalis, tetracycline, diazepam

6.75 If proprietary drugs must be referred to in print, capitalize the trade name or brand name:

In North America, Valium™ is probably one of the most widely prescribed drugs in the treatment of anxiety disorders.

Infectious Organisms

6.76 Capitalize and italicize names for infectious organisms (such as bacteria):

Bacillus, Staphylococci, Streptococci

When using the name as an adjective, set it in lowercase roman:

Staphylococci cause acute staphylococcal enterocolitus, an infection of the gastrointestinal tract.

6.77 Abbreviate members of the genus:

Escherichia coli becomes *E. coli* or *Esch. coli.*

Staphylococcus epidermidis becomes *S. epidermidis.*

Use the abbreviated term in most text:

Most *E. coli* strains ferment lactose readily.

Medical Equipment

6.78 Set terms for common medical equipment such as stethoscope and centrifuge in lowercase roman. As with most electronic equipment, terminology for advanced medical equipment involves many initialisms:

CT scan	computerized tomography (phonetically corrupted to CAT scan)
EEG	electroencephalogram

Medical Procedures and Treatments

6.79 Set terms for medical procedures and treatments in lowercase roman unless they include a proper name:

chemotherapy, appendectomy, Coombs' test, Credé's method

Vitamins and Minerals

6.80 Refer to vitamins by their name or letter convention. Names are lowercase; letter designations are uppercase. Distinguish the vitamin B complex by numeric suffixes. In typeset text, set the numeric suffix as a subscript: vitamin B_6. If subscripting is not available, the number may be set in line: vitamin B6. The following table lists vitamins in both forms; the alphabetic form is generally better known:

Some Common Vitamins and Their Designation

retinol	A	folic acid	B_c
thiamine	B_1	complex	B
riboflavin	B_2	ascorbic acid	C
niacin	B_3	calciferol	D
pyridoxine	B_6	riboflavin	G
cobalamin	B_{12}	biotin	H

Set the names of minerals and trace elements in lowercase roman:

calcium, phosphorous, potassium, iron

Scientific Terms and Symbols

6.81 Scientific terminology is well established because of a long history of scientific discovery and a tradition of published works. Some areas of scientific terminology are closely related to symbols and notations.

6.82 Science requires well-defined symbols to convey precise information. These symbols, often only a single character, can be grouped together in specific ways to describe a complex idea or operation (see also chapter 7).

Astronomical Terms

6.83 Many names come from Latin and Greek mythology. Often, entities are named for their discoverers.

Planets

6.84 The term *planet* generally refers to the nine bodies within our solar system. However, the general definition of a planet as a "small cold body orbiting a star" means that other planets can exist.

6.85 Capitalize the names of the nine planets of the solar system:

Mercury, Venus, Earth, Mars, Jupiter, Saturn, Uranus, Neptune, Pluto

Note the term *Earth* may be seen without an initial capital since it has become a common noun. However, for clarity and consistency, capitalization is recommended.

6.86 Capitalize the names of planet features:

Nix Olympia, Pavonia Mons, Plain of Chryse, Valles Marineres

Moons

6.87 Moons are planetary satellites. Earth's moon is referred to as *the Moon.*

All planets except Mercury and Venus have moons. In generic form, lowercase is used:

The moons of Saturn are . . .

All moons are named and the names are capitalized. For example:

the moons of Jupiter are Callisto, Europa, Ganymede, and Io

6.88 Capitalize features—craters, seas, mountains, radial ridges, rills, and valleys—on the Moon's surface:

Alpine Valley, Hadley Rill, Jura Mountains, Rook Mountains, Teneriffe Range

Some names derive from Latin or are formed in Latin style. Note *Mare* (sea) and plural form *Maria* (seas), and *Oceanus*:

Mare Nubium, Maria Crisium, Maria Smythii

Craters often have single names such as:

Copernicus, Lambert, Tycho

Comets

6.89 Comets are often named for their discoverers, hence treat the discoverer's name as a possessive adjective in the comet's name:

Halley's comet, Tycho Brahe's comet

A number of comets are not named for a discoverer. The usual form for these comets is the comet of 1843, the great comet of 1882. The term *great* is ascribed to a comet with a large tail that also approached close to Earth.

Meteors

6.90 Although individual meteors are unnamed, significant recurring meteor showers are named:

the Perseid, the Leonids

Note that the term *meteor shower* is omitted; it is understood from context.

Constellations

6.91 Capitalize the names of constellations (groups or configurations of stars). Many of these names come from mythology because their shape suggested a particular mythological figure:

Scorpius, Sagittarius, Capricornus, Ursa Major, Ursa Minor, Ophiuchus, Orion, Corona Borealis

Note the vernacular forms for some constellations:

Big Dipper, Little Dipper

Stars

6.92 Capitalize the names of stars:

Alpha Centauri, Arcturus, Polaris, Rigel, Sirius

6.93 Although many stars do not have individual names, there are conventions that uniquely identify them. For example, the traditional Bayer method for "naked eye" stars uses the Latin genitive form of the constellation name prefixed by a Greek identifier:

α-Centauri and β-Centauri

These identify the first and second stars in the constellation of Centaurus.

6.94 Newer systems identify fainter stars (not visible to the naked eye) and variable stars (which fluctuate in brightness). A common system uses roman capital letters. Historically, because the letters A and Q are already assigned to southern constellations, use R as the first letter:

1st star	R
2nd	S
. . .	
9th	Z

Thus:

U Geminorium is the fourth star in Gemini

R Scuti is the first star in Scutus

Using double letters in the range R to Z extends this system:

10th star	RR
11th	RS
. . . . . .	
18th	RZ
19th	SS
20th	ST
. . . . .	
54th	ZZ

Thus:

ST Cygni is the 20th star in Cygnus.

For further discrimination, use the double letters in the range AA to QZ:

55th star	AA
56th	AB
....	
334th	QZ

6.95 For new variables an open (unlimited) system of identification that reports stars as V335 Cygni, which indicates the 335th variable star in Cygnus, has been adopted.

6.96 Another common system of star identification, the Harvard observatory system, uses six digits to identify a star:

- The first four digits record the hours and minutes of ascension relative to star positions as of the year 1900.
- The next two digits describe the angle of declension in degrees.
- Roman digits indicate north, italic digits indicate south.

In this system, a designation such as Leonis 094211 indicates a star at an ascension of 9 hours 42 minutes and declension 11 North relative to the star Leonis (the first star in Leonis).

Classes of Stars

6.97 Stars are often classified in terms of their spectral radiation with the capital letters O, B, A, F, G, K, M, N, R, and S. A star may therefore be referred to as an M or a K star.

Novas

6.98 Novas are named in the Latin genitive form. Thus, a name such as Nova Persei indicates a nova in Perseus.

Galaxies

6.99 Galaxies have a common naming system and alphanumeric naming systems that are cataloged. Galaxies in the common naming system include Draco, Leo I, and Leo II systems.

6.100 Because of the immense number of objects in the universe, catalogs have been essential throughout the history of astronomy. The following catalogs provide the naming conventions in use:

Catalog	Abbreviation
Bonner Durchmusterung	BD
Cordoba Durchmusterung	CD
Flamsteed	
Henry Draper	HD
Index Catalog	IC
Messier	M
New General Catalog	NGC
Wolf	

6.101 Two catalogs of clusters, galaxies, and nebuli are still in common use: Messier (M) and the New General Catalog (NGC). Professional astronomers generally prefer Johan Dreyer's New General Catalog (NGC), which contains objects not cataloged by Messier. The Index Catalog (IC) supplements the NGC:

Comparison of Galactic Cataloging Systems

Messier Number	NGC Number	Name	Object Type
1	1952	Crab Nebula	supernova remnant
8	6523	Lagoon Nebula	nebula
31	224	Andromeda Galaxy	spiral galaxy
44	2632	Praesepe	open cluster
57	6720	Ring Nebula	planetary nebula
88	4486	Virgo A	elliptical galaxy

Modern catalogs cite objects in both the Messier and NGC systems. Objects traditionally named from other catalogs would appear as: 66 Cygni, 72 Ophiuchi (66 and 72 represent Flamsteed numbers).

Biological Terms

6.102 All life is cataloged in a hierarchy:

- Phylum
- Class
- Order
- Family
- Genus
- Species

6.103 Capitalize terms for phylum, class, subclass, order, family, and genus; set species in lowercase. For example, the common European squirrel would be classified as

phyla	**class**	**order**	**family**	**genus**	**species**
Vertebra	Mammalia	Rodentia	Sciuridae	Sciurus	vulgaris

Chemistry Terminology

Elements

6.104 Set the full names of chemical elements in lowercase. Represent elements by their chemical symbols, such as C for carbon and Au for gold. The symbols are uppercase, or uppercase and lowercase, roman.

6.105 Most element names are Latin. The more common elements—gold, silver, copper, tin, iron—are based on the vernacular. Note, however, that the symbols for these elements come from the Latin form of their names.

6.106 Chemical symbols are universally used in the equations of chemical reactions. They are never used in line in text.

6.107 The following list includes the full names of the chemical elements, as of August 1999, and their chemical symbols. Note that even though some of the elements are named for famous scientists or famous research institutes, the full names of the elements are not capitalized:

Symbol	Name	Symbol	Name	Symbol	Name
Ac	actinium	Ba	barium	Ce	cerium
Ag	silver	Be	beryllium	Cf	californium
Al	aluminum	Bh	bohrium	Cl	chlorine
Am	americanium	Bi	bismuth	Cm	curium
Ar	argon	Bk	berkelium	Co	cobalt
As	arsenic	Br	bromine	Cr	chromium
At	astatine	C	carbon	Cs	cesium
Au	gold	Ca	calcium	Cu	copper
B	boron	Cd	cadmium	Du	dubnium

Symbol	Name	Symbol	Name	Symbol	Name
Dy	dysprosium	Mo	molybdenum	Sb	antimony
Er	erbium	Mt	meitherium	Sc	scandium
Es	einsteinium	N	nitrogen	Se	selenium
Eu	europium	Na	sodium	Sg	seaborgium
F	fluorine	Nb	nobium	Si	silicon
Fe	iron	Nd	neodymium	Sm	samarium
Fm	fermium	Ne	neon	Sn	tin
Fr	francium	Ni	nickel	Sr	strontium
Ga	gallium	No	nobelium	Ta	tantalum
Gd	gadolinium	Np	neptunium	Tb	terbium
Ge	germanium	O	oxygen	Te	technetium
H	hydrogen	Os	osmium	Te	tellerium
He	helium	P	phosphorus	Th	thorium
Hf	hafnium	Pa	prothactinium	Ti	titanium
Hg	mercury	Pb	lead	Tl	tallium
Ho	holmium	Pd	palladium	Tm	thulium
Hs	hassium	Pm	promethium	U	uranium
I	iodine	Po	polonium	Uub	ununbium
In	indium	Pr	praesodymium	Uuh	ununhexium
Ir	iridium	Pt	platinum	Uun	ununnilium
K	potassium	Pu	plutonium	Uuo	ununoctium
Kr	krypton	Ra	radium	Uuq	ununquadium
La	lanthanum	Rb	rubidium	V	vanadium
Li	lithium	Re	rhenium	W	tungsten
Lu	lutecium	Rf	rutherfordium	Xe	xenon
Lr	laweencium	Rh	rhodium	Y	yttrium
Md	mendelevium	Rn	radon	Yb	ytterbium
Mg	magnesium	Ru	ruthenium	Zn	zinc
Mn	manganese	S	sulfur	Zr	zirconium

Chemical Compounds

6.108 Form the names of chemical compounds from their combining elements. Lowercase each word in the term for the compound. Do not hyphenate these terms:

copper sulphate

ferrous oxide (not iron oxide)

ferric oxide

6.109 Use the prefixes *bi, di, tri, tetra,* and *penta* to indicate chemical combinations in units of two, three, four, and five:

bi, di	2
tri	3
tetra	4
penta	5

Other prefixes include *meta,* meaning *above* or *beyond.* There is no numeric value associated with *meta.*

6.110 Lowercase the common names of compounds:

sulfuric acid, hydrochloric acid, prussic acid, caustic soda, potash, bleach

Chemical Notation

6.111 Chemical notation has a compact symbology and well-defined conventions. Since conventions change, follow the current ones. For instance, mass number once appeared at the upper right of the element symbol: U^{235} represented the fissile isotope of uranium-235. It is now written as ^{235}U.

6.112 In chemical notation, depending on the amount of information needed about an element, the symbol for the element could conceivably carry an attached notation at all four corners:

1. Indicate the mass number of an atom by a leading superscript:

   ^{1}H stands for mass-1 hydrogen (one proton in its nucleus)

   Mass-2 hydrogen, with one proton and one neutron in its nucleus, also exists in nature, and it would be indicated by

   ^{2}H (although it is preferably labeled as D, for deuterium)

   A third isotope of hydrogen, a mass-3 hydrogen atom (one proton and two neutrons in its nucleus), does not exist in nature but can be made in the laboratory; its designation is:

   ^{3}H (although normally it is labeled T, for tritium)

2. Reserve the leading subscript position for the element's atomic number. For hydrogen, this number is 1 (${}_{1}H$), which is usually omitted. The three hydrogen isotopes have the same atomic number (1, for the proton each has in its nucleus) but different mass numbers (they have 0, 1, or 2 neutrons in their nucleus). A few other examples of chemical element symbols with their atomic numbers are

   Carbon (${}_{6}C$)

   Sulfur (${}_{16}S$)

   Uranium (${}_{92}U$)

3. Use the trailing superscript position (upper right) for any of three different indicators:
   - Ionic charge, indicated by plus or minus signs (Na+, Ca2+, O3-)
   - Excited electronic state, indicated by an asterisk (He*, NO*)
   - Oxidation number, indicated by a roman numeral (PbII, PbIV)

4. Attach the trailing subscript 2 to the H (as in H_2O) to indicate the number of atoms in the molecule; the lack of a subscript on the O indicates there is only one oxygen atom.

Formulas

6.113 Chemical reactions are described by formulas written in chemical symbols for the reacting elements and compounds:

$$Cu + H_2SO_4 = CuSO_4 + H_2$$

Set the numeric suffixes (such as H_2) as subscripts; set numeric prefixes on the baseline (such as $2H_2$).

Geology Terms

6.114 Names for generic geological features are considered common nouns, hence set them in lowercase:

> anticline, fault, graben, igneous rocks, incline, magma, monocline, pyroclastic rocks, rift, strata

6.115 Names of minerals are lowercase, except for some vernacular forms that may contain proper names. One such example is *Blue John* (a specific type of fluorspar). Typical mineral names include

> calcite, feldspar, magnetite, mica, pyrite, quartz, silica

6.116 Geologic eras, periods, and revolutions are all proper nouns, hence they are capitalized:

- Eras—Precambrian, Paleozoic, Mesozoic, Cenozoic;
- Periods—Permian, Carboniferous, Devonian, Silurian, Ordovician, Cambrian;
- Revolutions—Alpine, Hercynian, Caledonian, Charnian, Karelian.

Physics Terminology

6.117 Like other scientific fields, physics relies on a standard set of symbols that include

A	Area	l	Wavelength
a	Attenuation of coefficient	*m*	Mass
c	Speed of light in a vacuum	*P*	Power
g	Shear strain	*r*	Radius
d	Diameter	*R*	Resistance (electrical)
d	Damping coefficient	r	Density (mass/unit volume)
E	Energy		
e	Linear strain, relative elongation	*t*	Time
F	Force	*T*	Temperature
g	Acceleration due to gravity	*V*	Voltage
h	Height	*W*	Work
I	Current (electrical)	*w*	Angular velocity
K	Bulk modulus of a material	(x, y, z)	Space coordinates

Laws, Principles, and Theories

6.118 Since many laws, principles, theories, equations, constants, and hypotheses are named for their originators, use capitalization and apostrophes:

> Bernoulli's Principle, Heisenberg's Uncertainty Principle, Newton's Laws of Motion, Newton's First Law

6.119 Other laws and principles, named for the topic or concept to which they relate, are set in lowercase:

> the laws of thermodynamics, the principle of conservation of energy, the second law of thermodynamics

Phenomena

6.120 Phenomena are named for their discoverer:

> Hall effect, van der Waal's force

Atomic and Nuclear Particles

6.121 In atomic and nuclear physics, identify types of radiation and particles with letters of the Greek alphabet:

> α-particle radiation (alpha-particle radiation)
>
> β-radiation (beta radiation)

6.122 Set the names for atomic and nuclear particles in lowercase roman. Use correct Latin grammatical forms:

> atom, nucleus, nuclei (plural), nucleons (collective term for protons and neutrons), proton, neutron, electron, positron, neutrino, antineutrino

6.123 Identify quantum numbers *j*, *l*, *m*, and *n* and set them in italic. If italic is not available, they can be set in roman without loss of meaning:

> The *l* quantum number determines the shape of the orbit.

6.124 Identify electron shells with the letters *k*, *l*, *m*, *n*, and *p*. (The letter *o* is omitted to prevent confusion with the digit *0*.) Set shell identifiers in italic:

> The *k* shell has only two available electron orbits.

6.125 Do not hyphenate quantum number and shell identifiers.

Technology Terms

6.126 High technology often creates new terminology to apply to new concepts, techniques, devices, and processes.

6.127 Major problems in technological terminology include

- Numbers of acronyms, abbreviations, and mnemonics;
- Multiple usage of abbreviations and acronyms;
- Obscure term origins; and
- Borrowing terms from other disciplines.

Each time a reader encounters poor terminology, comprehension declines and the reader's performance suffers. For example, consider the multiple meanings of the abbreviation *PCB*: polychlorinated biphenyls, printed circuit board, and process control block.

Similarly, the term *spool* occurs often in computing. While we write about *spool files*, we also use the term as a verb:

> The Print command *spools* the file to the printer.

The term *spool* was devised as an *initialism*: Slow Peripheral Operation On Line. But it became widely used and changed from an *initialism* to an acceptable word.

6.128 Professional organizations and standards organizations provide useful resources for selecting appropriate terminology and nomenclature. For information on these groups see also 13.6.

Computer Terms

Trends in Terminology

6.129 Trends in terminology are toward simplification. Such simplification combines words and uses lowercase letters.

6.130 Many of the terms used in computer technology are compound words, and they are commonly hyphenated:

> log-in, log-on, log-off; on-line, off-line, in-line; sign-in, sign-on, sign-off; up-load, down-load, cross-load

However, continued change has made many of these hyphenated words into composite terms. This method works well for simple familiar terms such as those just given:

> login, logon, logoff; online, offline, inline; signin, signon, signoff; upload, download, crossload

6.131 Be careful not to create awkward or incomprehensible terms. For instance, terms such as *cross-compilation* or *binary synchronous transmission* would not be combined.

Hardware Terminology

6.132 When used in a generic sense, computer hardware terms are set in lowercase roman. Such equipment terms include

central processor	high-speed laser printer	microcomputer
disk drive	ink-jet printer	minicomputer
flatbed scanner	laser printer	network router
hard disk	laser-jet printer	workstation

6.133 Identify specific computers by the manufacturer's name followed by the model designation:

Dell Precision™ Workstation 610	Data General™ AViiON AV 8700
Apple Power Macintosh™ G3	Compaq ProLiant™ 7000

Software Terminology

6.134 When writing in generic terms about computer software, use lowercase roman. Examples of such generic terms include

assembler	executive	operating system
compiler	interpreter	payroll program
cross-compiler	linking-loader	spreadsheet
database	loader	utility
decompiler	monitor	

Commercial Programs and Application Packages

6.135 Commercial software has legally registered names. The correct form for these names can often be found in trade journals. Manufacturers' literature offers a more definitive source. The manufacturer's name is not always an integral part of the software product name. Some well-known names for operating systems, for instance, include

Mac OS™ 9	Apple Macintosh operating system
OS/400™ Version 4	IBM OS/400™ operating system
Tru64 UNIX	Compaq™ UNIX operating system
UNIX	AT&T microcomputer operating system
Windows™ NT	Microsoft™ Windows operating system

6.136 It is common to encounter constructions such as *UNIX-based*:

The XYZ system will be UNIX-based.

Restructure such sentences as

The XYZ system will be based on the UNIX operating system

Personal Computer Software

6.137 Some of the better-known business application packages for personal computers, and their orthographic conventions, include

Excel™, Lotus Notes™, and PageMaker™

Mainframe Computer Software

6.138 Application software for mainframe computers uses many initialisms and acronyms as product names:

DCF	Document Composition Facility
GML™	Geometric Modeling Language
RPG/400™	IBM Programming Language

Releases and Versions of Products

6.139 A software release or version number frequently appears after the product name:

FileMaker™ Pro 5.1

Lotus SmartSuite™ 9.5

QuarkXpress™ 4.04

WordPerfect™ 9

Version numbers in product names do not always follow a simple progression:

Microsoft Office™ 95, Microsoft Office™ 97, Microsoft Office™ 2000

6.140 A common release or version convention for mainframe software takes the form xx.yy.zz:

	Release Level	Function
xx	feature	add new features
yy	enhancement	introduce new functions
zz	maintenance	fix software problems

For example, MML Rel 4.03.05 indicates that the MetaMarkup Language product's latest release is

- Feature 4
- Enhancement 3
- Maintenance 5

Computer Commands

6.141 Use the same typographic conventions found in the interface in documentation. For instance, if commands must be entered in uppercase, show them in uppercase in the documentation. This practice is particularly important in procedures and examples of user-system dialogue.

6.142 When referring to commands in a generic sense, capitalize them:

The Query Job command returns the status of the next job in the XYZ system queue.

Computer Messages

6.143 Set system messages—error messages, prompts, and information messages—in the same style as they appear on the computer display; screen and document text should be identical:

Please enter your PIN number and click: 'OK'

Software Development Documentation

6.144 Software documentation—design specifications, system and program descriptions, flowcharts, and similar information—describes the software's structure. In such documentation, use specific conventions to refer to programs, subroutines, modules, data entities, and states. Develop a style guide or adapt an existing one to promote consistency.

6.145 For graphical user interfaces select typographic conventions for user objects—buttons, dialog boxes, and menu items—and use those conventions throughout the product.

6.146 Be consistent with the verbs that describe user actions. For example, do not change "click on the **Open** button" to "click the **Open** button" or "select the **Open** button."

6.147 Set all common terms in lowercase:

buffer	method	queue
code	operand	routine
interrupt	operator	stack
list	process	thread
member	program	

6.148 When referring to these common terms by proper name, capitalize them:

> When an interrupt is received, control is transferred to the Clock subroutine. Bit 1 of the Program Status word is reset and Clock passes control to the Scan 1 program.

6.149 Use the solidus (/) in terms such as:

> input/output, read/write, and read/compare/write

6.150 Capitalize data names—record, field, and bit names:

Baud Rate field	Frames Received record	Status bit
Device Address field	Reset bit	X record

Programming Languages

6.151 There are many programming languages. Different languages work best for particular types of applications. Set names for programming languages uppercase:

> Pro*COBOL™, Pro*FORTRAN™

Few of these names are true acronyms.

6.152 Modern trends set programming languages in initial caps only, although some languages insist on all uppercase. The following list illustrates some mandatory capitalization:

Algol	Algorithmic language
Basic	Beginners' all-purpose symbolic instruction code
Visual C++	Systems programming language originally developed for the implementation of the UNIX operating system
Lisp™	Symbolics, Inc. list processing language
Turbo Pascal™	High-level programming language named for the mathematician Blaise Pascal
PRO*REXX™	CNS executive programming language

6.153 Most programming languages have *reserved* words that have a specific predefined meaning. Reserved words are set in text based on the conventions for the programming language. In a language such as FORTRAN, all reserved words are uppercase:

LET, PRINT, FOR, WHILE, GOTO

In other languages, such as Algol and PostScript™, reserved words are lowercase:

Algol—begin, end

PostScript—findfont, setfont, scale

Procedures

6.154 Precise terminology is particularly important in procedures. For example, consider the common term *login.* This term has several synonyms: *logon* and *signon* or *signin.* It is not too important which term is chosen as long as consistency is preserved. If *login* is used, then *logout* should be used. If *signon* is used, then *signoff* should be used.

6.155 Be consistent when using imperative verb forms such as enter, press, type, hit.

6.156 When writing procedure titles, use action titles rather than nominalizations. Thus, do not use such titles as

Printer option settings

System software initialization

Instead, use procedure titles that contain an action verb. Use the friendlier gerundive (*-ing*) form for documents such as user guides, and the imperative form for procedural instruction documents such as installation and troubleshooting manuals:

Imperative	Gerundive
Set printer options	Setting printer options
Initialize system software	Initializing system software

Electronics

6.157 Since electronics advances rapidly, terms often come into common usage before they appear in dictionaries. Useful sources for new terminology are trade journals and manufacturers' literature.

Circuit Technologies

6.158 The two genres of electronic circuits are analog (linear) and digital. Analog systems include audio systems, television and video systems, radar, and oscilloscopes. Digital systems include computers, microprocessors, computer terminals, and calculators. Digital technology has expanded into areas that have been traditionally analog, such as audio systems and telephone systems.

Passive Electronic Components

6.159 These electronic components do not perform active functions such as switching electronic signals. Passive components include

capacitors	inductors
coils	resistors
couplers	thermistors
filters	transformers

6.160 Normally, if a qualifier precedes these terms, neither term is capitalized. Such terms include

autotransformer	low-pass filter
band-pass filter	power transformer
ceramic capacitor	signal transformer
decoupling capacitor	smoothing capacitor
electrolytic capacitor	tantulum capacitor
film capacitor	twin-T filter
high-pass filter	variable capacitor
intermediate frequency transformer	variable resistor

If a proper name precedes any of these terms, capitalize that name:

Tchebychev filter

Active Electronic Components

6.161 Active components perform circuit functions such as amplifying and switching. Do not capitalize names for active components. If a qualifier that is not an acronym precedes these terms, capitalize neither term:

avalanche diode	silicon controlled rectifier
clamping diode	switching diode
diode	thyristor
field-effect transistor	transistor
junction transistor	travelling-wave tube
klystron	tunnel diode
magnetron	varactor diode

If a proper name precedes any of these terms, capitalize that name:

Zener diode, Gunn diode

If a mnemonic precedes any of these terms, capitalize the mnemonic:

PNP transistor

6.162 Common initialisms for active devices include:

Initialism	Device
CMOS	complementary metal-oxide-silicon field-effect transistor
FET	field-effect transistor
IGFET	insulated gate field-effect transistor
JFET	junction field-effect transistor
MESFET	metal-silicon field-effect transistor (no oxide)
MOSFET	metal-oxide-silicon field-effect transistor
NMOS or nMOS	N-channel metal-oxide-silicon field-effect transistor
PIN diode	P-type intrinsic N-type diode
PMOS or pMOS	P-channel metal-oxide-silicon field-effect transistor

Logic Circuits

6.163 Logic circuits are used extensively in digital systems such as microprocessors and computers. Acronyms identify a number of existing logic systems:

ECL	emitter-coupled logic
LSTTL	low-power Schottky TTL
TTL	transistor-transistor logic

6.164 Logic signals have two values based on symbolic logic: *true* and *false.* Capitalize these terms (TRUE). Reserve this convention for acronyms and signal mnemonics. In circuit descriptions and logic dictionaries, a typical sentence or statement in this style would be

> When RST is TRUE, the buffer contents are set to zero.

6.165 Another convention refers to logic signals as High and Low. Again, capitalize these terms (HIGH). The abbreviations Hi and Lo are acceptable in laboratory notes but not in formal writing. In circuit descriptions and logic dictionaries, a typical sentence or statement using this convention would be:

> When RST is HIGH, the buffer contents are set to zero.

6.166 Most logic elements have specific terminologies. The following list shows the accepted terms (and their capitalization) for these elements:

BCD decoder	full-adder
BCD-to-binary converter	JK flip-flop
Buffer	Latch
D flip-flop	XOR (Exclusive OR) gate
flip-flop	

Integrated Circuits

6.167 Integrated circuits (ICs) are complete circuits fabricated on a silicon chip. In contrast, circuits that use individual components are referred to as discrete component circuits. The terms used to describe the power, complexity, and type of integrated circuits are

ASTC	application-specific circuit
LSI	large-scale integration
VLSI	very large scale integrated circuits

The components fabricated in silicon integrated circuits include diodes, transistors, resistors, capacitors, and the logic elements mentioned in 6.163. The terminology for these devices is in 6.157–6.166.

Telecommunications

6.168 Modern telecommunication systems are computer-controlled. Many of the terms come from computer and electronic terminology. Since it is necessary to connect telecommunication products and systems, the industry follows international standards. For more information on these standards see 13.6.

Telecommunication Abbreviations and Acronyms

6.169 Telecommunications uses many abbreviations and acronyms. The following list offers a sample of typical acronyms:

Bell operating company	BOC
dynamic network architecture	DNA
integrated services digital network	ISDN
interexchange carrier	IEC
inter local access and transport areas	Inter-LATA
network control center	NCC
plain old telephone service	POTS
regional Bell operating company	RBOC
regional holding company	RHC

Transmission and Control Characters

6.170 The American Standard Code for Information Interchange (ASCII) Alphabet Number 5 defines the control characters for electronic transmission. Set the names for all such characters in uppercase:

ACK	Acknowledge
NAK	No acknowledge
WRU	Who are you?

Device States

6.171 Append the dynamic states of telecommunication devices to the device name:

DTE ready	Data circuit terminating equipment ready state
DTE wait	Data circuit terminating equipment wait state

Note the use of lowercase for the state (ready or wait).

6.172 If the state is used without the device name, capitalize the term:

Ready   Ready to Receive   Wait

7. Using Numbers and Symbols

Using Numerals

7.1 Scientific and technical writing depends significantly on numbers, symbols, units of measurement, and mathematical expressions. Ideas and processes must be recorded quantitatively with accuracy and precision. By using certain standard conventions, authors help their readers understand the technical material and concepts they present. This section offers guidance on such topics as

- Using numerals (7.1),
- Writing fractions and large numbers (7.11),
- Using words to express numbers (7.16),
- Using hyphens with numbers (7.20),
- Using scientific notation (7.27),
- Comparing powers of ten (7.31),
- Numbers in tables and graphs (7.36),
- Reporting significant digits (7.38),
- Units of measurement (7.41),
- Choosing appropriate symbols (7.49),
- Mathematical typography (7.54),
- Handling mathematical expressions (7.61).

Measurements

7.2 Express measurements as numerals:

7 meters	1/4 pound
8 by 12 inches	3 1/2 liters
1 1/2 miles	10 amperes
3 MeV	50 kG
1 Btu	17.5 seconds
8 cm	1/2 inch

But spell out isolated numbers zero to ten appearing in ordinary narrative text:

. . . a period of three years . . .

Mathematical Expressions

7.3 Use numerals instead of spelled-out numbers in mathematical expressions:

multiplied by 3	integrated from 0 to 1
divided by 14	*i*-values of 1, 2, and 3
a factor of 2	

Decimals

7.4 Put a zero before the decimal point in numbers less than 1, and omit zeros after the decimal point unless they are significant digits (see 7.38):

0.25 cm

specific gravity 0.9547

gauge height 10.0 feet

The leading zero can be omitted in a few traditional contexts:

.30 or .50 caliber guns

Percentage

7.5 Use numerals to express percentage:

12 percent	25.5%
0.5 percent	5 percentage points

Proportion

7.6 Use numerals to express ratios and proportions:

ratio of 1 to 4 1:2500

1/3/5 parts (by volume) of powder/alcohol/water

Time Spans

7.7 Use numerals to report time values (especially those reporting experimental results that happened over a specific time span):

6 hours 8 minutes 20 seconds

10 years 3 months 29 days

But:

four centuries, three decades, statistics of any one year

Dates and Clock Time

7.8 Use a consistent system for reporting times of events within a narrative:

4:30 P.M., 4:30 in the afternoon, 16:30 hours (24-hour clock)

10 o'clock or 10 P.M.; *not* 10 o'clock P.M. (redundant)

12 M. (noon), 12 P.M. (midnight)

It is natural to use local times for many purposes. However, if the time of an event has scientific significance, it should be reported in Universal Time (the standard time at longitude 0°, which is five hours later than U.S. Eastern Standard Time). Universal Time (UT), previously called Greenwich Mean Time (GMT), is always recorded on the 24-hour clock:

The tsunami struck at 13:24 UT (5:24 A.M. local time)

Or:

The tsunami struck at 5:24 A.M. PST (13:24 UT)

7.9 In accordance with the purpose of your writing, use a consistent system for reporting dates:

Date	Form
July 4, 2001 (*not* July 4th, 2001)	customary U.S.
4 July 2001	military
2001 July 4	scientific

The scientific form is preferable when reporting dates in data displays (see also 11.79).

7.10 Always use a four-digit format for years:

2001 July 4 (*not* 01 July 4)

It is not a good idea to use the informal format 01/02/03, since this form has different meanings in different countries.

Writing Fractions and Large Numbers

Writing Fractions and Mixed Numbers

7.11 Write fractions in the form ${}^{3}/_{7}$ rather than 3/7 if you can. Text fonts generally include a few common fractions such as ${}^{1}/_{4}$, ${}^{1}/_{2}$, and ${}^{3}/_{4}$. Other fractions can be written by setting the numerator and the denominator in a smaller point size and raising the numerator several points above the baseline.

7.12 Mixed numbers such as $4^{1}/_{4}$ are written with no space between the whole number and the fraction. If it is necessary to write mixed numbers with full-size numerals, use a hyphen between the whole number and the fraction:

16-1/3 pounds (*not* 16 1/3 pounds)

Writing Large Numbers

7.13 Conventions for writing large numbers vary by culture. In the United States, the custom is to use a period as the decimal marker ("decimal point") and to use separating commas in numbers of five or more digits on the left of the decimal marker to mark off the thousands, millions, and so on:

26,575.4515 3,775,000 3685.47785

However, in many countries a comma is used as the decimal marker and thin separating spaces (an en space) are used to mark off the thousands, millions, and so on. These separating spaces are also used to the right of the decimal marker to mark off groups of three digits:

26 575,451 5 3 775 000 3 685,477 85

The conflict in conventions causes numerical expressions such as 31,560 to be interpreted differently in different countries. To avoid this confusion, scientific journals often allow U.S. authors to use a period as the decimal marker but require that an en space and not a comma be used as a separator:

26 575.451 5 3 775 000 3 685.477 85

7.14 Separating commas or spaces are not used in four-digit whole numbers unless the numbers appear with other large numbers that are separated:

Acceptable	Not Acceptable
7,560	7560
890	890
18,365	18,365
1,230	1230
28,045	28,045

7.15 Use the words *million, billion* (note that in British usage *billion* means a million million; the American equivalent is a thousand million), and *trillion* instead of writing out the zeros:

sales of $6.3 billion (*not* $6,300,000,000 or 6.3 billion dollars)

population of more than 11 million (*not* 11,000,000)

Using Words to Express Numbers

7.16 When none of the preceding rules apply, use numerals for numbers larger than ten and spell out numbers ten or less:

They bought 50 computers.

We performed seven experiments.

The ship housed 152 crew members.

The target group had four members.

Always spell out round numbers in narrative text:

Thirty years ago One hundred people

Less than one million About a thousand dollars

If two numbers run together in narrative text, spell one out and use numerals for the other:

Fifteen 6-inch rods 12 ten-kilogram canisters

Spell out a number at the beginning of a sentence:

Seventeen students completed the examination.

Twenty-three 2-pound packages arrived at the lab.

7.17 Treat closely related numbers alike. If one is at the beginning of the sentence, spell them both out. If this seems too cumbersome, reword the sentence so the numbers do not appear at the beginning:

Fifty or sixty miles away is snowclad Mount Rainier.

Eleven and twenty-three players make up the respective teams.

But:

The two teams had 11 and 23 players, respectively.

7.18 A spelled-out number should not be repeated as a numeral except in legal documents:

The grace period extends for three (3) months.

7.19 Simple fractions that do not express units of measurement are generally spelled out, but mixed numbers are written with numerals:

one-thousandth

one-half of the samples

one-tenth of the cases

But:

$2^1/2$ times as much, $3^1/2$ cans of soup, $1^1/4$ inches

Using Hyphens with Numbers

7.20 Put a hyphen between numbers and words that form a unit modifier:

The rod has a 1-inch diameter (a diameter of 1 inch)

The 1-inch-diameter rod is heavy

6-foot-long board

five-member panel

6 percent–interest bonds (bonds yielding 6% interest)

11-fold, 150-fold

But:

threefold, tenfold

Note: This rule applies to spelled-out units but not to unit symbols. By international convention (see 7.47) a measurement is separated from its unit symbol by an en space rather than a hyphen:

35 mm camera (*not* 35-mm camera)

But use the hyphen if the unit designation is spelled out:

35-millimeter camera

7.21 When two or more hyphenated compounds in series have a common base element that is omitted in all but the first or last one of the series, retain the hyphen to indicate suspension (the suspense hyphen) (see also 4.102):

> Sections II-2, -3, and -7
>
> Serial numbers 14685-122, -129, and -137
>
> 2- or 3-inch pipe (*not* 2 or 3-inch pipe)
>
> 2-to-3- and 4-to-5-ton trucks
>
> 8-, 10-, and 16-foot boards
>
> 2-by-4-inch boards

But:

> boards 2 inches by 4 inches in cross section

7.22 Put a hyphen between the elements of spelled-out compound numbers from 21 through 99:

> twenty-one, thirty-seven, eighty-two, ninety-nine

7.23 Put a hyphen between the numerator and denominator of a spelled-out fraction, except when one or the other already contains a hyphen:

> two-thirds, one-thousandth, one three-thousandth
>
> two one-thousandths (or two thousandths)

7.24 For mixed numbers in which the fraction is written with full-size numerals (see 7.12), use a hyphen to connect the whole number and the fraction:

> 16-3/4 pounds (*not* 16 3/4 pounds)

7.25 A unit modifier that follows and refers back to the word or words modified (as in a parts list) takes a hyphen and is always written in the singular:

> motor: ac, three-phase, 60-cycle, 115-volt
> (*not:* motor: ac, three phases, 60 cycles, 115 volts)
>
> glass jars: 5-gallon, 2-gallon, 1-quart
>
> belts: 2-inch, $1\frac{1}{4}$-inch, $\frac{1}{4}$-inch

7.26 Do not hyphenate a modifier consisting of a number followed by a possessive noun:

> two months' layoff (*not* two-months' layoff)
>
> 13 weeks' pay

Using Scientific Notation

7.27 Numbers of all sizes, including very large and very small quantities, can be represented compactly as a small factor multiplied by an appropriate power of ten. This representation, called "scientific notation," simplifies calculations and saves space.

7.28 To represent a number in scientific notation

- Place a decimal point to the right of the first nonzero digit in the number,
- Delete all nonsignificant zeros (see 7.38), and
- Multiply the resulting number by the power of ten required to make the product equal to the original number.

Thus:

247,000,000 becomes 2.47 10^8 in scientific notation, and

0.000014647 becomes 1.4647 10^{-5}.

7.29 The exponent (the power of ten) can be negative or positive depending on whether the number being represented is smaller or larger than 1 (the number 1 is equal to ten to the zero power). A factor of 10^0 is customarily omitted. Various examples of small or large numbers and their equivalents in scientific notation include

27.5	2.75×10 (not 2.75×10^1)
1.375	1.375 (not 1.375×10^0)
37,040,000	3.704×10^7 (assuming the trailing zeros are not significant)
2,000,000	2×10^6 (not 2.0×10^6) (assuming the trailing zeros are not significant)
0.000715	7.15×10^{-4}
0.09004	9.004×10^{-2}
0.000000000145	1.45×10^{-10}
10.07	1.007×10

7.30 The factor preceding the power of ten, the "mantissa," is normally between one and ten, as in the preceding examples. However, there is an alternative convention of representing numbers in scientific notation so that the power of ten is a multiple of three. This practice fits the structure of the prefixes used in the International System of Units (see 7.41). When this convention is used, the mantissa will be between 1 and 999:

149.598×10^6 km (*not* 1.495 98 10^8 km)

Comparing Powers of Ten

7.31 When comparing the relative magnitudes of two numbers in the standard power-of-ten representation, remember that making an accurate comparison requires three steps:

1. Compare their factors and their exponents,
2. Combine the two comparisons, and
3. Record the resulting ratio.

7.32 To perform this process for the hypothetical numbers *a* and *b*, take the ratio of *b*'s factor to *a*'s factor and multiply it by the ratio of *b*'s exponent to *a*'s exponent. The result is the ratio of *b* to *a* (the desired comparison). For example, let *a* be 2.3×10^6 and let *b* be 6.9×10^3:

1. Divide the 6.9 factor of *b* by the 2.3 factor of *a* and get 3.

1a. Divide the 10^3 exponent of *b* by the 10^6 of *a* and get 10^{-3}.

(Remember, division of such numbers is done by subtracting the exponent in the denominator from the exponent in the numerator, i.e., $3 - 6 = -3$).

2. Multiply the 3 by the 10^{-3} and get 3×10^{-3} or 0.003.

3. Thus, the ratio of *b* to *a* is only 0.003; in other words, *b* is only 3/1000 as large as *a*.

7.33 Following are a few more examples of this comparison process:

Numbers	Factor Division	Exponent Division	Ratio
$b = 2.4$ $a = 1.6 \times 10^{-2}$	$2.4/1.6 = 1.5$	$1/10^{-2} = 10^2$	$b/a = 1.5 \times 10^2 = 150$
$b = 1.9 \times 10^2$ $a = 9.5 \times 10$	$1.9/9.5 = 0.2$	$10^2/10 = 10$	$b/a = 0.2 \times 10 = 2$
$b = 7.1 \times 10^{-5}$ $a = 1.42 \times 10^{-2}$	$7.1/1.42 = 5$	$10^{-5}/10^{-2} = 10^{-3}$	$b/a = 5 \times 10^{-3} = 0.005$

7.34 You can make the comparison obvious by putting the numbers to be compared in terms of the *same* power of ten, so that the mantissas alone reflect the comparative magnitudes of the numbers. Exhibit 7–1 illustrates how such a representation makes the relative magnitudes of the column entries obvious:

Exhibit 7–1: Comparison of Estimated Masses of Several Icebergs

Iceberg	Standard Representation Mass (kg)	Nonstandard Representation Mass (kg)
A	2.57×10^{10}	257×10^{8}
B	8.43×10^{8}	8.43×10^{8}
C	7.29×10^{9}	72.9×10^{8}
D	1.87×10^{11}	1870×10^{8}

The nonstandard scheme shown in Exhibit 7–1 is effective when the range of the numbers is only a few powers of ten. If the range is larger, say, six or more powers of ten, use the standard representation. For such wide-ranging numbers, the nonstandard representation of Exhibit 7–1 would cause some digits to extend inconveniently far to the left and right of the decimal point.

7.35 The idea used in the nonstandard representation in Exhibit 7–1—expressing all numbers to be compared in terms of the same power of ten—is commonly used in tables and graphs to avoid having to write out large numbers. For example, if a table contains a column of distances 3740, 3690, 3720, 3750, and 3710 expressed in meters, an editor might choose to represent the distances in kilometers instead—as 3.74, 3.69, 3.72, 3.75, and 3.71—to keep the numbers small.

Numbers in Tables and Graphs

7.36 Tables and graphs in engineering and scientific reports often change the original unit by some factor of ten in order to have convenient-sized numbers in the table columns or on the graph scales. Exhibit 7–2 illustrates the process with three presentations of the same data:

Exhibit 7–2: Tabular Data Presented as Powers of Ten

(Column 1) Yield Strength (kPa)	**(Column 2)** Yield Strength (10^3 kPa)	**(Column 3)** Yield Strength (MPa)
197,000	197	197
265,000	265	265
181,000	181	181
240,000	240	240

In column 1, the original values, the column head tells what the numbers in the column represent (yield strength) and the units of measurement (kPa, or kilopascals). But the numbers in that column are large.

The head in column 2 cites the unit of measurement as 10^3 kPa, or 1000 kPa. Since the unit of measurement is 1000 kPa and the first entry is 197 units, the first entry means 197,000 kPa.

Column 3 is really the same as column 2, but the units shown are megapascals (Mpa). Since 1 Mpa = 1000 kPa, the numerical entries in column 3 are the same as in column 2. Although column 3 is the most direct presentation of the data, column 2 will be preferable if yield strength is traditionally measured in kilopascals rather than megapascals.

7.37 Do not show the unit in column 2 as kPa 10^{-3}. Authors sometimes do this to indicate that the data shown are in kilopascals but have been multiplied by 10^{-3} (or divided by 1000). However, kPa 10^{-3} in the heading of column 2 would indicate the unit to be thousandths of kilopascals (or simply pascals), which is not what you intend.

Reporting Significant Digits

7.38 Account for all significant digits (or significant figures) when reporting measurements or when changing measurement units. Digits in a number expressing a measurement or used in a calculation are called "significant" when they indicate the number's accuracy. Scientists and engineers pay close attention to the accuracy of their numbers by expressing them in terms of significant digits, and they follow a well-established system of rules for achieving this accuracy:

1. All nonzero digits presented should be accurate: in a measurement recorded as 296.2 mg, the true value of the quantity should be at least 296.15 mg but less than 296.25 mg.

2. Zeros presented to the right of the decimal point should be accurate: in a measurement recorded as 296.00 mg, the true value of the quantity should be at least 295.995 mg but less than 296.005 mg.

3. Trailing zeros in a number with no decimal point are ambiguous: it is not clear whether the number 14,900 is accurate to the nearest 1, the nearest 10, or the nearest 100. To avoid this ambiguity, the measurement should be reported in scientific notation as 14.9×10^2, 14.90×10^2, or 14.900×10^2, whichever is correct.

4. When numbers are multiplied (or divided), the number of significant digits in the product or quotient is the minimum of the numbers of significant digits among the factors. For example, suppose a cylinder has measured height $h = 23.2$ mm and diameter $d = 5.6$ mm. The volume of the cylinder is

$$V = \tfrac{1}{4}\pi d^2 h \cong 5.7 \times 10^2 \text{ mm}^3,$$

not 571.4 mm^3. In this calculation, the mathematical constants 4 and π are precise, so they are considered to have an infinite number of significant digits. The height h has three significant digits, but the diameter d has only two. As a result, the volume has only two significant digits.

5. When numbers are added or subtracted, a slightly different rule applies. A sum is only as accurate as its least accurate summand. If three lengths are measured as 14.3 mm, 6.45 mm, and 9.3 mm, their sum is

$$(14.3 + 6.45 + 9.3) \text{ mm} \cong 30.1 \text{ mm},$$

not 30.05 mm. In this calculation, the three summands are accurate to the nearest 0.1 mm, so the sum has the same accuracy. (Notice that this accuracy gives the sum three significant digits even though one of the summands has only two.)

7.39 Similar considerations arise in putting numbers on the scale of a graphical plot. Although data accuracy is not involved, appearance and consistency, as well as avoiding excessive numerical precision, should be considered.

For example, to create a scale for time on the horizontal axis, start at zero on the left and proceed at 1-second intervals toward the right:

0, 1, 2, 3, 4

(*not* 0.0, 1.0, 2.0, 3.0, 4.0)

Since all the numbers are integers, no decimal points are needed. In particular, a decimal point should never be used with the zero at the zero position of a scale. To create a time scale at 0.5-second intervals, use:

0, 0.5, 1.0, 1.5, 2.0, 2.5

(*not* 0, 0.5, 1, 1.5, 2, 2.5)

Since it is necessary to use a decimal point and a 5 to represent the numbers at the half-second points, maintain a consistent appearance by using a decimal point and a zero to represent those numbers at the one-second points.

Orders of Magnitude

7.40 Two quantities are of the same "order of magnitude" if their ratio is less than ten. An order of magnitude larger means ten times as large, but two orders of magnitude larger means $10 \cdot 10 = 100$ times larger, not 20 times as large. If A is said to be three orders of magnitude larger than B, this means A is 10^3 or *one thousand times* the size of B.

Since your readers may not be familiar with this concept, when writing for a general audience it might be better to describe A as 1000 times as large as B rather than writing that A is three orders of magnitude larger. If it seems desirable to make comparisons in terms of orders of magnitude, define the concept for general readers.

Units of Measurement

Choosing a System of Units

7.41 The International System of Units—usually called the SI from its French name *Système Internationale*—provides a coherent system of units of measurement for scientific and technical work.

Most scientific and technical journals now require the use of SI units (with a few well-defined exceptions in each scientific field). The official U.S. policy, expressed in the Metric Conversion Act (1975) and the Omnibus Trade and Competitiveness Act (1986) (see also 13.7), is that the SI units are the "preferred system of weights and measures for U.S. trade and commerce." Metric conversion policies in all the engineering disciplines require or strongly encourage the use of SI units. As a result, scientific and technical authors should use SI units of measurement except when there is an obvious need to use other units.

The SI units form a small subset of the metric system. Since most readers in the United States are more familiar with traditional English units, it is still advisable to use these traditional units in writing for a general audience.

7.42 Always report measurements and observations in the units in which they were actually made. If they were made in traditional units, report them in those units:

The low temperature that morning was 5 °F (–15 °C).

The window was 8 ft 3 in (2.51 m) from the ground.

Similarly, use traditional units to describe the dimensions of objects built in those units:

Six-foot pole

14 ten-inch rods

7.43 When you use traditional units because it seems necessary in order to communicate with your intended readers, contribute to metric education by supplying the SI equivalents:

The meteor dug a crater 25 feet (7.6 meters) deep.

The pressure is about 36 psi (or 250 kilopascals).

The National Institute for Standards and Technology's (NIST) *Guide for the Use of the International System of Units* includes a precise and comprehensive table for converting practically any measurement into SI units. This table is reprinted in the *CRC Handbook of Chemistry and Physics* (see also 13.7).

In stating unit equivalences, pay careful attention to significant digits (see 7.38). The length of a "six-foot pole" is probably accurate within 0.1 foot, perhaps within 0.01 foot, so write

Acceptable	Not Acceptable
The pole's length is 6 feet (1.8 m).	The pole's length is 6 feet (1.8288 m).
The pole's length is 6 feet (183 cm).	
Add 4 oz (110 g) of sugar.	Add 4 oz (113.4 g) of sugar.

Using SI Units and Symbols

7.44 *The International System of Units* and the *Guide for the Use of the International System of Units* provide a strict protocol for the use of SI units and their symbols (see also 13.7). The remaining sections under this heading summarize these international rules.

7.45 There are 28 SI units, including seven independently defined "base units" and 21 "derived units" defined in terms of the base units. A number of additional units are not SI units but are acceptable for use with SI units. The following tables list the units in these three categories (see 7.47).

Exhibit 7–3: International System of Units—Base Units

Quantity	Unit	Symbol
length	meter	m
mass	kilogram	kg
time	second	s
electric current	ampere	A
temperature	kelvin	K
amount of substance	mole	mol
luminous intensity	candela	cd

Notes: The symbol for the second is s, not sec. The temperature unit is the kelvin, not "degree Kelvin." The unit of luminous intensity is the candela, not the candle.

Exhibit 7–4: International System of Units—Derived Units

Quantity	Unit	Symbol	Definition
plane angle	radian	rad	m/m
solid angle	steradian	sr	m/m
frequency	hertz	Hz	s^{-1}
force	newton	N	$m \cdot kg/s^2$
pressure	pascal	Pa	N/m^2
energy, work	joule	J	N·m
power	watt	W	J/s
electric charge	coulomb	C	A·s
electric potential	volt	V	W/A
capacitance	farad	F	C/V
electric resistance	ohm	Ω	V/A
electric conductance	siemens	S	A/V
magnetic flux	weber	Wb	V·s
magnetic flux density	tesla	T	Wb/m^2
inductance	henry	H	Wb/A
Celsius temperature	degree Celsius	°C	
luminous flux	lumen	lm	cd · sr
illuminance	lux	lx	lm/m^2
(radio)activity	becquerel	Bq	s^{-1}

Other units acceptable for use with SI units include the

- Customary larger units of time: minute (min), hour (h), day (d), week (wk), and year (yr or a);
- Units of angle measure: degree (°), arcminute (′), and arcsecond (″);
- Liter (L) for liquid volume and capacity, the hectare (ha) for land area, and the tonne or metric ton (t) for large masses;
- Bit (b) and the byte (B);
- Standard logarithmic units such as the octave and the decibel (dB); and
- Electronvolt (eV) and unified atomic mass unit (u).

7.46 The SI prefixes for multiples and submultiples of units—for example, kilo- (k), mega- (M), milli- (m), and micro- (μ)—make it easy to select a properly sized unit for the data under consideration. In reporting the mass of very small objects, for example, using the prefix μ- replaces a mass measurement of 2.4×10^{-6} g (0.0000024 g) by the much simpler 2.4 μg.

Exhibit 7–5: Acceptable International System of Units Prefixes

Multiple	Prefix	Symbol	Multiple	Prefix	Symbol
10^{1}	deka	da	10^{-1}	deci	d
10^{2}	hecto	h	10^{-2}	centi	c
10^{3}	kilo	k	10^{-3}	milli	m
10^{6}	mega	M	10^{-6}	micro	μ
10^{9}	giga	G	10^{-9}	nano	n
10^{12}	tera	T	10^{-12}	pico	p
10^{15}	peta	P	10^{-15}	femto	f
10^{18}	exa	E	10^{-18}	atto	a
10^{21}	zetta	Z	10^{-21}	zepto	z
10^{24}	yotta	Y	10^{-24}	yocto	y

Multiple prefixes are not allowed: write 5 pF, not 5 $\mu\mu$F.

To avoid common errors, notice that the symbol for kilo- is the lower case k-, not the capital K-, and that da-, not dk-, is the symbol for deka-.

If the Greek letter μ is not available, mc- can be used as a symbol for micro-. However, always use the official symbol μ if you can.

7.47 The symbols for the SI units are defined as "mathematical symbols," not as abbreviations. Accordingly, they follow the rules for mathematical symbols rather than the normal grammatical rules for abbreviations.

1. Never follow unit symbols with a period, unless they fall at the end of a sentence, and no *s* is ever added to indicate a plural: 75 cm (*not* 75 cm. or 75 cms).
2. Print unit symbols in roman (upright) type, even if the surrounding text is in italic.
3. Separate unit symbols from the number they follow by an en space, not a hyphen:

   35 mm camera (*not* 35mm or 35-mm)   15 °C (*not* 15°C)

4. In general, set unit symbols in lowercase. Capitalize the first (or only) letter of a symbol if, and only if, a symbol comes from a proper name (such as N for the newton). There are two exceptions to this rule. First, use the capital L rather than the lowercase l as the symbol for the liter. Second, the capital B is used as the symbol for both the bel and the byte.

Acceptable	Unacceptable
50 km	50 Km or 50 KM
10 A	10 a or 10 amp
15 kJ	15 KJ
20 mL	20 ml
512 kB	512 kb

5. Separate compound unit symbols, such as kilowatt hour, with an en space or a raised dot indicating multiplication:

   1200 kW h or 1200 kW·h (*not* 1200 kWh)

6. Do not use the traditional English unit abbreviation *p* for *per* when units are combined by division. The slash (solidus) character / can be used, but only once (see also 4.106):

Acceptable	Unacceptable
15 m/s	
15 m · s^{-1}	
85 km/h	85 kph
	85 kmph
25 cm/s^2	25 cm/s/s

Using Traditional Units

7.48 In various situations (see 7.42–7.43) it may be necessary or desirable to use units other than SI units. Apply the same style rules to traditional units as to the SI units, both to maintain consistency within a document and to improve clarity. Applying the SI style rules to traditional units means following the typographical rules of the previous sections. It also means avoiding the use of *p* for *per* and not using such traditional English abbreviations as *sq. ft.* and *cu. yd.:*

Acceptable	Unacceptable
45 mi/h	45 mph
23.7 ft^2	23.7 sq. ft.
2 ft 4 in	2′ 4″
1.56 Mgal/d	1.56 MGD
15 ft^3/s	15 cfs
1100 r/min	1100 rpm

Notice how following the SI rules consistently takes little, if any, additional space. It also leads to greater clarity in the presentation of data by not requiring readers to be familiar with special abbreviations such as psi, MGD, or cfs.

Choosing Appropriate Symbols

7.49 Scientists and engineers represent many quantities with symbols. The symbols are usually letters, either roman (i.e., English) or Greek. Most are a single letter, sometimes with attached subscripts or superscripts. The symbols for chemical elements, such as Fe and Pb, and a few engineering symbols, such as the Prandtl number (Pr) and the Nusselt number (Nu), contain more than one letter.

7.50 In algebra the variables in an equation are usually written as *x* or *y*. A scientist or engineer will label variables with a letter having a definite mnemonic connection to the name of the object or concept for which it stands. Velocity, for example, will probably be labeled *v*. If a problem involves the velocities of two objects, perhaps a train and an automobile, the scientist distinguishes them by adding appropriate subscripts, such as v_t and v_a. Italic type is used for such symbols (see 7.58).

7.51 In selecting a symbol for vehicle weight, one might be tempted to select *VW*. This would be a bad symbol choice for two reasons:

1. Scientifically knowledgeable readers seeing the *VW* in the equation would assume it represented two symbols, *V* and *W*, multiplied together, and
2. Scientists normally choose the *quantity of interest*, in this case weight, as the mnemonic basis for a one-letter symbol, and use a subscript qualifier if need be.

A conventional choice for a symbol for vehicle weight would be w_v (*w* for weight, the quantity of interest, and *v* for the vehicle whose weight it is).

7.52 Use standard symbols as much as possible. Most quantities of interest in science have been assigned traditional symbols, and failing to use these symbols can confuse readers. The *CRC Handbook of Chemistry and Physics*, published annually, contains a comprehensive list of standard symbols for quantities from nearly all fields of science (see also 13.7).

Mathematical Symbols

7.53 Use standard mathematical symbols for reporting mathematics. There are hundreds of conventional mathematical symbols; see, for example, the *McGraw-Hill Dictionary of Mathematics* (1997), pages 284–291 (see also 13.7). Following are the most important conventional symbols in six areas of mathematics.

Arithmetic, Algebra, Number Theory

$+$	Plus (positive)
$-$	Minus (negative)
$\pm$	Plus or minus (positive or negative)
$\cdot$	Multiplied by (used between numbers; see also 7.65)
$\times$	Multiplied by (used between complicated expressions, or for the cross product of vectors)
ab	*a* times *b* (*a* multiplied by *b*)
a/b	*a* divided by *b* (ratio of *a* to *b*)
$=$	Equals
$\equiv$	Is identically equal to
$\neq$	Is not equal to
$\cong$	Is approximately equal to
$>$	Is greater than

Symbol	Meaning
$<$	Is less than
$\geq$	Is greater than or equal to
$\leq$	Is less than or equal to
a^n	$a \cdot a \cdot a \cdot a \ldots$ to n factors
$a^{1/2}$	The positive square root of a, for positive a
$a^{1/n}$	The nth root of a, usually the principal nth root
a^0	The number 1 (for $a \neq 0$); by definition
a^{-n}	The reciprocal of a^n $(1/a^n)$
$a^{m/n}$	The nth root of a^m
e	The base of the system of natural logarithms ($e \cong 2.71828\propto$)
$\log_a x$	Logarithm (to the base a) of x
$\log a$	Common logarithm (to the base 10) of a
$\ln a$	Natural logarithm (to the base e) of a
$\exp x$	e^x, where e is the base of the system of natural logarithms
$a \propto b$	a varies directly as b
i	The imaginary unit (square root of -1)
$n!$	$1 \cdot 2 \cdot 3 \propto n$ (n factorial)
$f(x)$	A function f whose argument is the variable x
$\lvert z \rvert$	Absolute value of z (numerical value of z, without regard to sign)
$\mathrm{Re}(z)$	Real part of z (where z is a complex number, i.e., has a real and an imaginary part)
$\mathrm{Im}(z)$	Imaginary part of z (where z is a complex number)
$\mathbf{i}, \mathbf{j}, \mathbf{k}$	Unit vectors along the coordinate axes
$\mathbf{a} \cdot \mathbf{b}$	Dot product (scalar product) of the vectors **a** and **b**
$\mathbf{a} \times \mathbf{b}$	Cross product (vector product) of the vectors **a** and **b**

Trigonometric and Hyperbolic Functions

Symbol	Meaning
sin	Sine
cos	Cosine

tan	Tangent (sin/cos)
cot	Cotangent (1/tan)
sec	Secant (1/cos)
csc	Cosecant (1/sin)
$\sin^{-1} x$	The principal value of an angle whose sine is x (for x real)
$\sin^2 x$	$(\sin x)^2$, and similarly for $\cos^2 x$, $\tan^2 x$, and so on.
sinh	Hyperbolic sine
cosh	Hyperbolic cosine
tanh	Hyperbolic tangent

Elementary and Analytic Geometry

$\perp$	Perpendicular (is perpendicular to)
$\parallel$	Parallel (is parallel to)
π	Ratio of a circle's circumference to its diameter ($\pi \cong 3.14159\propto$)
(x, y)	Rectangular coordinates of a point in a plane
(x, y, z)	Rectangular coordinates of a point in space
(r, θ)	Polar coordinates
(r, θ, z)	Cylindrical coordinates

Calculus and Analysis

Σ	Summation sign, applied to an expression to indicate successive addition (requires specification of the addition to be done)
Σ_1^n or $\sum_{i=1}^{n}$	Sum to n terms, one for each positive integer from 1 to n
$\sum_{i=1}^{\infty} x_i$	The sum of the infinite series $x_1 + x_2 + x_3 + \ldots$
Π	Product sign, applied to an expression to indicate successive multiplication (requires specification of the multiplication to be done)
Π_1^n or $\prod_{i=1}^{n}$	Product of n terms, one for each positive integer from 1 to n
$\lim_{x \to a} y = b$	The limit of y as x approaches a is b, for $y = f(x)$
Δy	An increment of y

Symbol	Meaning
dy	Differential of y
dy/dx	Derivative of y with respect to x, where $y = f(x)$
$\partial y/\partial x$	Partial derivative of y with respect to x, where $y = f(x)$
$d^n y/dx^n$	The nth derivative of y with respect to x, where $y = f(x)$
D_x	The operator d/dx
$\int f(x)dx$	The indefinite integral of $f(x)$ with respect to x
$\int_a^b f(x)dx$	The definite integral of $f(x)$ between the limits a and b

Logic and Set Theory

Symbol	Meaning
$x \in M$	The point x belongs to the set M
$M = N$	The sets M and N are equal
$M \subseteq N$	Each point of M belongs to N (M is a subset of N)
$M \subset N$	Each point of M belongs to N and N contains a point not in M (M is a proper subset of N)
$M \cup N$	The intersection of M and N
$M \cap N$	The union of M and N
$\varnothing$	Empty set

Statistics

Symbol	Meaning
χ^2	The chi-square distribution, $\chi^2(n) = \sum_{i=1}^{n} u_i^2$, where n is the number of independent observations $x_1, \ldots, x_n$ from a normal distribution $N(x, s^2)$, and the u_i are the standardized variables, $u_i = (x_i - x)/s$.
P.E.	Probable error (same as probable deviation)
s	Standard deviation (from a sample)
σ_x	Standard deviation of the population of x
t	Student's t-test statistic
V	Coefficient of variation
Q_1	First quartile
Q_3	Third quartile
$E(x)$	Expectation of x
$P(x_i)$	Probability that x assumes the value x_i

Mathematical Typography

7.54 In scientific and technical works, mathematical symbols and quantities are presented in particular ways, with distinctive typefaces. These conventions have two benefits. They set off the mathematical material from the accompanying text, improving the visual effect. In addition, they speed the experienced reader's understanding of the mathematics (see also 12.26).

Nearly all scientifically trained readers recognize the typographic coding of mathematical material, but few are thoroughly familiar with all the coding rules. Fortunately, nearly all word processing software packages now come with equation editors that make it possible to set mathematical expressions with almost the same level of quality and precision as commercial printing. As a result, authors have available—and they should use—the tools they need to set mathematics clearly and accurately.

7.55 Setting mathematical expressions properly requires an extended set of characters, including:

- Standard, or roman, typeface (a, b, c, A, B, C);
- Italic (*a*, *b*, *c*, *A*, *B*, *C*);
- Boldface (**a**, **b**, **c**, **A**, **B**, **C**);
- Greek letters, both uppercase and lowercase (π, ρ, Π, P, Σ); and
- A variety of special symbols (∂, ∇, $\varnothing$, $\in$, $\equiv$, $\oplus$).

Fortunately, all the standard word processing software packages come with fonts including all these symbols and many others.

7.56 Take particular care to choose the correct symbol rather than an approximation. For example, distinguish clearly between

- Zero (0) and the capital letter O;
- Hyphen (-) and the minus sign (−);
- Letter x and the multiplication sign (×);
- "Straight quotes," apostrophe ('), and the prime symbol (′); and
- Superscript zero (^{0}) and the degree symbol (°).

Use of Roman Type

7.57 Always use roman (upright) type for

- Numerals: $45x$, $f(3.68)$, $\pi/6$;
- Abbreviations of standard functions and operators: arg, av, cos, cosh, cot, det, erf, exp, Im, lim, ln, log, max, min, mod, Re, Res, sin, sinh, tan, tanh, tr;
- Unit symbols and their prefixes: ft, Hz, kg, mL, rad;
- Chemical symbols: B, Be, C, F, He, N, Pb, U;
- Standard abbreviations: abs, approx, diam, tot.

Use of Italic Type

7.58 Use italic type for

- Ordinary or scalar variables such as *n*, *x*, *y*;
- The constant *e* as the base of the natural logarithm system; and
- Functions defined in the work, such as $f(x)$.

The following examples illustrate how to combine roman and italic type in mathematical expressions:

$$x + 14 \qquad f(x) = x + 3a \qquad \exp x = e^x \qquad \sin\,(t - \pi/2)$$

Use of Greek Letters

7.59 Use lowercase Greek letters instead of italic letters for scalar variables. An exception is the letter π, which is reserved for the constant 3.14159∝. Since the lowercase Greek letters are written with a slant (α, δ, θ, ψ), they fit well with variables set in italic:

$$\sin\,(n\theta + \pi/4) \qquad f(x) = \rho x + \mu \qquad \xi y^2 + \psi x^2$$

Use capital Greek letters for the names of operators or functions; in particular, the capital sigma (Σ) traditionally indicates a summation, the capital pi (Π) indicates a product, and the capital delta (Δ) indicates a difference.

Use of Boldface

7.60 Boldface roman type is used for vectors, which are quantities having both a scalar magnitude and a spatial direction. This convention makes it easy to distinguish between scalar and vector quantities in an expression:

$$\mathbf{F} = m\mathbf{a} \qquad \mathbf{a} \times \mathbf{b} \qquad \overline{\mathbf{F}}\Delta t = \Delta\mathbf{p}$$

Handling Mathematical Expressions

7.61 This section explains the physical arrangement of mathematical expressions:

- Horizontal spacing of elements (7.62),
- Placement of equations (7.67),
- Breaking equations (7.72),
- Miscellaneous typographic details (7.77).

Most contemporary word processors include an equation editor capable of automatically formatting mathematical expressions.

Horizontal Spacing of Elements

7.62 Use one-space (an en space) separation on either side of mathematical symbols +, −, ±, and the like, between two quantities. Exceptions: Closed spacing—no space—is used in subscripts and superscripts, and may be used in other places where compactness is desired—as in tables:

$xy < p$	29 ± 0.4	$(x + y) + (u + y)$
$a + b = c - d$	$3.4 \times 10^{n-2}$	$r > r_0$

7.63 Use an en space to separate the trigonometric and hyperbolic function abbreviations from characters on either side:

$\sin \omega t$	$2 \cos \theta$	$i \cosh y$
$\tanh^{-1} y/x$	$r \tan \phi \sinh x$	$\log p$
$\sin^2 p$	$2\pi\mathrm{e}\ \mathrm{xp}\ (x^2/2 + y)$	$\log_{10} R$

7.64 Use an en space to separate integral, summation, continued product signs, the abbreviation lim, and other calculus and analysis symbols from characters on either side. When these signs carry limits extending to right or left, the limit (rather than the sign) is used as the starting point for counting the one-space separation:

$$\frac{1}{\pi}\int xdx \qquad p(x)\sum_{n=0}^{p+q+r} R_n X^{n-2} \qquad 2x\sum_{n=0}^{k}(y - y_n) \qquad \int_{a+b}^{c} t^2 dt$$

7.65 Use the raised dot character (·) between *unenclosed* fractions multiplied together:

$$\frac{pg}{xy} \cdot \frac{s}{t} \qquad \frac{1}{2} \cdot \frac{2}{3} \cdot \frac{3}{4}$$

But:

$$\left(\frac{x-y}{2}\right)\left(\frac{x+y}{2}\right)$$

7.66 In the following situations, no space is used between

- Quantities multiplied without a multiplication symbol:

  $2ab \qquad (p-q)(r-s) \qquad Y_aY_bZ$

- The symbol for a function and its argument (independent variable):

  $\phi(x) + \theta(y) + \tau(z) \qquad f(z) = 2z$

- A differential (derivative) operator and the variable to which it is applied:

  $$\frac{\partial t}{\partial x} \qquad \frac{d^2y}{dx^2}$$

- A symbol and its subscript or superscript, or between elements of the subscript or superscript:

  $B_k \qquad re^{i\omega t} \qquad x^{2+b}$

- A signed quantity and its sign:

  $\pm 0.5 \qquad -3/2 \qquad -ab$

  $\sim 7.5 \qquad <3 \qquad -\cos t$

Placement of Equations

7.67 The space above and below displayed equations (those set apart from text) should be double the space used between lines in the text. This rule applies to the space between the equation and text, and the space between one equation and another. The uppermost character in an equation is the starting point for counting space above; the lowermost character is the starting point for counting space below.

7.68 For horizontal placement of equations, centering is the standard practice if it can be done easily and conveniently. If centering is not convenient, however, choose some scheme, such as using a two-paragraph indent for all equations that will fit on a single line with that indent, and no indent (flush left) for all longer equations.

7.69 Do not begin an equation to the left of the left margin. Instead, break the equation and use two or more lines as required.

7.70 When two or more similar short equations are written one after the other, align them on their equal signs:

$$\begin{aligned}
x(y) &= y^2+3y+2 \\
p(x,y) &= \sin(x+y) \\
p(x_0,y_0) &= 2\sin x_0 \cos y_0 \\
q(x,y) &= \cos(x+y) \\
q(x_0,y_0) &= \cos^2 x_0 - \sin^2 y_0 \\
l_2(z) &= \frac{25}{\pi^4} l_0 \int_1^\infty \frac{n^3}{e^n-1}\,dn + \sum_{i=1}^{7} P_i(z)
\end{aligned}$$

7.71 When equations need to be identified, place the identifying numbers at the right or left margins:

$$(3) \qquad f(x) = 2\tan\left(3x - \frac{\pi}{r}\right)$$

Breaking Equations

7.72 Equations are commonly broken just before the equals sign, with the first line centered and the second line (including the equals sign) indented from the first:

$$\int_0^1 \left(q_n - \frac{n}{r} f_n\right)^2 r\,dr + 2n \int_0^1 q_n f_n\,dr - 3\left(n^2+r^2\right)\int_0^1 \left(4f_n - \frac{6}{q_n-1}\right) dr$$
$$= \int_0^1 \left(f_n - \frac{n}{r} q_n\right)^2 r\,dr + nf_n^2 \qquad (1)$$

7.73 The second most preferred place to break an equation is just before a plus or minus sign not enclosed within parentheses, brackets, or the like. The plus or minus sign is moved to the next line with the remainder of the equation, and is indented with respect to the equals sign or placed so as to end near the right margin:

$$r(x, y, z) + q(x, y, z) + f(x, y, z) = (x^2 + y^2 + z^2)^{1/2} \times (x - y + z)^2$$
$$+ [f(x, y, z) - 3x^2]. \qquad (4)$$

7.74 The next best place to break an equation is between parentheses or brackets indicating multiplication of two terms. A multiplication sign is brought down to the next line to precede the remainder of the equation. Thus the equation above may be written

$$r(x, y, z) + q(x, y, z) + f(x, y, z) = (x^2 + y^2 + z^2)^{1/2}$$
$$\times (x - y + z)^2 + [f(x, y, z) - 3x^2]. \qquad (5)$$

7.75 The same general rules apply to breaking equations in the line of text. When such an equation cannot be broken except at a place that leaves the line of text awkwardly short or long, it is advisable to set the entire equation apart from the text as a displayed equation.

7.76 When an equation is so long that it requires three or more lines, observe the following rules:

1. Start the first line at the left margin,
2. End the last line at the right margin, and
3. Indent the intermediate lines so that the equation is displayed from upper left to lower right:

$$h \int_{u_k}^{u_{k+1}} \left[y_n + \Delta_1 y_n u + \frac{\Delta_2 y_n}{2} \left(u^2 + u \right) \right.$$
$$\left. + \frac{\Delta_3 y_n}{6} \left(u^3 + 3u^2 + 2u \right) du \right]$$
$$= h \left[y_n u + \Delta_1 y_n \frac{u^2}{2} + \frac{\Delta_2 y_n}{2} \cdot \frac{u^3}{3} + \frac{u^2}{2} \right.$$
$$\left. + \frac{\Delta_3 y_n}{6} \cdot \frac{u^4}{4} + u^3 + u^2 \right]. \qquad (6)$$

Miscellaneous Typographic Details

(*Note:* Most of the rules in this section will be applied automatically by the equation editor routines of word processing software. However, you will need to pay careful attention to them if you type mathematical expressions without using an equation editor.)

7.77 Use a minus sign whose length matches the length of the plus sign. In most text fonts this will not be the hyphen character; hyphens (-) are usually much shorter than plus signs (+). In most fonts, the en dash character (–) makes a good minus sign.

7.78 The large integral and summation signs are the ones normally used in displayed equations; the small signs are appropriate for use in the line of text.

7.79 Use a size for parentheses, brackets, braces, and similar marks of enclosure in keeping with the height of the expression enclosed. The small size is satisfactory for one-line expressions having either simple subscripts or superscripts. Built-up fractions and other tall expressions require larger sizes. Parentheses, brackets, braces, and the like should always be centered vertically on the main line of the expression they enclose. (In most cases, setting large parentheses, brackets, or braces requires use of an equation editor.)

7.80 The level of subscripts or superscripts outside marks of enclosure should be on the same level as or outside the level of the most extreme subscripts and superscripts inside. Outside subscripts should be as low as or lower than the lowest inside subscript, and outside superscripts should be as high as or higher than the highest inside superscript. When no inside superscript exists, put the outside superscript on the level where the inside one would be if there was one:

$$(a^2 + b^2 + c^2)^3 \qquad (x + y)^3 \qquad \left[x_0 (a_1 + b_1)\right]_{av}$$

$$\left[\left(\frac{x_1}{a_1} + \frac{x_2}{a_2} + \cdots + \frac{x_n}{a_n}\right)_{x_i = y_i}\right]^3$$

7.81 In general, set the superscript directly above the subscript, which gives a neat and compact appearance. If this arrangement is not possible with the equipment you have (and most word-processing equipment cannot do it), then use other arrangements:

General Rule	Typical Exception
$a_1^2 + b_1^2 = c_1^2$	$a_1{}^2 + b_1{}^2 = c_1{}^2$
$L^{-2}\,(x+y)^2_{\max}$	$L^{-2}\left[(x+y)^2\right]_{\max}$
$x_i' + y_i' = z_i'$	$x'_{\,i} + y'_{\,i} = z'_{\,i}$

Punctuating Mathematics

7.82 In general, the rules of punctuation used for ordinary text are also used for text including mathematical expressions. In applying the rules of punctuation, remember that an equation, such as $f(x) = 2x + 7$, is a clause. It has a subject, $f(x)$, a verb, =, and an object, $2x + 7$. The same is true for other statements of mathematical relationship, such as $x < 5$, or $y \in B$. Expressions of this kind are treated in a sentence just like any other dependent clause:

> Whenever $x < 5$, $f(x) = (5 - x)x^2$.
>
> The equation holds if $x = 3$, $x = 4$, or $x > 6$, but not otherwise.
>
> If $x \in A$ and $x \in B$, then $x \in A \cap B$.

7.83 Many authors (and some editors) seem to believe that the last line before a displayed equation must end in a colon. In fact, an equation is displayed only because of its size, or to increase its visibility. Grammatically, it acts just as a short, nondisplayed equation would act. A colon is needed only if it would be needed even if the equation were not displayed. Furthermore, a displayed equation should be followed by a period, comma, or semicolon, if that is what the grammatical situation requires. The following examples are typical:

> Provided $x \in A \cap B \subseteq G$,
>
> $$h(x) = F(x + 0.5) + \frac{q(x)}{4} + \frac{3\pi}{4}\,.$$
>
> Thus we observe that the height, h, is given by
>
> $$h = ut\,(\text{m}), \qquad (1)$$
>
> and the volume, V, by
>
> $$V = \pi r^2 ut\ (\text{cm}), \qquad (2)$$
>
> where u is upward velocity (m/s), t is time (s), and r is radius (m).

7.84 The last example includes a special feature of mathematical writing, the *where* list: a list of definitions of variables placed after an equation. If a *where* list is short, it can be run into the text as shown above. If it is longer, something like the following example is advisable:

The force of gravitational attraction between two objects a and b is

$$F = km_a m_b / r^2 ,$$

where

$$\begin{aligned} F &= \text{force (newtons)} \\ k &= \text{gravitational constant } (6.673^{-11} \times \text{N} \cdot \text{m}^2/\text{kg}^2), \\ m_a &= \text{mass of object } a \text{ (kilograms)}, \\ m_b &= \text{mass of object } b \text{ (kilograms)}, \\ r &= \text{distance between } a \text{ and } b \text{ (meters)} \end{aligned}$$

In this example, notice the *where* is placed below the equation and flush left; the symbol definitions are indented and aligned on the equal signs; the definitions are in the same order in which the symbols appear in the equation; appropriate punctuation appears at the end of each line; and units of measure are included for each quantity.

8. Using Quotations, Citations, and References

8.1 To establish a sense of both ethics and credibility, authors should always recognize the work of others that has been incorporated into their text, illustrations, or data displays. Virtually all professions, professional organizations, and publishing agencies provide guidance on recognizing the contributions of others (see 13.8).

8.2 To acknowledge sources of quoted materials within your document, use brief bibliographical notes (citations). At the end of your document you may provide a full bibliography or reference list that describes to the reader exactly where quotations and other source materials can be found. You may also include explanatory endnotes or footnotes.

When quoting and citing outside sources, writers need to be aware of current copyright conventions. For example, the Digital Millennium Copyright Act has created far broader and, at the same time, more specific requirements for electronic publishing (see 13.8).

8.3 To assist authors in understanding these issues, this chapter considers

- Legal guidelines (8.4);
- Quotations (8.7); and
- Citations, notes, and references (8.86).

Legal Guidelines

8.4 Legal issues for the scientific or technical writer can be summed up quite simply; if you are not certain whether to cite a source, then cite it.

8.5 If you have quoted so much from a source that you are not certain whether you need permission, get permission. In doubtful instances, check with company lawyers, the publisher, or supervisory personnel.

8.6 For every quotation, determine whether

- A source must be cited, and if so, the correct attribution for the source;
- The quote has been accurately reproduced, paraphrased, or summarized, to faithfully retain the source's meaning;
- The copyright holder needs to grant permission and if so, whether you have gotten it.

Quotations

8.7 Quotations are words, ideas, objects, or data—including data displays; illustrations; and online, audio, and video materials—found in another source and used in your printed or online documents. The following section offers guidance with

- Quotation usage (8.8),
- Types of quotations (8.17),
- Properly setting quotations in text (8.38),
- Revising quotations (8.46),
- Capitalization (8.66), and
- Punctuation (8.72).

Quotation Usage

8.8 In general, three considerations govern the decision to use a quotation:

- Quoting an established source lends credibility to your position.
- Quotes add emphasis or interest by isolating a specific part of a text.
- Quotations can help create a well-constructed argument.

8.9 In contrast, a similar set of considerations can help authors determine when not to use a quotation:

- Do not quote if it offers no special information or nothing is gained by directly quoting someone.
- Do not quote when a paraphrase or summary can provide the same information more concisely.
- If you use so many quotes that the quotations, rather than your own writing, carry the message, you should be aware of the legal issues involved with overusing quotations (see 8.4).

How to Use Quotations

8.10 Follow these guidelines when using quotations:

- Make quotations as brief as possible.
- Use only as much of the quotation as necessary to convey its sense.
- Form sentences so that quotes fit naturally into the logical, grammatical, and syntactic flow of the writing.

8.11 Limit lengthy quotations as much as possible, although you may need them on rare occasions, such as when the quotation source is unavailable to most readers. In these cases, put the information in a separate appendix and use standard footnote techniques to refer to it.

When to Cite Quotations

8.12 Cite both the source of an idea or a particular expression of an idea. The following situations demand citation:

- Direct quotations—When another's words, data, or ideas are used directly;
- Indirect quotations—Paraphrasing or summarizing another's words or ideas, especially important to attribute if the source provides a first exposure to an idea;
- Dialog—The exact words used by someone during an interview;
- Little-known facts—Facts or data, such as a report on new research, that are commonly unavailable.

8.13 Technical authors should reference how their work is related to supporting information. Acknowledging these sources is an important part of writing for scientific and technical audiences.

8.14 Properly citing sources satisfies legal requirements, strengthens the author's argument, maintains intellectual integrity, and provides a valuable resource for the reader.

When Not to Cite a Source

8.15 Although authors should properly reference supporting material, sources need not be credited at all times. Writing that is too densely documented can be difficult to read. In such cases, rather than documenting every sentence in a passage, group references and document the entire passage as a unit.

8.16 Ideas and information available from many sources are considered common knowledge and need not be documented unless you are directly quoting the sources. Do not cite

- Commonly known facts—George Washington was the first president.
- Facts or beliefs commonly available from many sources—Mount Everest is the highest mountain in the world.
- Proverbial, biblical, or well-known literary expressions—Love of money is the root of all evil.

Types of Quotations

8.17 Direct quotations faithfully reproduce the source's exact words or data. Use direct quotations when there might be a question about the original statement or a representation of it. Use direct quotations to reproduce words or data exactly as they appear in their source.

8.18 Indirect quotations are useful in representing the sense of a long quotation, when the original is unnecessarily long, or the exact wording is not as important as the sense it conveys. Use indirect quotations to convey only the main idea or ideas behind a lengthy external source. An indirect quotation calls less attention to itself.

Types of Direct Quotations

8.19 Run-in quotations are usually shorter than three or four lines and are set directly within the text.

8.20 Block quotations are set off from the main text if their length exceeds four lines.

8.21 Nonstandard quotations—epigraphs, oral sources, correspondence, footnotes, and foreign language—require special techniques (see 8.24).

Guidelines for Direct Quotations

8.22 Keep them brief. Use only as much of the quotation as necessary to convey the sense of the quotation.

8.23 Form sentences so that quotes fit naturally into the logical, grammatical, and syntactical flow of the text:

> John Carroll tells us that users often miss important information in manuals because they "come to the learning task with personal concerns. . . . They skip critical information if it doesn't address those concerns."

Guidelines for Nonstandard Direct Quotations

8.24 This section provides guidance on the use of nonstandard direct quotations including

- Epigraphs (8.25),
- Quoting oral sources (8.27),
- Quoting from correspondence (8.31),
- Quoting within footnotes (8.33), and
- Foreign-language quotations (8.34).

Epigraphs

8.25 Epigraphs are interesting, relevant quotations placed at the beginning of a book or each chapter of a book. Use the following guidelines for epigraphs:

- Do not use quotation marks.
- Use a "display" typeface. This can be as simple as using an italic font several points larger than the body text.
- If the source needs to be cited, it should be right-aligned below the quote, usually in a slightly smaller typeface.

8.26 An epigraph might look like this:

> *Small opportunities are often the beginning of great enterprise.*
>
> —Demosthenes

Quoting Oral Sources

8.27 Treat quotes from speeches or conversations like quotations from written sources. Cite the date, time, and location of the speech or conversation accurately. Also, get the speaker's permission to use quotations from the speech.

8.28 Direct and indirect dialogue are similar to direct and indirect quotes (see 8.17–8.19). Direct dialogue reproduces the dialogue verbatim; indirect dialogue paraphrases the speech.

8.29 Direct dialogue does not require that the speaker be identified for each utterance. If only two people are speaking, then it is sufficient to indicate the change of speaker by starting each person's words on a new line and by beginning and ending each speaker's words with double quotation marks:

> Bob asked, "What are you doing?"
>
> "Starting the experiment," Tom replied.
>
> "But isn't the flask supposed to be empty?"
>
> "No! It has to have distilled water in it."
>
> "I think you better check that procedure list again."

8.30 Question and Answer format (Q & A) is sometimes used as an instructional device to convey technical information. For this kind of Q & A, do not use quotation marks:

> *Why should I send in my registration card?*
>
> Registered owners receive immediate notice of updates, as well as a special price on upgrades.
>
> *Are there any other benefits?*
>
> Yes. You will also be notified of new products and receive other special offers from time to time.

Quoting from Correspondence

8.31 Correspondence includes letters or memoranda written by one person and sent to one or more other people. The rules for including quotations from correspondence are the same as for including quotations from other printed sources, such as books.

8.32 If parts of the correspondence such as the salutation are included, reproduce them exactly as you would a block quotation.

Quoting within Footnotes

8.33 To include a quotation within a footnote, follow the same rules for quoting and citing as if the quotation were within the body text. If the quotation is longer than two or three sentences, consider including the quotation within the primary text as part of the discussion rather than as a footnote.

Foreign-Language Quotations (see also chapter 2)

8.34 In general, treat foreign-language quotations in the same manner as other quotations:

- Accurately reproduce the quotation, including special typographic conventions such as non-English characters and punctuation.
- Except for rare circumstances, always provide an accurate translation. The entire point may be lost by including a foreign quotation that the reader does not understand.
- Put the translation within square brackets like other quotation additions. Or, put the translation in a footnote, especially if additional explanation is needed.
- Provide your own translation only if no other is available.
- Never retranslate a quotation that has already been translated. Find the original and use it.
- Provide definitions of specialized or controversial terms.

Guidelines for Indirect Quotes

8.35 Use indirect quotations to convey only the main idea or ideas behind a lengthy external source.

8.36 Any summary or paraphrase must accurately represent the source's ideas and arguments. Summarizing or paraphrasing does not eliminate the responsibility for accuracy.

8.37 Do not simply rearrange the words in the original quote. To paraphrase successfully put the idea in your own words:

> *Skipping.* New users often seem to feel that what cannot be used can be skipped. One person we observed dismissed several pages of explanation in a training manual, commenting, "This is just information," as she flipped past it. People come to the learning task with personal concerns that influence their use of training materials. They skip critical material if it doesn't address their concerns. They browse ahead until they find an interesting topic and ignore its prerequisites.

This passage can be successfully paraphrased as

> Readers use documents the way they want, not the way the author intends.

Properly Setting Quotations in Text

8.38 Quotations should fit naturally, grammatically, and logically into the flow of the writing. When using a quotation that is not a complete sentence, either use enough of the quotation to make it sound like a complete sentence or incorporate it into the sentence in such a way that the sentence and the quotation form a single, natural-sounding sentence.

Guidelines for Setting a Quotation within the Text

8.39 To set quotations properly within a text

- Introduce your quotation to identify the source and show the reader the relevance of the quotation in relation to the subject.
- Place the introduction to suit the style of the surrounding text.

    As Carl Nagas once begged, “Stop, in the name of science.”

    “Stop,” Carl Nagas once begged, “in the name of science.”

    “Stop, in the name of science,” Carl Nagas once begged.

- Try to avoid overly formal introductions; those that use “said” or “the following quotation” are poor style.
- More formal introductions may be necessary for longer quotations.

    The warning at the end of his article is clear: “In today’s litigious society it is foreseeable that trainers could be sued. A lawyer wanted to get a trade article that stated that inadequate training was being given to restaurant staff admitted as evidence in a recent court case.”

Introducing Block (Set-off) Quotations

8.40 As with run-in quotations, the introduction for a block quotation depends on the quotation’s context. In general, use a complete sentence to introduce a block quotation that starts with a complete sentence, and use an incomplete sentence to introduce a block quotation that starts with an incomplete sentence.

8.41 Quotations introduced with a complete sentence can use either an informal sentence that ends with a period or a formal sentence that ends with a colon.

Informal Introduction

8.42 Use an informal introduction when the exact content of the passage is not important:

> In *The Mythical Man-Month,* Frederick Brooks describes the biggest fallacy of all when scheduling.
>
> > All programmers are optimists. Perhaps programming attracts those who believe in happy endings. Perhaps the nitty frustrations drive away all but those who habitually focus on the end goal. Perhaps it is merely that computers are young, programmers are younger, and the young are always optimists. But however the selection process works, the result is indisputable: "This time it will surely run," or "I just found the last bug."
>
> So the first false assumption that underlies the scheduling of systems programming is that all will go well.

Formal Introduction

8.43 Use a more formal introduction when the exact content of a passage is important:

> In *The Mythical Man-Month,* Frederick Brooks lists three common causes of scheduling disasters:
>
> > First, our estimating techniques are poorly developed. More seriously, they reflect a false assumption, that is, that all will go well.
> >
> > Second, our estimating techniques falsely confuse effort with progress, hiding the assumption that men and months are interchangeable.
> >
> > Third, because we are uncertain of our estimates, software managers often lack flexibility.

8.44 If the block quotation begins with an incomplete sentence, introduce it in the same manner as a run-in quotation—form the introduction so that the text flows smoothly from the introduction into the quotation. The incomplete sentence in the quotation should complete the initial sentence:

> In the third chapter, Brooks describes a method for completing large jobs efficiently based on a team that is
>
> > organized like a surgical team rather than a hog-butchering team. That is, instead of each member cutting away at the problem, one does the cutting and the others give support that enhances effectiveness and productivity.

Introducing Indirect Quotations

8.45 Usually, an indirect quotation needs no specific introduction. The purpose of an introduction is chiefly to identify the source to the reader. This information should be naturally included within the summary or paraphrase:

> Brooks makes it obvious in chapter 2 that the major reason for missing schedules is the lack of accurate scheduling.

Revising Quotations

8.46 Revise a direct quotation to

- Maintain the grammatical sense of the writing,
- Clarify the quote within the context of the writing,
- Keep the quotation as brief as possible,
- Omit extra or distracting words or phrases.

8.47 In revising a quotation, remember the following guidelines:

- Do not change the meaning of the quotation. Intentionally altering a quotation to change its meaning can be construed as libel;
- Treat the source fairly;
- Do not point out such revisions as

  Matching the text's capitalization by changing the first letter's case (unless writing a legal document or matching a publisher's style);

  Changing commas and periods at the end of a quotation to match the punctuation needs of the sentence that includes the quotation;

  Changing all caps to small caps, italic to underline, and similar typographical items, as long as the intended meaning is retained.

Adding Material to Quotations

8.48 Add material to a direct quotation only to clarify something that might not be apparent to the reader without being able to see the entire quotation in its original context. Enclose additions within a quotation in square brackets [like this].

8.49 If a misspelling involves the omission of a letter, add the letter in the appropriate place within square brackets. If the misspelling is obvious, either add the correct letter within square brackets or use [*sic*]:

> He failed to use the cor[r]ect address.

8.50 *Sic*, Latin for *thus* or *so*, indicates that an obvious error has been faithfully reproduced from the original. Misspellings, erroneous data, or apparent errors in logic can be indicated with [*sic*]:

> Jackson intended to oriant [*sic*] the learner to the system.

Note that, stylistically, [*sic*] can convey more than the fact that a quote has been accurately reproduced. In some cases, especially where a quotation has been included so that it can be refuted, using [*sic*] has negative implications.

8.51 When the correct spelling, number, date, or other error is not obvious, put the correct information in square brackets:

> Galacia, a region of central Europe, includes the north slope of the Carpathian Mountains and the valleys of the upper Vistula, Diester [Dniester], Bug, and Seret Rivers.

8.52 Use square brackets [ ] with nothing between them to indicate information missing from the source, for instance, if the original is unintelligible or incomplete. This practice often occurs when the source is damaged and unreadable.

8.53 Put clarifying notes in brackets. Such explanatory text might include the source of italics, explanation of a pronoun reference, translations, and other clarifications:

> According to several theorists, one of the most frequent errors made by instructors is expecting too much commitment *before* [emphasis in original] providing reinforcers. In volunteer organizations, consistent, frequent, timely, and *appropriate* [emphasis mine] reinforcement significantly influences commitment.

> When he [Tom Smithson] was a student, they say he was very wild.

> These questions [motivation planning questions] apply to planning for meetings, instructional sessions, projects, and other activities.

Omitting Parts of Block (Set-off) Quotations

8.54 When an incomplete sentence introduces the block quotation, the first sentence of the quotation, normally another incomplete sentence, finishes the introduction. In this case, do not begin the quotation with ellipses:

> Frederick Brooks says that the biggest scheduling fallacy is that
>
> > all programmers are optimists. Perhaps this modern sorcery attracts those who believe in happy endings and fairy godmothers.

8.55 When a complete sentence introduces the block quotation, the quotation normally begins with a complete sentence. In this case, do not use ellipses:

> In *The Mythical Man-Month*, Frederick Brooks describes the biggest fallacy of all when scheduling.
>
> > All programmers are optimists. Perhaps this modern sorcery attracts those who believe in happy endings and fairy godmothers.

8.56 When a complete sentence introduces the block quotation and the quotation begins with an incomplete sentence, precede the first sentence with ellipses:

> In *The Mythical Man-Month*, Frederick Brooks describes the biggest fallacy of all when scheduling.
>
> > ... programmers are optimists. Perhaps this modern sorcery attracts those who believe in happy endings and fairy godmothers.

8.57 If text has been removed from within the quotation or at the end of the quotation, follow the same guidelines as for run-in quotations:

> In *The Mythical Man-Month*, Frederick Brooks describes the biggest fallacy of all when scheduling.
>
> > All programmers are optimists. Perhaps this modern sorcery attracts those who believe in happy endings.... Perhaps the nitty frustrations drive away all but those who habitually focus on the end goal. Perhaps it is merely that ... the young are always optimists.

8.58 Setting off a quotation allows the writer to reproduce a long, continuous piece of text. However, if the resulting block quotation needs considerable editing, either paraphrase or rewrite the quote to incorporate it into the primary text as a run-in quotation.

Omitting Parts of Run-in Quotations

8.59 The treatment of a run-in quotation depends on how much of the quotation is used and what part of it is omitted. To indicate an omission, use ellipses—three periods with a space between the periods and the surrounding text.

8.60 When the quotation is an obviously incomplete sentence, no ellipses are needed:

> Jackson warns that, partly due to "today's litigious society," trainers have to worry about being sued.

8.61 When the quotation is a sentence with the beginning omitted, do not use ellipses. Simply treat the quotation's initial punctuation as part of your sentence:

> At the end of his article he warns that "it is foreseeable that trainers could be sued. A lawyer wanted to get a trade article that stated that inadequate training was being given to restaurant staff admitted as evidence in a recent court case."

8.62 When the quotation is a sentence with the end omitted, only use ellipses if the omission makes the ending grammatically incorrect, or to intentionally call attention to the fact that the original text does not end there:

> The warning at the end of his article clearly places the blame: "In today's litigious society . . ."

8.63 Use ellipses to indicate an omission when the quotation is a sentence with text removed from the middle.

> The warning at the end of his article is clear: "In today's litigious society it is foreseeable that trainers could be sued. A lawyer wanted to get a trade article . . . admitted as evidence in a recent court case."

8.64 When the quotation is longer than a sentence, and text is omitted between sentences or at the end of the first sentence, include enough of the text to form complete sentences and use a four-dot ellipsis to indicate the missing text. In fact, a four-dot ellipsis is actually a period followed by an ellipsis:

> Frederick Brooks identifies the biggest scheduling fallacy as the fact that "all programmers are optimists. . . ." The result is indisputable: "This time it will surely run," or "I just found the last bug."

8.65 If the sentence preceding the omission in a multisentence quotation requires punctuation other than a period, use the punctuation and a standard ellipsis.

> The article once again raised the question: "Should we certify technical writers? . . . We must answer the other questions before we can answer this one."

Capitalization

8.66 Change the initial capitalization of the quotation to make it grammatically correct with the remainder of the text. Base that decision on how the quotation would appear if it were your own words rather than a quotation, and whether the quotation can be read as a part of main text or as a separate sentence.

8.67 Changing capitalization from the original does not normally have to be indicated in any way. For legal documents, for some publishers, and in cases where the meaning may be affected if capitalization is altered, a change must be indicated by putting the changed letter in square brackets (see 8.48).

Capitalizing Run-in Quotations

8.68 Lowercase run-in quotations that are intended to be read as part of a sentence:

> Emerson might just as well have been talking about useless software interface features when he said that "a foolish consistency is the hobgoblin of little minds."

8.69 Capitalize a quotation that should stand out as a separate sentence:

> Some software interface features are created simply to be consistent with another feature. But, as Emerson once said, "A foolish consistency is the hobgoblin of little minds."

Capitalizing Block Quotations

8.70 Because block quotations are reproduced exactly as they appear in the original and generally are not part of the primary text, the capitalization of the original should not be changed.

8.71 Block quotations that begin with an incomplete sentence follow the same rules as run-in quotations; capitalize according to how the quotation fits into the sentence that introduces it:

> In the third chapter, Brooks describes a method for doing large jobs efficiently. The method is based on a team that is
>
> > organized like a surgical team rather than a hog-butchering team. That is, instead of each member cutting away at the problem, one does the cutting and the others give him every support that will enhance his effectiveness and productivity.

Punctuation

8.72 In addition to square brackets (see 8.48) and ellipses (see 8.59), quoted text uses a series of conventions to control the placement of other punctuation in relation to quotation marks.

Quotation Marks

8.73 Place double quotation marks at the beginning and end of simple quotations:

> "New users often seem to feel that what cannot be used can be skipped."

8.74 Start and end run-in or nested quotations with double quotation marks. Use single quotation marks for nested quotations:

> As Fredd says, "The paper confirms that she said 'I think the robber called it a cannon.' "

8.75 If a quotation has more than one quotation nested within it, reconsider whether the quotation is obvious enough to be useful to the reader.

Quotation Marks in Block (Set-off) Quotations

8.76 Do not use quotation marks to start or end a block direct quotation, unless it is part of the style used by the publisher or the person for whom the writing is intended.

8.77 If a block quotation includes a quotation, use double quotation marks and alternate with single quotation marks as needed for nested quotations:

> John Carroll addresses the issue of why users often miss important information.
>
>> "New users often seem to feel that what cannot be used can be skipped. One person we observed reacted dismissively to several pages of explanation in a training manual, commenting, 'This is just information,' as she flipped past it."

8.78 Punctuation that is part of a quotation should be retained and included within the quotation marks. However, commas and periods at the end of a quotation should be changed to fit the sentence's grammar.

Commas and Periods

8.79 Use a comma to introduce a quotation unless the quotation is a predicate nominative or a restrictive appositive:

> Emerson might just as well have been talking about useless software interface features when he said that "a foolish consistency is the hobgoblin of little minds."

8.80 Commas and periods should always fall within the quotation marks (see also 4.117):

> Some software interface features are implemented simply because of consistency with another feature. But, as Emerson once said, "A foolish consistency is the hobgoblin of little minds."

8.81 Change commas and periods at the end of a quote to match the context of the sentence containing the quote:

> "I refuse to testify," he said.
>
> He said, "I refuse to testify."

Other Punctuation and Quotations

8.82 Punctuation marks other than commas and periods that are part of the quotation go within the quotation marks:

> "I refuse to testify!" he said.
>
> He said, "I refuse to testify!"

8.83 Punctuation marks other than commas and periods that are not part of the quotation go outside the quotation marks:

> He said, "I refuse to testify!"; the prosecuting attorney heard him!

8.84 When the quotation ends a sentence and the sentence has punctuation other than a period, put that punctuation outside the quotation marks:

> What does he mean by "I refuse to testify"?

8.85 If a quotation falls at the end of such a sentence and the quotation has special punctuation, use both the original punctuation and the punctuation needed for the sentence:

> Did he say "Do I refuse to testify?"?
>
> Did he say "I refuse to testify!"?

Citations, Notes, and References

8.86 Scientific and technical authors document their primary source material in many ways, depending on the context in which they distribute or publish their work. Three preliminary decisions can help writers organize and document supporting resources:

- Selecting an appropriate style guide (8.88),
- Using a supportive database program (8.90), and
- Understanding the basic aspects of documentation (8.92).

8.87 Writers should acknowledge sources by using the appropriate citation method for their audience:

- In-text citations—Full or abbreviated references in the text, illustration, or data display (8.95);
- Footnotes—Traditional, scholarly referencing practice (8.101);
- Endnotes—Full references compiled after the text (8.107);
- Reference lists—Separate and complete listing of all relevant or useful resources (may be combined with any of the above techniques) (8.117).

Selecting an Appropriate Style Guide

8.88 Choose the reference method most appropriate to the context in which the work will be published. Many professional organizations and technical publishers require specific styles for documenting sources:

Exhibit 8–1: Style Guides for Specific Professions

Professional Field	**Style Guide** (see 13.8 for complete citations)
Humanities	*MLA Handbook for Writers of Research Papers* *MLA Style Manual and Guide to Scholarly Publishing*
Psychology, social sciences, and some technical fields	*The Publication Manual of the American Psychological Association*
Mathematics	*AMS Author Handbook*
Mathematics and the natural sciences	*The CBE Manual for Authors, Editors, and Publishers*
Medical	*The AMA Manual of Style*
Physics	*Style Manual for Guidance in the Preparation of Papers*
General publishing	*The Chicago Manual of Style* *Words into Type*

8.89 Most publishers will provide style sheets or the name of a standard reference work for you to follow. If you have not been provided a specific set of guidelines, you should select a format related to the professional field and use it consistently throughout your work.

Using a Supportive Database Program

8.90 The serious scholar or research writer should take advantage of one of the excellent bibliography software applications now available. For anyone engaged in research, these tools provide enormous advantages over other methods of tracking research materials by providing the ability to

- Create multiple, special-topic bibliographies;
- Merge two or more bibliographic databases;
- Transform a bibliography into a variety of traditional formats—*APA*, *MLA*, and many others—or including your own custom formats;
- Maintain consistent bibliographic databases for a lifetime of research;
- Search bibliographic databases for key terms;
- Acquire and automatically incorporate bibliographical records from other researchers or research organizations.

8.91 Generally, these applications, including EndNote™ and ProCite™, offer a choice of on-screen forms to record each source: books, book chapters, electronic sources, and so on. These forms contain fields for different kinds of identifying information—in most cases, more than many researchers would normally use. Most of these applications will also provide the means to create your own specialized forms for unusual sources.

Understanding the Basic Aspects of Documentation

8.92 To identify source material properly, you have to provide enough information so interested readers can readily locate the original works. Although the conventions vary, a proper citation of source material appearing in books includes

- Author(s) names (first and middle initials are preferred to full names in technical documents),
- Book title (or, in the case of a compilation, the chapter or paper title, followed by the book title),
- Place of publication,
- Publisher,
- Publication year,
- Page number or range of pages.

8.93 In addition, other information such as the editor or translator, the volume and number, and the series or edition may be included.

8.94 Document source material appearing in journals or periodicals by including:

- Author(s) names (first and middle initials are preferred to full names in technical journals),
- Title of the article or paper,
- Name of the journal or periodical in which the work appears,
- Volume and/or issue number,
- Date of publication,
- Page numbers or range of pages.

In-Text Citations

8.95 An in-text citation, or "author-date citation," acknowledges a reference immediately after it appears. Place the citation in parentheses. The in-text citation typically has only two parts: the author(s) last name(s) and the publication date. If the work is a book or a long article, include a chapter or page number. The example shows an abbreviated in-text citation; the last number is a page reference:

> The structure of a computer tutorial must consider student motivation (Alessi and Trollip 1985, 281). Furthermore, the success rate for such programs . . .

This example—and most of the examples in this chapter—illustrates a general convention. Be aware of the conventions in your field and recognize that styles vary considerably.

Complete Reference in Text

8.96 In a work with only a few sources or in a less formal document (such as a magazine article), it may be appropriate to include the entire reference in the text. Place the necessary information within parentheses and treat it as part of the text:

> Adults typically read texts to learn how to perform an action (S. T. Kerr, "Instructional Text: The Transition from Page to Screen," *Visible Language* 20 [1986], 368).

As the example shows, the reference overwhelms the text in which it appears. Therefore, use this format when there are only a few references.

Exhibit 8–2: Sample Citation Variations for Major Professional Style Guides

Modern Language Association	American Psychological Association	Chicago Manual of Style
	Author Unnamed in Text	
(Smith 444)	(Smith, 1955, p. 444)	(Smith 444)
	Author Named in Text	
(444)	(p. 444)	(p. 444)
	Corporate Author Named in Text	
The information viewers receive from the screen cannot be predicted by their apparent attention (Acme Limited 444).	The information viewers receive from the screen cannot be predicted by their apparent attention (Acme Limited, 1955).	The information viewers receive from the screen cannot be predicted by their apparent attention (Acme Limited 1955).
	Two or Three Authors	
(Smith and Jones 444) (Smith, Jones, and Morgan 444)	2 authors in APA (Smith & Jones, 1955)	(Smith and Jones 444) (Smith, Jones, and Morgan 444)
	Three or More Authors	
(Smith et al. 1955) Or: (Smith, Jones, Morgan, and Helmreich 444)	3–5 authors in APA Smith et al. (1955) 6+ authors in APA (Smith et al., 1955)	More than 3 authors: (Smith et al. 1955) Or: (Smith, Jones, Morgan, and Helmreich 444)
	An Author with Two or More Cited Publications	
After the author's surname abbreviate the title; then add page numbers. A limited environment differs from an enriched environment in both number of personnel and amount of equipment (Kirsch, Studies 55–85).	Add a letter to the date only after alphabetizing the references. A limited environment differs from an enriched environment in both number of personnel and amount of equipment (Kirsch, 1988b, pp. 55–85).	

Illustrations and Data Displays as In-Text Notes

8.97 For illustrations and data displays (see also chapter 11) place the reference at their foot. For multipage illustrations or data displays, place the reference at the foot of the first page. In ruled tables or figures, enclose the reference in a box:

Mid	15.0	14.0	1.5	1.5
Stern	32.0	29.3	2.5	2.4
SOURCE: Put Attribution Here.				

8.98 Introduce references with SOURCE: or SOURCES: in emphatic type (conventionally cap and small caps).

8.99 If the illustration or data display is reproduced directly from another publication, follow the citation with the words "reprinted with permission of the publisher" in parentheses:

> SOURCE: John Paulson, "Tensile Strength Loss due to Radiation," *Metallurgy Review*, September 1986, p. 55 (reprinted with permission of the publisher).

8.100 If the publisher requires that the citation give credit with some other wording or format, honor the publisher's request.

Footnotes

8.101 Footnotes are the traditional and accepted method for citing and documenting sources in many disciplines.

8.102 Place footnotes at the bottom of the page on which the reference occurs. A superscripted number (or less commonly, a bracketed number) follows the reference in the text. Also, place the number at the bottom, or foot, of the page, followed by the author's name and the rest of the necessary bibliographic information, as in an endnote. Use the same format (either superscript or bracket) for the note number at the foot of the page as for the note number in the text.

8.103 Begin numbering footnotes with one (1) for each chapter or section of a work. Although in many older works, the first footnote on each page is assigned the number 1, this practice requires continual renumbering when the work is edited and paginated.

8.104 When citing the same work in two or more footnotes, follow the guidelines suggested in the section on endnotes (see 8.107).

8.105 Footnotes place references close to the text they support. This placement eliminates flipping from the text to a list of references at the end of the book or chapter.

8.106 However, footnotes have several disadvantages. They

- Interrupt readers by directing their attention to the note at the bottom of the page,
- Can create typographically untidy pages,
- Do not provide a collected resource—as endnotes do—unless they are accompanied by a bibliography.

Endnotes

8.107 Endnotes are similar to footnotes, except endnotes are grouped together at the end of the entire work or at the end of individual sections, rather than at the bottom of the page.

8.108 In some cases, endnotes and footnotes may be substituted for a full reference list. Consult your publisher to see if this practice is acceptable.

8.109 Place endnote numbers at the end of a sentence (or clause), so that they do not interrupt the text. Use a small superscripted number placed after any punctuation to refer to a specific endnote. Endnote numbers begin at one (1) and continue sequentially within chapters:

> A computer tutorial's structure must also consider the students' motivation.[4]

8.110 When enclosed in square brackets or parentheses, the number is placed before a punctuation mark:

> A computer tutorial's structure must also consider the students' motivation [4].

8.111 Create a numbered list, entitled "Notes" or "Endnotes," to provide the necessary bibliographic information. If superscripted endnote numbers appear in the text, punctuate them with a period—not superscripted—in the notes:

> 4. S. M. Alessi and S. R. Trollip, 1985, *Computer-Based Instruction: Methods and Development* (Englewood Cliffs, NJ: Prentice-Hall), 281.

8.112 If the endnote numbers appear bracketed in the text, bracket them in the notes:

> [4] S. M. Alessi and S. R. Trollip, 1985, *Computer-Based Instruction: Methods and Development* (Englewood Cliffs, NJ: Prentice-Hall), 281.

8.113 If a work is cited more than once, treat subsequent citations in one of two ways:

- If the work cited is the same as that in the immediately preceding note, use the expression "ibid." ("in the same place") followed by the page number, instead of a complete citation.
- If the work has been cited several notes earlier, use only the author's last name plus the expression "op. cit." ("in the work cited") or "loc. cit." ("in the location cited") followed by the page number.

8.114 Endnotes containing ibid., op. cit., or loc. cit. can make it difficult to find the correct citation. Thus, in contemporary practice, references to previously cited works are recorded by a shortened form consisting of the author's last name, key words of the title, and the page number. Here, for example, is the first reference to a work:

> 3. C. D. Perkins and R. E. Hage, 1949, *Airplane Performance, Stability, and Control* (New York: John Wiley & Sons, Inc.), 148.

Here is a subsequent citation for the same work:

> 6. Perkins and Hage, *Airplane Performance*, 236.

8.115 Endnotes group many references together; therefore, they are a valuable resource to those readers who are interested in learning more about a subject. On the other hand, endnotes interrupt readers by forcing them to flip back and forth between the text and the end of the book or chapter.

8.116 Because endnotes are grouped together, editing the text and updating the notes list are easier than updating footnotes distributed throughout the work. Editing a work containing numbered notes, either footnotes or endnotes, requires extensive renumbering. This task can be time-consuming, although many modern word processing programs automatically renumber footnotes and endnotes during editing (see 8.90).

Reference Lists

8.117 Provide a reference list at the end of the article or chapter to supplement textual notes; the reference list gives complete citation for all works cited.

8.118 At the end of the article or book, compile a single list of all sources cited in textual notes. In accordance with the type of document and context of publication, title this list "References," "Works Cited," "Works Consulted," or the like. Each entry in the list should include all the bibliographic information about the source:

Alessi, S. M., and S. R. Trollip. 1985. *Computer-Based Instruction: Methods and Development.* Englewood Cliffs, NJ: Prentice-Hall.

8.119 Organize the list in alphabetical order by author or, if no author or editor is listed, by sponsoring organization (see 8.123).

8.120 Give special attention to multiple authors, multiple publications in a given year, and multiple works in a single citation (see Exhibit 8–2).

Bibliography

8.121 A bibliography may be a list of works cited, consulted, or of interest to the subject; that is, it may be an extension of the reference list or a subset or a completely different set of documents. The reader usually assumes that the compiler of a bibliography has listed the most important works, if not all of them. Thus, a bibliography has a broader purpose than a reference list or a list of works cited.

8.122 A bibliography can be used alone or with footnotes or endnotes. Since footnotes deprive the reader of a collected group of references (see 8.106), a bibliography is especially convenient in addition to footnotes.

8.123 Arrange bibliographies alphabetically by the last name of the principal author or editor. In works where there is no explicit author or editor, list the entry alphabetically by sponsoring organization:

Ohio University. 1973. *Receivership Skills: Restructuring Our Principal Justifications.* Athens, Ohio: Ohio University Broadcast Research Center.

8.124 Divide a long bibliography into sections by subject matter or by document type. Provide headings that describe the organizational method. Dividing a bibliography by subject matter, for instance, produces:

Computer-Assisted Training

Keller, J. M. 1983. "Motivational Design of Instruction." In C. M. Reigeluth (ed.). *Instructional Design Theories and Models: An Overview of the Current Status.* Hillsdale, NJ: Lawrence Erlbaum.

Stein, J. S., and M. C. Linn. 1985. "Capitalizing on Computer-Based Interactive Feedback: An Investigation of Rocky's Boots." In M. Chen and W. Paisley (eds.). *Children and Microcomputers.* Beverly Hills, CA: Sage Publications.

Video Disk Training

Manning, D., P. Balson, D. Ebner, and F. Brooks. "Student Acceptance of Videodisc-Based Programs for Paramedical Training." *T.H.E. Journal* 11 (November 1983): 105–8.

8.125 Organization by document type is a characteristic of dissertations and of academic bibliographies. Here is an example organized by document type:

Books

Dynkin, E. B. 1982. *Markov Processes and Related Problems of Analysis.* Cambridge: Cambridge University Press.

Iosifescu, M. 1980. *Finite Markov Processes and Their Applications.* Chichester: John Wiley & Sons, Ltd.

Journals

Biggins, J. D. 1987. "A Note on Repeated Sequences in Markov Chains." *Advances in Applied Probability* 19, 739–42.

Citing Electronic Sources

8.126 While most major academic and professional organizations have standards or guidelines for electronic citations, such guidelines are relatively new, and they vary in style and format. This section presents examples from some of the most prominent organizations with the caveat that there is, currently, no universal standard for citing electronic sources. Furthermore, the continuing changes to the online environment itself and the invention of new media affect the appropriateness of any recommendations.

8.127 Because conventions and guidelines are far from stable, the examples in this section are subject to change. If you are writing for a particular publication or organization that uses a specific format, be sure to get the most current copy of its standards or guidelines. Otherwise, select one style and use it consistently throughout your work. Just keep in mind that you must include enough information for your reader to locate the document you are referencing.

8.128 Since information on the Internet can appear and disappear with no notice or warning, electronic citations tend to be less reliable than citations for hard-copy documentation such as books and journal articles. Because of this, you should try to cite only locations that you know to be relatively stable. The types of electronic sources you may encounter include

- Websites on the World Wide Web
- Electronic mail
- Gopher sites
- FTP archives
- Online magazines or newspapers
- Online databases
- Graphics (e.g., .gif, .jpeg)
- PostScript files
- PDF files
- Audio/sound files
- Video files
- Listserv messages
- IRC, MOOs, MUDs
- Telnet sites
- Usenet (newsgroups)

8.129 The key to effective electronic citation, as with any type of citation, is providing an accurate and complete path to the original source document. It is advisable to either download onto a disk the online document you are citing or print off a paper copy for your records. Be sure to date the copy and add any other notes that will help you identify it in the future.

8.130 Because of the variety and complexity of electronic documents, the electronic citation must include as much detailed information as possible so that your readers can locate it. For example, your record should include

- Title of the online document and/or the title of the site on which the document appears. If the title is not obvious, try viewing the source code and looking for the <title> tag; or if the document has a subject line and no title, use the subject line.
- Author, editor (maintainer), translator, or sponsoring organization;
- Number of pages, sections, or any other apparent organizational unit;
- Date, if provided (often "last date changed")—include the month, day, and year;
- Date you accessed the information included in parentheses at the end of the citation (13 Jan. 2000);
- URL (uniform resource locator) or other identifier (see also 4.107);

In addition, if available, you should also include

- Information on purchasing a print version;
- Version, volume, or issue number;
- Name of the online service if the document is accessible only through that service.

8.131 The primary difference between conventional citations for print documents and electronic citations from the World Wide Web is the inclusion and method of presenting the URL. The style is similar but not consistent among the various standards and guidelines. For URLs that take up more than one line, break the URL at a slash (/), and continue on the next line. Note that the URLs in the citations below are examples only; they are not, necessarily, actual locations on the World Wide Web or in other online databases.

Modern Language Association (MLA)

Bradford, Roark. *John Henry.* 1931. 12 February 2000 <ftp://ftp.ebooks.org/etext/fiction/southern/bradford.txt>.

CNN Interactive. 19 January 2000. Cable News Network. 19 January 2000 <http://www.cnn.com/>.

Coastal Writers Project. Ed. Thomas Shields. January 2000. East Carolina U. 26 June 2000 <http://www.ecu.edu/~shieldst/coastal_writers >.

Jewett, Sarah Orne. *Country of the Pointed Firs.* Boston, 1896. 30 January 2000 <gopher://gopher.vt.edu:10010/02/73/1>.

International Standard Organization ISO 690-2

Dunken, Robert. The Most Frequently Asked Questions about NASA Earth Observatory. In *The NASA Homepage* [online]. 12 February 1999. [cited 2 January 2000]. Available from: <http://www.nasa.gov/EOS>.

Journal of Technology and Computers in Education [online]. Blacksburg (Va.): Virginia Polytechnic Institute and State University, 1999– [cited 15 January 2000]. Semiannual. Available from Internet: <gopher://borg.lib.vt.edu:70/1/jte>. ISSN 1045-1064.

Mathematical Markup Language Home Page [online]. W3C [World Wide Web Consortium]. User Interface Domain. 20 December 1999; 13:02:41 [cited 2 February 1999]. Available from: <http://www.w3.org/MML/>.

Snebur, Belinda. Re: Sextants as Required Sailing Equipment. In USENET newsgroup: *sailing.lit.training.navigation* [online]. 29 December 1999; 09:07:11 [cited 4 January 2000; 13:03 EST]. Message-ID: uni-stuttgart.de. Available from Internet.

American Psychological Association (APA)

Databases accessed from the World Wide Web:

Sackler, L. B. (2000). Researchers can make sense of the Net. *San Francisco Business Journal, 21*(3), pp. 12+. Retrieved January 2, 2000 from DIALOG database (Masterfile) on the World Wide Web: http://www.dialog.com

Other online databases:

Boldt, P. D. (2000). The manager goes to clown school: An experiment in humanistic education, 1991–1995. *The Harvard Business Review*, 164, 10+. Retrieved [month day, year] from DIALOG online database (#74, IAC Business A.R.T.S., Item 15004275)

CD-ROM databases:

Federal Bureau of Investigation. (2000, March). *Encryption: Impact on law enforcement.* Washington, DC: Author. Retrieved from SIRS database (SIRS Government Reporter, CD-ROM, Fall 1998 release)

E-mail (APA recommends citing e-mail communications as personal communications):

D. G. Lawrence (personal communication, 16 May 2000)

Chicago Manual of Style

Brenda F. Knowles, review of *The Priestess of Love*, ed. Carol F. Morgan, in *Welsh Academy Medieval Review* [electronic journal] (Cambridge: Cambridge University Press, 1999– [cited 15 March 2000]), file no. 87.1.3; available from listserv@cc.welshacad.edu;Internet.

"The Development of Christian Architecture: The Roman Church," in ArchiDocs: Primary Architectural Documents from Renaissance Europe [database onlilne] (Greenville, NC: East Carolina University, 1996– [cited 10 April 2000]); available from http://www.ecu.edu/archidon/sbook.html#church.

Sidney Churchill, e-mail to author, 12 November 1999.

American Chemical Society (ACS)

Northern Illinois University Chemistry Gopher Site. gopher://gopher.hackberry.chem.niu.edu (accessed January 2000).

Society for Technical Communication Home Page. http://www.stc-va.org (accessed January 2000).

American Medical Association (AMA)

Albany Center for Medical Education. The ACME Forensic Biochemistry Page. Available at: http://webmed.edu/acme/biochem/home.htm. Accessed August 2, 2000.

Simpson J. A., Wooldridge J. W. Relationship of Niacin to MTBE Levels. *JAMA* [serial online]. 2000;292:2150-2179. Available from American Medical Association, Chicago. Accessed January 14, 2000.

9. Creating Indexes

9.1 Indexing scientific and technical documents has two phases:

- Preparation: setting up the indexing environment (9.7), scheduling (9.15), and determining index organization (9.26) and format (9.40);
- Index processing: marking entries (9.55), filing and sorting (9.56), using indexing software (9.59), writing and editing index entries (9.82), and final proofreading (9.115).

9.2 This chapter will assist individuals with little or no indexing experience in both phases. In addition, it provides a useful refresher for the experienced indexer.

9.3 Standards and indexing organizations provide guidance for developing useful indexes (see also 13.9). All of these groups offer differing opinions concerning such aspects of indexing as sorting words and individual characters, punctuating and capitalizing indexes, and treating symbols. Everyone agrees, however, that once an indexing system has been established, consistency in its implementation improves an index's effectiveness.

Key Indexing Terms

9.4 Five terms are important in discussing indexes:

- Entry—a single index record, consisting of a heading, a locator and/or a cross-reference, and any subheadings together with their locators and cross-references.

- Heading—a word or phrase chosen to represent a document's item or concept in the index.
- Subheading—a word or phrase subordinate to the heading.
- Locator—an identifier that points to information in a document or collection of documents. Although generally represented as a page number, a locator might also be a section or paragraph number.
- Cross-reference—a pointer from one heading or any of its subheadings to another heading. A *see* cross-reference points to a heading that collects all relevant information; a *see also* cross-reference points to another heading that contains additional information (see 9.48, 9.96).

Need for Indexes in Technical Material

9.5 An index identifies and locates relevant information within the material being indexed. The index functions as an organizer and subject finder, not as a duplicate table of contents. The indexer analyzes the subject matter and produces a systematic series of headings that gather the references related to a specific subject in one place, pointing out the relationships among them.

The Goals of a Good Index

9.6 The best indexes help the reader find information with the least effort in the shortest time. Good indexes have five characteristics:

- Complete: If the author emphasizes a particular fact in the text, the index should provide similar emphasis for that fact. Depth is important; index every important point in the text.
- Accurate: Information should be located where the index entries indicate it can be found. Names, dates, and sequences must be spelled correctly, be in a logical order, and cross-referenced right. Edit the index with the same care given to its text.
- Concise: Entries should be clear and logical. The more concise the index entries, the more usable the index will be. For example, *Steel* is general; *Steel, annual U.S. production* is more exact.
- Consistent: Spell the word *color* or *colour*, but never use both. If the person's name is *Jane Smyth*, do not index her as *Smythe* or *Smith*.
- Impartial and objective: The indexer's work has no place for bias or prejudice. The indexer must be objective, faithfully representing the author's presentation, using the author's terminology.

Planning an Index

9.7 In planning an index consider

- Who will use the index,
- Who should create it,
- How long it should be,
- How much time it should take, and
- What tools are required.

The Audience

9.8 An indexer should consider the potential audience's age, education, and expectations, and how they will use the index. While preparing the index, take the user's point of view.

9.9 Consider the life span of the index and the need to revise it over time. The index must accommodate revisions, additions, and deletions. Plan the index not only for today's readers, but also tomorrow's.

9.10 Examine indexes in works on similar subjects to see if your index includes material that someone searching through more than one source would expect to find.

The Indexer

9.11 A good indexer has

- A knowledge of index preparation,
- The ability to work with care and patience,
- An awareness of reader expectations, and
- The ability to analyze technical content.

9.12 Typically, the author knows both the material and the audience, and is a likely candidate for preparing the index. However, the author's closeness to the topic might mean a loss of objectivity, and the author may not know enough about indexing to prepare a good one. On the other hand, a professional indexer brings both objectivity and experience to the task and can add significant value to the finished project. Still, authors of technical documentation routinely prepare indexes as part of their assigned workload.

9.13 An indexer should work with an unabridged dictionary, a technical dictionary and thesaurus for the subject being indexed, if they are available, and any applicable style guides or specifications. Thesauri are available for many technical disciplines, and are valuable for reconciling inconsistencies in terminology in a project involving multiple authors, indexers, or volumes of material (see also 13.9).

9.14 If more than one indexer works on a project, one person should assume the role of index manager, charged with

- Maintaining index consistency,
- Assuring compliance with relevant specifications, and
- Maintaining the indexer's style guide (see 9.21–9.25).

Planning the Index Length

9.15 Index length varies with the document's length and complexity. The indexer must consider project realities—how much space and preparation time can be devoted to creating an index? Space and time often conspire against the indexer, making it all but impossible to develop an index of sufficient depth to meet the users' needs.

9.16 It is difficult to offer an ideal index length. Physical constraints like page size, type size, and number of columns per page affect the index—but the most important constraint is the complexity of content. A short index, occupying 2 percent of the total number of pages, might be adequate for an elementary-level text. On the other hand, in a serious scientific or technical work, the index can account for 15 percent of the total document length. In some standard desk references, such as *Roget's International Thesaurus* and *Bartlett's Familiar Quotations*, the index occupies over 60 percent of the total number of pages.

9.17 A publication's size determines an index's potential depth. If space is limited, the indexer must estimate the number of possible entries within that limit. To calculate this limitation, multiply the number of columns per index page by the number of lines per column to derive the number of entries per page; then, multiply that figure by the number of pages. From this total, subtract 20–25 percent for turnover lines, cross-references, and separators between letter groupings.

Preparation Time and Indexing

9.18 The index is often a "rush" job. Final copy of the source is essential to ensure accuracy, but final page proofs arrive late in the publication cycle.

Even indexers with access to the latest versions of source documents know that electronic publishing makes it possible for authors to edit text up to the last possible minute. Such editorial changes introduce new content or alter pagination, which can affect the accuracy of locators.

9.19 Just as it is difficult to estimate index length, it is also difficult to estimate development time. An indexer might be able to complete 8–12 pages per hour while developing a light—one or two terms per page—index. However, in dense technical material yielding ten or more terms per page, that same indexer might complete only two pages per hour. As an initial estimate, assume an indexer can process four to six pages of source material per hour. Indexing is more akin to an editor's pace than a writer's.

9.20 While indexing, plan on spending about 40 percent of the time studying the text and marking and sorting entries, and 60 percent analyzing and editing. This includes the time spent editing both interim and final drafts of the index.

Indexer's Style Guide and Keyword List

9.21 A publishing house will usually supply the specifications for any index it is having developed, but such specifications might not be available in all organizations. Sustained indexing efforts can benefit from the use of an in-house style guide, developed to take the guesswork out of indexing and ensure consistency across projects. The style guide, an "evolving" document that bridges an organization's editorial and indexing process, has two main sections:

- A guide to house conventions for format, design, and mechanics;
- A similar guide to index content.

9.22 The first section of the indexer's style guide might

- Identify any thesaurus or reference works used to support indexing efforts;
- Document in-house templates, style sheets, master documents, or document-type definitions that control index preparation and production;
- Detail typographic and formatting considerations; and
- Log indexing decisions and their rationale.

9.23 The second part of the style guide lists keywords selected from words and phrases in the material being indexed. Using such a list ensures index consistency by serving as the local authority. Outside sources—industries, professional societies, and trade groups—publish authorized lists of keywords that may augment this local list. For additional information on well-known sources of subject authority lists see 13.9.

9.24 A working keyword list should include abbreviations and acronyms and their expanded forms, the structure of synonymous or hierarchical relationships, and the use of organization-specific terms such as product names or trademarks. Listing preferred terms with related terms and cross-references should help resolve questions about standard meaning and usage, and control the introduction of new terminology by correlating obsolete terms with current ones.

9.25 Keyword lists can be formal or informal, as fits the need, and flexible, to change over time to fit an organization's specific objectives. The key is to build both a useful keyword list and a method for developing such lists that will help create better indexes.

Filing Index Entries

9.26 A fundamental indexing decision is deciding how entries will be filed or arranged. The indexer must organize the index alphabetically—either in word-by-word or letter-by-letter order—or by subject.

Alphabetic versus Subject Indexes

9.27 Indexes arranged in alphabetic order dominate in technical publishing because their creation is more straightforward than that of subject indexes, and because an alphabetized, indented index offers three essential usability features that assist the user in retrieving information quickly:

- The vertical, visual orientation allows rapid scanning;
- The alphabet provides a predictable organization method;
- The alphabet lends itself to cross-referencing and makes movement within the index easier for the user.

9.28 Although alphabetic indexing is recommended, there are some useful alternatives for special situations. A subject-oriented classified index is often appropriate in disciplines such as law or the physical sciences.

9.29 However, the classified index introduces its own special problems:

- Creating a subject index requires a level of topical expertise generally beyond most indexers;
- Subject indexes are subjective, and one person's logic can seem complicated and unusable to others;
- Revising or replacing an unstable classification scheme could create a burdensome task for an indexer editing an index prepared with the original scheme.

9.30 A few cases employ a limited, alternative arrangement of subheadings within the headings' alphabetic arrangement. For example, under a heading for *geologic time periods*, the time periods themselves might be listed chronologically. Similarly, under a heading encompassing a series of related numeric codes, perhaps software error codes, the codes themselves might be indexed as subheadings in numeric order.

9.31 Some situations might use a second index to gather acronyms and abbreviations, taxonomic names, significant dates, symbols, or titles; or numeric identifiers for legal cases, parts, patents, or publications.

Alphabetic Arrangements

9.32 Before the widespread use of computer software as an indexing aid, indexing rules were often complicated and inconsistent. However, the current trend treats any letter or number as a graphic symbol, each with a specific value in the sorting order. The older "as if" and "sounds like" rules are slowly passing from use.

9.33 An index can be alphabetized using the computer's strict character-by-character ASCII (American Standard Code for Information Interchange) sort order, but this process may be more trouble than it is worth. More realistic approaches are the word-by-word or letter-by-letter arrangements.

9.34 Any computer program, from operating system utilities to word processors and spreadsheets, using the ASCII character set can produce a reliable, predictable ASCII sort. However, this capability provides a mixed blessing for the indexer because every character has a sort value, and some are not index-friendly. For example, all uppercase letters (ASCII 65-90) are sorted before any lower case letters (ASCII 97-122). Thus, an ASCII sort would "alphabetize" both *Above* and *Zebra* before *about*. In addition, characters typically ignored in indexing, such as the hyphen (ASCII 45) or apostrophe (ASCII 33), are instead sorted in ASCII order. For these reasons, ASCII sorting is of little value to indexers when compared with word-by-word and letter-by-letter sorts.

9.35 In the word-by-word system, start by alphabetizing by the entry's first word. If two entries have identical first words, alphabetize by the second word, then the third, and so on to the first comma, semicolon, or period. The space is critical: in the word-by-word system, "nothing comes before something," that is, letters followed by a space precede the same letters followed by more letters. *New York*, for example, precedes *Newfoundland*. Library catalogs and most major encyclopedias use the word-by-word system.

9.36 The letter-by-letter system considers every letter up to the first significant comma, semicolon, or period. It "ignores" spaces, so *Newfoundland* precedes *New York*. Most dictionaries use the letter-by-letter method. Alphabetization starts over when letter-by-letter encounters a comma or period, meaning *New, York* (if that were a valid entry) would precede *New York.*

9.37 Both systems ignore certain characters in sorting—the apostrophe, the dash or hyphen, diacritical marks (an important feature of foreign-language text), diagonal slashes, quotation marks, and any punctuation that is part of the entry itself. However, since no international standard governs how to sort these characters, national standards bodies sometimes treat them in slightly different ways that create small differences in the sort.

9.38 A comparison of an unmodified ASCII, word-by-word, and letter-by-letter sort illustrates how different methods treat the same list of words:

ASCII Sort	Word-by-Word	Letter-by-Letter
Liquid helium	Liquid helium	Liquid, Karen
Liquid lattices	Liquid, Karen	Liquidambar
Liquid, Karen	Liquid lattices	Liquid-fueled rockets
Liquid-fueled rockets	Liquidambar	Liquid helium
Liquid/solid reactions	Liquid-fueled rockets	Liquidity
Liquidambar	Liquidity	Liquid lattices
Liquidity	Liquid/solid reactions	Liquid/solid reactions

Although both the word-by-word and letter-by-letter alphabetic systems are in widespread use, be consistent after choosing a system and identify which is in use in an introductory note.

General Rules for Alphabetizing

9.39 Both the word-by-word and letter-by-letter arrangements follow these general rules. For additional information on applying these rules to specific cases see 9.82.

- Ignore apostrophes, diacritical marks, diagonal slashes, hyphens, and quotation marks when alphabetizing in either system. Ignore, for example, the hyphen in *Gell-Mann* and the apostrophe in *Pickett's Charge.*
- Stop alphabetizing at the first significant punctuation—a period, comma, or semicolon. However, ignore these characters when they are part of the term being indexed, such as commas in book titles or chemical formulas.

- In the letter-by-letter style, alphabetize words that precede a comma or period before any identical words followed by additional words with no intervening punctuation. Thus *directory, root* precedes *directory hierarchy*. This is not an issue in the word-by-word style, which uses the space as the key stop point.
- Some indexers ignore relational words such as articles, conjunctions, or prepositions, and alphabetize instead by the key word in the noun phrase. This practice is not recommended. Rather, focus on writing index entries that do not need or include relational words. If such entries are written, however, alphabetize relational words like any others.

Index Format and Design

9.40 The publisher or in-house specifications, not the indexer, often control format and design. These good-practice recommendations direct indexers or organizations in developing their own specifications and style. Format considerations include indentation practices (9.41), page layout (9.42), typography (9.47), and punctuation (9.52).

Indented or Paragraph Format

9.41 The two basic index formats are indented (line-by-line) and paragraph (run-in). The indented format, recommended for technical material, provides a well-organized vertical display of entries that the user can scan quickly and easily. The paragraph style, though harder to read, is more compact.

Indented

Maps 136–48
  blowups 142
  creating new 138
  deleting 143–45
  display options 136–39, 141
  exposing and hiding 141, 143–44
  printing 147

Paragraph

Maps 136–48; blowups 142; creating
  new 138; deleting 143–45; display
  options 136–39, 41; exposing and
  hiding 141, 143–44; printing 147

This text considers only the recommended indented format, which is more common in technical and scientific works. The sample index (9.53) combines format, layout, and typographic examples.

Layout Considerations

9.42 The indented index layout uses a hanging indent style, in which

- The heading is set flush against the left margin.
- Each subheading begins on a new line, indented one em. Sub-subheadings are indented two ems. Avoid using more than three heading levels.
- "Turnover lines," any line indented beneath another to accommodate words that do not fit on the preceding line, are indented one em deeper than the deepest subheading indentation.
- Locators follow the entry, in some indexes separated by a comma and in others by a one-em space (see 9.41, 9.53).

9.43 Set indexes in two columns with the right margin unjustified. Using more than two columns, if the page size seems to accommodate them, might create narrow columns and many turnover lines. This layout adds unnecessary length to the index and slows the users' scanning rate.

9.44 The index's first page should follow the convention used elsewhere in the document for chapter or major section breaks. It should include "Index" as the title, as well as an introductory note, if appropriate, that spans all columns. Also, follow document conventions for running headers and footers.

9.45 A blank line is sufficient for separating alphanumeric groups within the index.

9.46 If an entry continues from one page to the next, repeat both the heading and the subheading carrying over, and include some form of "continued" as a visual aid for the user.

Type Choices

9.47 Use the same typeface as the text for the index. Typeset the index in a type size two sizes smaller than the text. Thus, if the text is set in 12-point type, the index would be set in 10-point. Avoid using a type size smaller than 7-point (see also 12.32).

9.48 Italicize the cross-referencing terms *see* and *see also*, as well as any phrase representing an inclusive class of terms: *See also names of participating institutions.* Finally, a term that appears in italic type in the text, such as a Latin taxonomic name, should appear in italics in the index (see also 6.69, 7.58).

9.49 If a heading lists a series of locators, the principal locator (e.g., the one representing that heading's definition) may appear in distinctive type, using italic or boldface type.

9.50 Indexers disagree on capitalization, leaving it to individual indexers to choose a consistent scheme. Some indexers capitalize the first word of a heading and set other entry words, and the initial word in subheadings, in lowercase. Others set the initial word of all headings and subheadings in lowercase; and still others set all initial words in uppercase. Any words or abbreviations normally set in lowercase letters retain that practice as a heading or subheading, such as *p-n junctions* or *f-stop*. Similarly, any terms normally capitalized, such as a proper name or acronym, retain that form. Finally, italicized terms, such as a taxonomic term (see also 6.69), appear in italics in the index.

9.51 Be careful indexing commercial and trademarked names, which often feature unusual combinations of alphanumeric characters and symbols. Use the spelling provided by the author; discrepancies can be resolved by researching the company or standard references.

Punctuation Considerations

9.52 Indented indexes use straightforward punctuation practices and less punctuation in order to simplify the visual image:

- A comma marks an inverted phrase so that the keyword comes first, and separates locators within an entry from one another. You may use a comma to separate an entry from its locator if the entry ends with a numeral, which could itself be mistaken for a locator.
- A one-em space separates an entry from its locators or cross-references. Some index specifications do use a comma for this separation; either method is acceptable if applied consistently.
- A semicolon separates the parts of a multiple cross-reference.
- An en dash delimits page ranges: 89–104.
- Parentheses enclose qualifying information, such as distinguishing between homographs.
- Do not punctuate the end of an entry.

Format and Design Example

9.53 This sample index illustrates this section's recommendations. An em space separates the entry from its locators or cross-references. Note that turnover lines are indented one em further than the sub-subheadings, as in the Cassiopeia A entry:

Cameras, electronic 492, 499
Carbon flash 559
Cassiopeia A (3C461) *See* Radio
sources; Supernovae
Catalogs
3C catalog of radio sources, revised
630
bright nebulae 74
dark nebulae 74, 120
Lundmark 121
Lynds 123
Schoenberg 123
Herschel's (of nebulae) 99
planetary nebulae 484–85, 563
Perek-Kohoutek 484, 563
Voronstov-Velyaminov 484
reflection nebulae 74
Central stars *See* Planetary nebulae
Cerenkov radiation 668, 709
Chemical composition 71 *See also*
Abundances; *names of*
individual elements
of stars and nebulae 405
Clark effect 756, 758, 765

Indexing Procedure

9.54 This section considers

- An overview of the indexing process (9.55),
- Using the computer in index preparation (9.59),
- Writing and editing index entries (9.82), and
- Final proofreading (9.115).

An Overview of the Indexing Procedure

9.55 When the preparatory indexing decisions have been made, begin the indexing procedure by following these steps. Remember that, like many publishing tasks, these steps are iterative rather than strictly serial:

1. Skim through the manuscript to examine its organization and main ideas. Identify important topics to index by studying the table of contents, front matter, abstracts and summaries, opening paragraphs of sections, and figure and table captions.
2. Progress through the document to identify and mark the terms and topics that will appear as index entries. Some indexers complete this task on paper, others enter terms directly into the computer. If working on paper, highlight the headings in the text and write the appropriate modifier (heading, subheading, etc.) in the margin, preceded by a colon or some other indicative punctuation mark. Finally, enter all potential entries into the computer.
3. Analyze the potential entries to organize them into coherent headings and subheadings and identify the relationships among headings.
4. Arrange the potential entries, either alphabetically or by subject, to view the draft index.
5. Refine the index to provide multiple access points for a given topic by double-posting entries (9.89), rearranging word order to create a new heading, and creating entries for synonyms and acronyms (9.93 and 9.102).
6. Create cross-references (9.96).
7. Edit the index in comparison to the source document. At the content level, consolidate similar headings, clarify heading/subheading relationships, and delete unimportant headings. At the copyedit level, check the consistency of all spelling and capitalization and verify the accuracy of all page locators.
8. Arrange the entries in final order.
9. Complete the index by writing an introductory note and proofreading the final version.

Determining What to Index

9.56 Reading every indexable page—introductory matter, abstracts and executive summaries, all chapters or sections, conclusions, and appendixes—is the best way to capture all index entries. Footnotes and endnotes can be indexed, but they can represent tangential material. The proportion of content allotted to each topic determines approximately how many entries that topic will contribute to the index (9.15).

9.57 Index the presence of data displays (tables, charts, and diagrams) and illustrations. Base the entry on the caption. Such nontextual elements can be identified in an index with an italicized abbreviation—*il.* for illustration, *tab.* for table, and so on (9.47, see also 10.62).

Matter to Omit

9.58 Some items need not be indexed:

- Title page material
- Table of contents
- Lists of figures, tables, equations
- Trivial or passing mentions of people or places
- Dedications
- Authors or titles cited as references and listed in the bibliography
- Section summaries or abstracts
- Review questions
- Bibliographic entries
- Advertising or promotional material
- Glossary terms

Using the Computer in Index Preparation

9.59 Determining what concepts need indexing and imposing a logical organization that makes these concepts into a cohesive index require intellectual decisions beyond the capability of any software. However, the computer provides a powerful tool for marking, collecting, and sorting potential index entries. This section examines four types of computer-assisted indexing:

- Indexing with word processors (9.61);
- Stand-alone indexing composition programs (9.67);
- Markup languages (9.71);
- High-end publishing packages (9.74); and
- In addition, it provides a summary of recommendations (9.76).

9.60 Currently, computerized indexing offers two options. In the first, the indexer works with a word processor or indexing software separate from the text. The indexer enters all indexing data including locators in a separate file, perhaps with another program. In the second, the indexer uses more powerful indexing features to embed indexing data directly in the source file (the same file that contains the text being indexed), and generates indexes automatically from these data.

Word Processors and Alphabetized Lists

9.61 The most basic way to use a word processor as an indexing program is to create an alphabetized list:

1. Create a template, or "style sheet," based on the organization's index specifications.
2. Open a new file, and process the paginated source document (9.55), typing entries and locators into this file.
3. Use the sort feature to produce an ASCII-sorted list, and edit the list to order it word-by-word or letter-by-letter.
4. Use the word processor's built-in search-and-replace and spell-checking features to make the index consistent. Locators require careful verification.

9.62 This process can provide simple indexes for short documents, or longer documents with a repetitive structure that concentrates on objects, such as a product catalog, rather than more complex concepts. But this process uses only the most basic aspects of computer-aided indexing.

Word Processors and Embedded Indexing

9.63 In a more powerful approach, contemporary word processors use data inserted (embedded) into the text stream to create indexes. Using built-in features, word processors can index words and phrases selected from the text, or from terms entered directly into the text stream using index "field codes" to delimit an index entry. The program collects the marked entries and their locators and creates an index in a predefined, though locally programmable, format that can satisfy a variety of index specifications.

9.64 In some word processors, the indexing function uses an on-screen data entry form that contains heading, subheading, cross-referencing, and typographic instructions, and controls the index's content and appearance.

9.65 Embedding introduces the idea of reusing index data. After inserting the index field code, the text may be moved within a document or to another document; however, as long as the code stays with the text, the index function will display the correct locator in the index.

9.66 Despite the advantages offered by word processors' built-in indexing function, they have their flaws as indexing tools:

- Indexers cannot control the sorting algorithm. Word processors generally rely on a variation of the ASCII sort (9.33–9.38) and are unable to alphabetize using the word-by-word or letter-by-letter methods. This limitation creates the need for additional editing.
- The indexing function may not be able to understand and process multipart locators, such as a two-part page number 4.4 or IV–4.
- Last-minute source document changes can affect the index. Every text change requires regenerating the index to correct locators. A text change made after finalizing the index can seriously impair overall index accuracy.
- Word processors are proprietary programs that rely on proprietary file formats, which may be incompatible with other systems and programs.
- The word processor may limit the number of index entries allowed.

Stand-Alone Index Composition Programs

9.67 Stand-alone indexing programs answer the needs of the indexer who does not have access to documents in their electronic form. In some respects, they function like a specialized word processor that produces a sorted index, in a predetermined format, as the indexer enters headings, subheadings, and locators. They provide an excellent solution when the source files are unavailable. Their flexibility and ease of use make them the preferred choice of professional indexers, who usually work from page proofs.

9.68 Once the indexer enters the terms, the index composition program fully automates the procedure. The indexer marks potential entries on the page proofs and then types them into a formatted data file. The program automatically sorts the index data file and applies typographic, layout, and punctuation rules.

9.69 Although program features vary, stand-alone index composition programs typically have three general classes of capabilities:

- Core features—copying previous entries, flipping entries (transposing heading and subheading), search and replace, spell checking, and cross-reference verification.
- Sort options—either word-by-word or letter-by-letter sorting, choices for handling relational words, and the ability to produce a listing of entries in page order (a valuable editing function).
- Formatting features—ability to use indented or paragraph style, page layout control, and automatic punctuation. Most programs allow user-created style sheets and the automatic conversion of the output to a preferred typesetting, word processing, or database format.

9.70 Stand-alone index composition programs also have limitations:

- They sort word-by-word or letter-by-letter based on subtle differences in filing standards and conventions for such sorts; in word-by-word, for example, the hyphen may be ignored in one implementation and treated like a space in another. Since such differences affect the sort order, check the process used by your software, especially if you routinely merge index entries created in some other way or with some other program.
- Since stand-alone indexing programs, such as CINDEX, process unavailable source files, editing those files can continue during indexing. If editing the source file changes locators, the stand-alone program, unlike a word processor with embedded indexing, has no knowledge of this event; the indexer must manually update the index data.

Markup Languages and Indexes

9.71 Many commercial and corporate publishers create and store their source documents with a markup language—a set of characters or symbols that defines a document's logical structure—such as Standard Generalized Markup Language (SGML). If an organization uses SGML for publishing, it makes sense to create indexes by embedding an SGML tag to identify index entries in similarly tagged texts. Some stand-alone indexing programs can produce SGML-tagged output as one of their standard formats.

9.72 Although markup languages provide flexibility and offer a good opportunity for reusing content, creating indexes with them can be difficult:

- Even with a structured editing tool, indexing with tags can require a lot of typing.
- SGML has a demanding syntax that may require multiple edits to correct errors.

9.73 Since using a markup language requires software expertise, many indexers would rather concentrate on writing good index entries than learning the details of a markup language application. But difficult or not, markup languages can create versatile, accurate embedded indexes with a high degree of reuse, without the need for additional editing.

Indexing with High-End Publishing Programs

9.74 High-end publishing programs, such as Adobe FrameMaker™, use the embedded indexing model outlined in the word processing section (9.63) to collect index codes from source documents to generate an index file. Each subsequent index processing overwrites the existing index data file and produces a new index with a single command. Like the stand-alone indexing programs, these programs have more indexing features than word processors (9.66); they can

- Interpret multipart locators,
- Identify unresolved cross-references,
- Alphabetize in either the word-by-word or letter-by-letter methods, and
- Perform special sort orders.

9.75 The indexer must have direct access to the source files in order to embed the indexing data within them or edit the index entries later. The ability to create a "one-click link" from the index entry back to its location in the source file makes indexing easier.

Computer Indexing Conclusions

9.76 Selecting a computer indexing approach depends on the indexer's access to the document being indexed in its electronic form:

- If the indexer has no access and must rely on page proofs or Portable Document Format (.pdf) files, a stand-alone index composition program is a good choice. Such programs provide both ease of use and flexibility in alphabetic arrangement and formatting; and, with the index always visible on the screen, the indexer can edit entries at any time.
- If the indexer can access the document in its electronic form and can embed indexing data directly within it, the word processor's indexing features or the more powerful high-end publishing packages might be the better choice.
- Organizations already invested in systems based on high-end publishing packages or markup languages might extend the use of these technologies to include indexing.

9.77 A word processor or high-end publishing program with an indexing function

- Saves time for authors/indexers by allowing them to index while writing,
- Eliminates learning time by allowing writers to index entries rather than learning software, and
- Reduces the need to resolve potential incompatibility between a word processor and an indexing program.

9.78 In any of the embedded indexing models, the publishing process views the index data file as an output file, and changing the index entails changing the index entry, or tag, in the source document. The embedding process has several disadvantages:

- The index is not visible as the indexer enters or edits entries.

- Although the index file can be edited, any changes are overwritten the next time the index is processed. Thus, any change requires reprocessing the index.
- Any change in the source document affects the index, and every change to the source requires that the index be regenerated.

9.79 Embedded indexing, the index-as-you-go method, allows indexers to produce and edit interim drafts, which can save substantial time at a project's end. Indexers can use these drafts to check for inconsistencies in terminology, punctuation, and capitalization, and to identify imbalances between a topic's importance in the text and its representation in the index.

9.80 The kinds of documents produced and the publishing systems already in place could well dictate the author's indexing tool selection. For documents that will be indexed

- Only once, will not be revised, or have a short shelf life, the stand-alone approach works well;
- With every new document version, or have a long shelf life, the embedded approach is appropriate.

9.81 It is no longer simply a matter of whether or not the indexer should use a computer-based indexing method; rather it is a choice of which approach to take. That decision depends on more than features; it depends on how indexing fits into the publishing practice within the indexer's organization, as well as the long-term use of the document.

Writing and Editing Index Entries

9.82 Index entries should be concise and meaningful. This section shows how to

- Prepare good headings and subheadings (9.85),
- Work with synonyms and cross-references (9.93),
- Develop useful locators (9.100), and
- Handle special cases (9.101).

9.83 Indexing language consists of both vocabulary and syntax—*information science* is quite different from *information, science*—based on concrete nouns or noun phrases. Apply these stylistic guidelines in index entries:

1. Use the author's preference in terminology to determine synonym choice, spelling, and capitalization. Use the forms that appear most often in the text. Resolve discrepancies by consulting a standard reference work or specialized thesaurus (see also 13.9).

2. Use natural word order. Until recently, it was common to invert noun phrases to place the keyword first. However, indexers recommend natural word order, because it most closely resembles the way people think and are likely to search. Thus, index *Lake Michigan* and not *Michigan, Lake.*

3. The same term may have different meanings in different technical fields. *Nesting*, for instance, means something quite different to an ornithologist and to a computer programmer. To differentiate between such homographs, use separate headings with qualifiers in parentheses:

Filters (electrical)	Condensers (electrical)
Filters (mechanical)	Condensers (steam)

9.84 Two rules govern the use of singular versus plural number in index entries:

- Use the singular for processes (fermentation), properties (conductivity), concepts (relativity), and collective or mass objects (oxygen, water).
- Use the plural for countable objects and classes of objects (trees, planets).

If the singular and plural forms of a word have different meanings, index both. *Paper* is a mass item, *papers* is a class of object, including bond papers, coated papers, and so on.

Headings and Subheadings

9.85 Make headings general, subheadings specific. For instance, the heading *printing processes* is a general classification. The subordinate subheadings divide this general heading into its specific subclasses: *gravure, digital, lithography,* and so on. Similarly, both *diesel engine* and *gasoline engine* are subheadings of *internal combustion engine.*

9.86 Make each entry a heading rather than a subheading, unless it

- Begins with the same word or words as the heading, thus denoting a close relationship, and
- Is ambiguous unless read in the context of the main heading. *Annual U.S. production* cannot stand alone without a heading—*Steel*—to provide context.

9.87 A common entry or subentry error is to assume that if a series of entries all begin with the same word, then that word is the heading and all subheadings should be constructed from it. This error introduces needless levels of subheadings while reducing index clarity. Rather than using *Data* as a heading in the following example, the correct technique—creating repeated headings with *Data* as the initial word—emphasizes the different meaning of each entry:

Data dictionary

Data entry

Data modeling

Data processing

Data structures

9.88 Use no more than two subheading levels. Additional subheadings might indicate a flaw in heading choice.

9.89 Headings and subheadings can sometimes be transposed, or flipped—each term can serve as a subheading of the other. An abbreviated example of double-posting is

CGI applications
  Perl programs in 181–84

Perl programming
  CGI applications 181–84

9.90 In addition to a logical relationship, headings and subheadings have a grammatical relationship. Subheadings can include a relational word such as an article, preposition, or conjunction for clarity. If a subheading includes such words, sort them like any other word. Keep in mind, however, that it is better to write subheadings that are free of relational words; include them only to avoid ambiguity. Observe that in the following example, the second entry eliminates all the prepositions with no loss of index clarity:

Semilogarithmic charts 85–90
  characteristics of 85–87
  for portraying spatial data 108–90
  interpretation of 87–91
  statistics applications of 91–95

Semilogarithmic charts 85–90
  characteristics 85–87
  interpretation 87–91
  portraying spatial data 108–90
  statistics applications 91–95

9.91 Avoid underanalyzing headings, typically indicated by a long string of undifferentiated locators. Study the heading again to divide it into meaningful subheadings. Try to include no more than four undifferentiated locators with an entry.

9.92 In contrast, a long list of subheadings, each with only one locator, might indicate overanalysis. In this case, the narrow focus of the locators provides an additional clue, and the subheadings can be consolidated under a single subheading:

Over-analyzed	**Consolidated**
Maps	Maps
borders 137	display characteristics 136–38
icons used 138	
projection system 136	
styles 136	
varying size of 137	

Synonyms

9.93 Consolidate synonyms by

- Grouping those terms that have nearly the same meaning, and
- Deciding which term should be the primary entry and which should be cross-referenced to it.

For instance, in indexing a book on *printing technology*, choose one term—flying, misting, spitting, or spraying—as the primary heading and cross-reference the others to it.

9.94 Follow the author's preference and choose as the main entry the form used most often in the document. If the author does not show a preference, use the most familiar form as the main entry.

9.95 Synonyms present problems in disciplines where terminology changes frequently over time. New terms replace obsolete terms, as in the progression from *conscription* to *draft* to *selective service.* People, places, and entities all change names. Seemingly opposite terms can have the same meaning: *developing countries* and *underdeveloped countries.* In cases such as these, make the current usage the primary term. If necessary, index any outdated terms and cross-reference to the current one (see also 13.9):

Burkina Faso *See also* Upper Volta for references prior to November 1984

Upper Volta *See also* Burkina Faso for references more recent than November 1984

Cross-References

9.96 Use cross-references to provide guides to related information. There are two types of cross-references—*see* and *see also.*

9.97 In the *see* cross-reference, all the information related to a term is included within another term. If a *see* reference is used, try to provide at least two locators:

Reference from/to	Example
Full name to acronym or abbreviation	Extensible Markup Language *See* XML
Replaced name to current name	Leningrad *See* St. Petersburg
Given name to pseudonym	Munro, H. H. *See* Saki
Synonym to entry with information	Security *See* Authorization
Popular to official term	Table salt *See* Sodium chloride
Acronym to expanded form	XML *See* Extensible Markup Language

9.98 The *see also* cross-reference directs the reader to an entry that contains additional related information. *See also* cross-references often point from the general to the specific, as well as among topics with a close association. In the third example, the cross-reference implies a relationship of antonyms:

Access devices 224–30 *See also* Access servers; FRADs; Traffic concentrators

Caxton, William 89 *See also* Photography; Wolff, Francis

Linear recursion 7–4 *See also* Nonlinear recursion

Place the cross-reference immediately after the heading or subheading it augments, rather than after the last subentry.

9.99 In cross-referencing, avoid

- Creating a faulty logical relationship between entries,
- Blind *see* references pointing to nonexistent headings, and
- Circular searches.

For instance, save readers from the frustration of this circular reference:

Inert gases 150, 177 *See also* Neon

Neon *See* Inert gases

Rules Governing Index Locators

9.100 Record and verify the accuracy of every locator based on these guidelines:

- List locators in ascending order.
- Supply complete page numbers: 96–98, not 96–8. Lengthy case locator numbers can be elided: 102026–29.
- Record every page: 73, 75–76.
- Use en dashes, not hyphens, in specifying locator ranges. Use the hyphen to separate parts of a multipart locator.
- Collect entry repetitions that are in close proximity as a single page range for the entry: 96–120.
- Use type variation, for example, using boldface to distinguish a principal locator in a string of locators.
- Index footnotes with a lowercase n: 54n; if the footnote appears on a page with other footnotes, specify the indexed footnote by number, in italics: 54n*2*.
- Use a colon to separate the nonpage locators—paragraph numbers, abstracts, legal cases, and patents—from the page in a multipart locator in which one part is not a page number. For example, use the colon to denote a volume: chapter-page locator: 5:17–10.
- Use a slide or frame number to index a nontext item in a presentation.

Rules Governing Special Cases

9.101 This section describes some special cases often encountered in scientific and technical indexing, and offers guidelines for dealing with them.

Abbreviations and Acronyms

9.102 Until recently, indexers avoided indexing abbreviations and acronyms directly, using instead the full spelling, with the abbreviation or acronym in parentheses, for example, *Seasonal Affective Disorder* (*SAD*). However, current practice recommends filing abbreviations and acronyms as given, especially in scientific and technical writing because of the extensive use of professional acronyms.

9.103 Indexing by acronym or by its expanded form depends on reader familiarity with these terms. If an acronym is more familiar than the full name to an audience, use it. It is doubtful, for example, that an audience of telecommunications professionals would look for information in an index under the expanded forms of *ATM*, *IP*, or *ISDN*. Despite this potential familiarity, include a *see* reference to the expanded term for new users who may not know the abbreviation.

Accents, Apostrophes, and Diacritical Marks

9.104 Ignore accents, apostrophes, and diacritical marks in alphabetizing—but not when printing. Do not use the Americanized spellings of ä, ö, and ü (ae, oe, ue); retain specialized marks and treat them as if they were unmarked letters. If there are many such terms, follow the custom of the native language. For example, in Swedish, the å, ä, and ö follow the z in the dictionary. Be sure to note any such alphabetizing practices in the index introduction (see also chapter 2).

Foreign Terms

9.105 Do not translate foreign phrases, especially in such specialties as legal or medical indexing. Since such translations have no meaning to professionals, index the term as readers expect to see it (see chapter also 2).

Latin Names

9.106 The scientific names of plants and animals should remain untranslated. They should be italicized in their Latin form, with the first word always capitalized and the second word never capitalized. In some cases, a well-known translated name can be provided as a qualifier:

Portulaca grandiflora (moss rose) 99

Entities and Alphabetic Ordering

9.107 Although inverting a heading to place a keyword first in an entry is generally discouraged, entity names are an exception. Alphabetize business or institutional names, book and periodical titles, and other entities by the first keyword, with any leading articles inverted; for example, *John Hancock Insurance Company, the*, and *Wall Street Journal, the.* It may be necessary to invert some names to avoid ambiguity: *Commerce, Department of.*

Proper Names

9.108 Follow the author's preference in treating proper names. Names should appear in the index as they appear in the text. To index names, use the following guidelines:

- Use the personal name that is best known: *Eliot, George*, and not *Evans, Mary Ann.*
- Use the familiar form of a personal name: *Poe, Edgar Allan*, and not *Poe, E. A.*

- Index by the first part of a compound surname: *Bulwer* in *Bulwer-Lytton.*
- Index by the first part of a compound entity name: *Hewlett* in *Hewlett-Packard Company.*
- Use the full form of a geographic place name rather than its inverted form: *West Virginia* instead of *Virginia, West.* Also, retain articles that are an integral part of a place name: *Los Angeles, El Salvador.*
- Use the last or most current name of a geographic place name: *Istanbul,* not *Constantinople,* with a cross-reference to the superseded name if necessary.

Numerals

9.109 Index numerals as is, in their own grouping preceding the alphabetic entries. Avoid the convention of treating numerals as if they were spelled out.

9.110 Some cases dictate the limited use of serial order over alphanumeric order; an example is within a series of related items with numerals in their names. To avoid placing the subheading *Route 128* before *Route 66,* use a numeric order to list the highways as readers would expect to find them. The same is true with roman numerals.

9.111 In biology and chemistry, numerals in the names of chemical compounds are disregarded when alphabetizing unless the numerals provide the only differentiation between two compounds. This keeps like compounds together and avoids the confusion that would result from indexing all compounds that happen to start with 3- consecutively.

Symbols

9.112 Symbols, which appear routinely in disciplines such as mathematics, astronomy, and chemistry, can be indexed separately, before numerals. Another approach might be to arrange them in a classified order recognized by the audience. For example, in an astronomy publication, planetary symbols can be grouped together, ordered by their distance from the sun. A discipline-specific dictionary or thesaurus should suggest a symbolic order (see also 13.9). If the text uses many symbols, consider placing them in a secondary index. List them where it makes the most sense for the intended audience and explain that placement in the introductory note.

9.113 Indexers also use various approaches for Greek letters. Like other symbols, they can be either grouped separately or listed in a separate index.

Some sciences present special problems; biology and chemistry sometimes ignore Greek letters in alphabetizing, as with the *gamma* in γFe_2O_3. Other fields spell out Greek letters to alphabetize them; this practice produces *chi square distribution* instead of χ^2 *distribution* in the index. Since these represent exceptional cases, indexers are best guided by reviewing standard references in the discipline.

Including an Introductory Note

9.114 An introductory note helps the reader use the index easier. It should

- Explain the alphabetic arrangement used.
- Include instructions for interpreting special typography.
- Indicate abbreviations used for denoting examples, illustrations, or data displays.
- Describe how symbols and numerals are treated.
- Describe how to interpret foreign characters.

Final Proofreading

9.115 After completing the index, make a final proofreading review before sending it to the editor:

1. Verify the ordering and consistency of headings and subheadings.
2. Check for consistent spelling, capitalization, and punctuation.
3. Check that no single heading or subheading has more than four undifferentiated locators.
4. Verify the accuracy of all locators.
5. Make sure all cross-references identify headings correctly and are neither blind nor circular.
6. Delete repetitious, extraneous, or trivial entries.

10. Creating Nontextual Information

10.1 Typically, illustrations can be used in a variety of documentation types, from job training materials to proposals, brochures, and pamphlets. Illustrations can

- Add information,
- Help explain content,
- Visually confirm information in text,
- Motivate readers and make reading more interesting,
- Provide visual relief for a page heavy with text, and
- Sometimes replace text.

10.2 Using illustrations will help readers

- Perform procedures,
- Visualize mechanisms and processes, and
- Remember what they have read.

10.3 Illustrations will be especially helpful when users

- Have limited reading skills,
- May not have on-the-job time to read, or
- Speak a language other than English.

10.4 This section provides guidelines on

- Electronically acquiring, processing, and printing or displaying illustrations (10.5);
- Selecting illustration types—photographs, renderings, line drawings, infographics, and icons (10.18);
- Preparing illustrations for foreign audiences (10.29);
- Constructing illustrations (10.37); and
- Special considerations for producing illustrations (10.42).

Electronically Acquiring, Processing, and Printing or Displaying Illustrations

10.5 All photographs, as well as paintings and shaded or washed drawings, are *continuous tone* images. That is, they contain a tonal range of colors or gray shades. Traditional, offset production processes these illustrations with *halftone screens* to allow them to be printed. New, electronic versions of similar processes provide other methods for achieving the same results. On the basis of these newer processes, this section offers guidelines for

- Acquiring (10.6),
- Processing (10.12), and
- Printing or displaying illustrations (10.14).

Acquiring Illustrations

10.6 All illustrations can still be acquired by traditional means: conducting photography sessions, purchasing stock photographs, or using original or purchased artwork. However, even traditional elements will be processed in electronic form at some stage. Three input devices can help acquire illustrations: digital still or video cameras; flatbed, drum, or three-dimensional scanners, or graphic drawing tablets. Typically, these devices have a considerable range of capabilities. Digital cameras, for instance, can image specific areas (e.g., 640 × 480 to 1080 × 1024 pixels). Within that range they also have a number of limitations. The former image area can only optimally create a 3-by-5-inch final image; the latter a 5-by-7-inch. Currently, they share a bit depth of 24 bits. Scanners have similar capabilities and limitations.

10.7 Since these capabilities will influence and control the quality of published illustrations, one needs to recognize the interrelationship among input devices for acquiring illustrations, processing hardware and software, and the capabilities of potential output devices. Two measures provide clues for assessing potential illustration quality and both illustration and equipment compatibility: bit depth and resolution. Bit depth simply refers to the amount of information a device can store and process for each specific point (pixel or picture element) in an illustration.

Bit Depth Examples

Number of Bits	Numeric Expression	Number of Colors or Shades
1	2^1	Two, black and white
4	2^4	64
8	2^8	256
24	2^{24}	16+ million

A 24-bit (red-green-blue) image has three color channels with 8 bits for each channel:

- (R)ed 256 colors
- (B)lue 256 colors
- (G)reen 256 colors

A 32-bit CMYK, four-color (cyan-magenta-yellow-black) image, has three color channels with 8 bits for each channel, and a fourth mask, black, channel:

- (R)ed 256 colors
- (B)lue 256 colors
- (G)reen 256 colors
- (K)ey 256 colors

Bit depth provides evidence for users about the amount of information input devices can acquire. Digital and video cameras, scanners, monitors, and software for manipulating illustrations all have their own bit depths.

10.8 The second measure that controls quality is resolution. Like bit depth, this measure also has considerable variability:

Exhibit 10–1: Some Typical Input Device Resolutions

Input Device	Resolution	Bit Depth
Low-end digital camera	320 × 240 ppi[1] 1600 × 1000	24
High-end digital camera	1010 × 1068 ppi 7000 × 7000	36 48
Low-end video	160 × 100 ppi 640 × 450	30
Low-end scanner	600 × 1000 dpi[2]	36
High-end scanner	14,000 dpi	48
Three-dimensional scanner	752 × 582 ppi	36
Graphic tablet	800 × 600 ppi 1024 × 768	

Notes: [1] ppi, pixels (picture elements) per inch; [2] dpi, dots per inch.

10.9 While these differences in the amount of data an input device can acquire (bit depth and resolution) may seem problematic, most acquisition software levels these differences automatically by a process called "dithering," literally the removal of some information. Despite this ability, it is best to remember that one needs to acquire the best possible image, or illustration, in order to create a useful, and expected, final image. Of these input devices, only scanners currently offer any form of sophisticated resolution control.

10.10 Scanner resolution for acquiring a continuous tone image depends on the frequency of the halftone screen and the size of the final image. The basic formula suggests scanning images at 1.5 to 2 times the frequency of the halftone screen to be used to print that image. For the most part, lower values—1.5 to 1.7—can produce acceptable results.

10.11 For continuous tone illustrations that will be cropped or resized during processing, calculate scan resolutions by

1. Multiplying the halftone screen frequency by some value between 1.5 and 2; and
2. Multiplying that result by the ratio of the final to the original illustration's dimensions.

For example, to scan a 2-by-3-inch original to produce a 6-by-9-inch final image to be printed with an 85-lpi screen, the calculations would be

- 85 lpi × 2 = 170
- 170 × 3 = 510 dpi scan resolution.

Exhibit 10–2: Scanning Recommendations for Same-Size Continuous Tone Illustrations

Projected Final Document	Printer Resolution (dpi)	Halftone Screen Frequency (lpi)	Recommended Scanner (dpi/bits)	Recommended Scan Resolution (dpi)
Text only	300–600	65 (coarse, used for newsletters and other short-life documents)	Flatbed, 300 dpi, 24-bit color	98–130
		85 (average, used for newsprint)		108–170
1-color, quick print, or color offset	600–1000	100	Flatbed, 600 dpi, 30-bit color	150–200
	1000–2540	133 (high-quality, generally used for 4-color reproduction)		200–266
Color offset	2540	150	Flatbed, 600 dpi, 30-bit color	225–300
		177 (very fine, used for reproducing fine-art illustrations, sometimes used for annual reports)		265–355
High-quality color offset	2540	200	Flatbed, 600–1000 dpi, 36- to 48-bit color Or Film scanner or low-end drum scanner, 4000 dpi, 36- to 48-bit color	300–400

One obvious caution is that you need to experiment with the range of 1.5 to 2 to determine which produces the best results for your needs. You should also read your local printer's manual and discuss your needs with your typesetter and/or printer.

Processing Illustrations

10.12 Once an illustration has been acquired as a digital image, it can be displayed on a monitor and processed by graphic software. With the possible exception of some high-resolution monitors and sophisticated software, most of this work is completed on the desktop with relatively low resolution monitors. Two monitor resolutions are in common use: 72 and 96 ppi. Like scanners and digital cameras, they rely on bit depth as an indicator of their ability to display a range of colors. Thus, an 8-bit display can show 256 colors, a 24-bit display shows over 16 million.

10.13 Bit depth also plays a role in how a graphic's software processes information. Like other computerized processes, such software can read and process only specific bit depths. For example, you may be using a program that can process 32 bits. Such a program can process an 8-bit grayscale image, a 24-bit RGB image, or a 32-bit CMYK image. But it may also be able to process as much as a 64-bit image if it has the ability to dither a file (level or reduce its color information). All of these considerations influence productivity and final image quality.

Printing or Displaying Illustrations

10.14 Although most technical and scientific information continues to be printed as paper documents, a considerable amount of that same information has become electronic text–designed for presentation on a computer monitor. Although computers can now accomplish much of the prepress process, printed documents still rely on traditional press production techniques. This means, for example, that a full range of halftone screens is available; and these documents can be printed at a variety of appropriate resolutions.

10.15 In contrast, those documents that will appear on computer displays impose new conventions on information design. Displayed information cannot achieve the same resolutions as printed information on the typical 72–96 ppi computer display. A further, and possibly more dramatic, limitation is the download time for graphics over a modem. Given such limitations, information designers must create graphics and text that still convey their intended meanings within the confines of this new medium.

Exhibit 10–3: Download Time for Sample Displayed Documents

	Modem Speed (seconds)			
File Characteristics	**14.4**	**28.8**	**56.8**	**T1**
15 k file with 1 GIF black-and-white photograph	10	5	2	1
26 k text-only file	18	9	4	1
160 k text with 7 GIF files of black-and-white and color photographs	113	56	28	1

10.16 One significant difference between paper-based documentation and displayed documents is that the latter can also display dynamic images: self-running slide presentations, animations, and video. If your organization decides to use any of these techniques to offer technical or scientific information on a computer display, consult with your designers and programmers.

10.17 Before incorporating dynamic images into a displayed document, consider your rationale for employing these techniques. If you are trying to simulate a phenomenon or the actions of a mechanical process, any dynamic technique may contribute to your audience's understanding of the topic. However, you should consider that adding any dynamic technique does require additional software, and sometimes hardware, for your audience to access such documents (see also 12.103).

Selecting Illustrations

10.18 When selecting an illustration technique:

To Show	Use	See
Exactly how something looks to help readers recognize it or to prove it is real	Photographs	10.19
Something that does not exist now but will in the future, or something that is difficult or impossible to photograph	Renderings	10.20
Specific features or characteristics of a complex object How to assemble, install, use, or maintain a product How something works	Line drawings	10.22
Quantitative or tutorial information combined with suggestive or supportive graphics	Infographics	10.26
A picture that represents objects, actions, or locations in a form that can be quickly understood	Symbols and icons	10.27

Photographs

10.19 Use photographs to show exactly how something looks to help readers recognize it or to prove it is real. Photos also can motivate readers by making documents more interesting, attractive, and accessible:

- Create or purchase original photographs, or acquire them using the techniques discussed in 10.6 and following, above;
- Examine proofs (electronic displays, transparency slides, or contact sheets) to select the best photos based on subject matter, tonal range, and other quality measures typically used by your organization;
- Have prints made or acquire them in some electronic manner for computer treatment;
- Examine the prints, or electronic files, with the artist to decide on retouching and cropping;
- Mark callouts (if required) on photocopies of the prints. To do so, leave a generous (one inch) border on each photocopy. Mark the callouts legibly near the features to be labeled, and draw a line from the callout to the appropriate feature.

Renderings

10.20 Use renderings to show something that does not exist now but will in the future, or something that is difficult or impossible to photograph. Some renderings are so realistic that they can be mistaken for photographs.

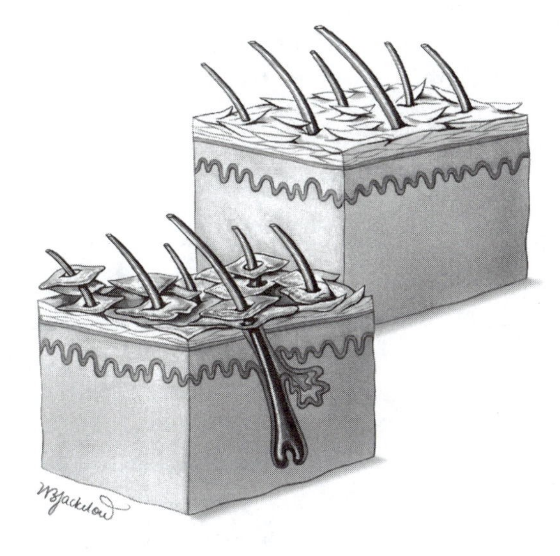

10.21 Provide the artist with as much information as possible to create a rendering. Engineering design drawings often prove helpful. In addition, you may wish to arrange a meeting among the artist, marketing, and engineering or scientific personnel.

Line Drawings

10.22 Use line drawings to show specific features or characteristics of a complex object. For instance, a one-dimensional line drawing might be included in a product description to illustrate the size of an object. More detailed line drawings are generally part of the design and production process. Such drawings must meet specific tolerances and production criteria because they support the production process:

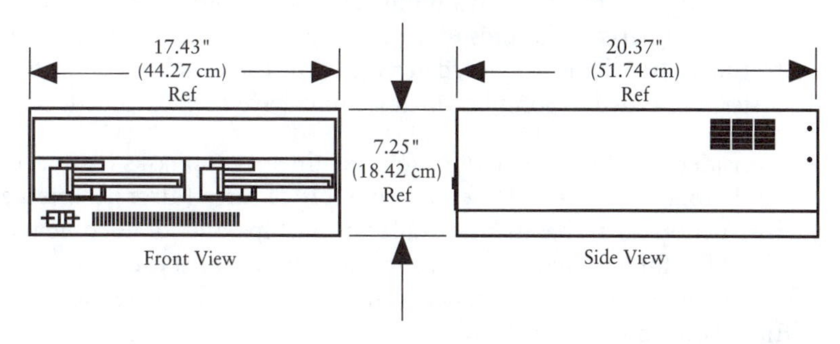

A similar illustration can show how something works. The following figure, for example, integrates text and a line drawing to help the reader perform a procedure properly:

Single Style Illustration

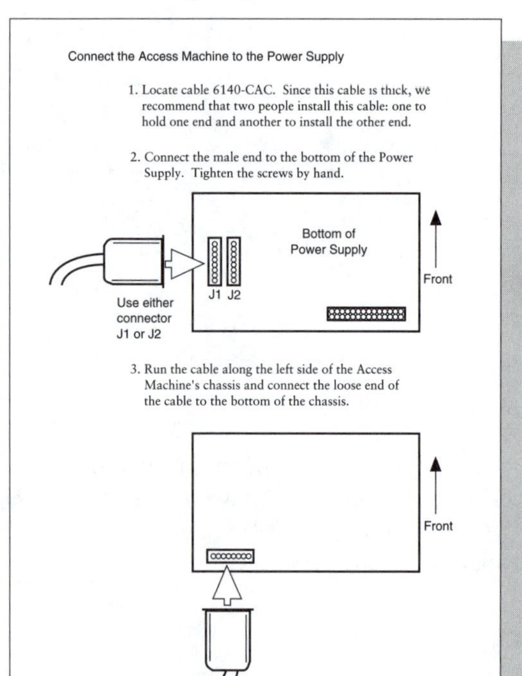

10.23 Do not mix illustration types in procedures. Using one style (photos, line drawings, etc.) allows the reader to recognize the object and its parts from step to step and between procedures.

10.24 Excluding needless detail from line drawings allows readers to focus on important parts or features they must identify or manipulate. The illustration at 10.22, for instance, details only the cable and its connector; the fasteners needed to complete the assembly have been excluded.

10.25 Consider wordless instructions for machinery setup tasks (connecting a cable to a computer), for emergency procedures (aircraft emergency instruction cards) based on spatial relationships, and for documents destined for foreign audiences. Techniques similar to those used in visual dictionaries for foreign languages should also be considered for international documents (see 13.10).

Infographics

10.26 Infographics combine suggestive or supportive illustrations with quantitative data or tutorial information. They require the same accuracy and ethical treatment as any data display (see also chapter 11). Although this class of illustrations may be viewed as inappropriate for scientific and technical information, it is often used to create interest. For example, popular articles in lay publications often illustrate scientific content with infographics:

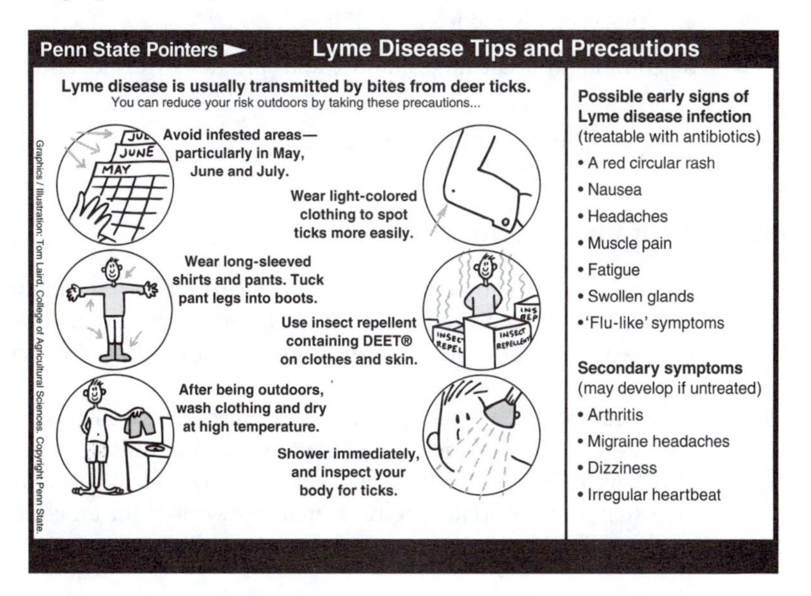

Symbols and Icons

10.27 Use standard symbols for specific disciplines that recognize these symbols as accepted graphic shorthand: chemistry, electrical engineering, geology, mathematics, and so on. (Some corporations distinguish glyphs from symbols. Glyphs refer to graphic symbols that have universality; warning and caution glyphs fall into this category.) Some symbols, such as international signage for personal services, seem to have gained universal recognition.

10.28 In contrast to symbols, icons tend to be arbitrary and cultural-specific. For example, the American rural mailbox as an icon for electronic mail has little meaning in most of the world. If accompanied by explanatory terms, icons seem to function better. The most commonly known icons in scientific and technical documentation are those approved by established standards organizations (see also 13.10).

Preparing Illustrations for Foreign Audiences

10.29 Documents destined for foreign audiences require some additional planning. Typically, an information designer must consider:

- Preparing Simplified English or translatable text (see also chapter 2);
- Planning for cultural differences (10.30);
- Providing for text expansion (10.33);
- Using symbols and icons (10.35);
- Using appropriate time, number, and currency styles (see also chapter 7); and
- Printing in ISO formats (see also chapter 12).

Planning for Cultural Differences

10.30 While it is difficult to plan for every cultural difference, information developers should recognize that many such differences exist. For example, representations of people and professional relationships differ culturally.

10.31 Other common Western representations may even be illegal: some countries do not allow depictions of government buildings. Clothing styles may have religious significance. Even relatively typical visual metaphors are often inappropriate: the rural mailbox symbol for an electronic (e-mail) mailbox has no counterpart in many cultures.

10.32 Finally, illustrations of hands are likely to offend since some combination of fingers has an offensive meaning somewhere; similarly, colors have specific cultural meanings. The bibliography for this section provides some useful pointers for understanding these issues (see also 13.10).

Providing for Text Expansion

10.33 Translation typically produces text that is from 20 to 30 percent longer than the English original. While text expansion influences page count and page breaks, it has a more drastic impact on illustrations. If an illustration, of any kind, has its text incorporated into the artwork, translating that text may alter the relationship between text and illustration.

10.34 To compensate for text expansion in illustrations, use callout numbers in the illustration and provide a summary of those callouts:

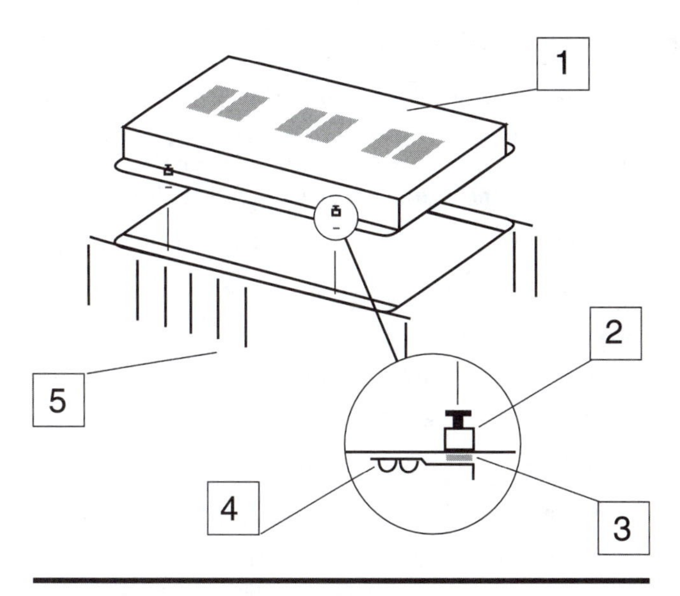

1. Fan Assembly
2. Locking Pin
3. Grounded Washer
4. Channel Connector
5. Drum Assembly

Using Symbols and Icons

10.35 Symbols and icons can provide commonality across cultures. Symbols, especially those based on international standards, provide excellent examples.

10.36 Icons can achieve similar results provided they do not rely on cultural recognition. If they do require cultural knowledge, then they should be accompanied by explanatory text.

Constructing Illustrations

10.37 The following sections discuss the aspects of the creation process important to providing functional, technically accurate illustrations:

- Gathering source information (10.38),
- Working with the artist (10.39), and
- Printing final illustrations (10.41).

Gathering Source Information

10.38 To use artist-created illustrations, provide technically accurate source material. The source material quality varies with professional relationships and the illustration's subject. Generally, appropriate source information can include documents from engineering, research, or marketing departments:

- Engineering drawings (schematics or blueprints),
- Prototype object or product,
- Artist's rendering,
- Hand sketch of an object or device,
- Photograph of an object or device,
- Scientific or engineering reports,
- Specifications,
- Engineering design drawings, and/or
- The object or device itself.

Working with the Artist

10.39 If the artist needs additional information, the author must obtain it. For instance, if the illustrations depict hardware, provide the actual device for the artist as reference material.

10.40 Review the artist's work at least once for

- Technical accuracy,
- Typographic errors,
- Correct typographic conventions,
- Terminology-text consistency,
- Accurate use of abbreviations,
- Correct specialized symbols and signs,
- Appropriate line weights for specific uses,
- Leader arrows correctly positioned,
- Obvious physical size cues, and
- Accurate object orientation.

Consider any suggestions a knowledgeable artist makes for presenting information.

Printing Illustrations

10.41 In printing illustrations on an offset press, review a blueline, or similar proofing copy, with the illustrator. Do not make substantive changes at this point; however, check for:

- Illustration accuracy,
- Proper screen placement,
- Consistency of screen percentages, and
- Missing type.

Special Considerations for Producing Illustrations

10.42 Some aspects of illustration production may require more planning. These issues include

- Type legibility for illustration text (10.43),
- Positioning illustrations (10.46),
- Selecting views (10.49),
- Including a hand (10.50),
- Highlighting (10.51),
- Numbering illustrations (10.56), and
- Labeling illustrations (10.61).

Type Legibility for Illustration Text

10.43 Use the same size and style typeface as the body text for illustration captions, subtitles, and callouts—but practical constraints may require that they be smaller.

10.44 Use upper/lowercase in figure captions. All caps are harder to read and should be used only occasionally for emphasis or for acronyms and abbreviations.

10.45 Set illustration notes in smaller or less emphatic type than captions, subtitles, or callouts.

Positioning Illustrations

10.46 When placing illustrations on the page, keep them within the page's grid structure and close to their reference in the text.

10.47 Many companies use a standard grid structure to control their documentation (see also 12.48). Placing the illustration within the grid structure enhances the document's organization and contributes to clean copy; however,

- If positioned improperly, the illustration will appear detached or randomly placed (floating) on the page; and
- If too little or too much space surrounds the illustration, the page will look cluttered or the illustration will float.

10.48 If the reader needs to see an illustration in order to understand the text, place the illustration as close as possible to the first reference to it. Illustration references mean the place in the text where the reader is told to look at the illustration.

Selecting Views

10.49 In documenting hardware, the view of the device depends on the illustration's purpose and the reader's needs. Alternatives to the typical front view include top view, side view, bottom view, partial view, and cutaway view. When constructing the initial print package, check with an artist and consider the view (or views) that will be most useful to readers.

Including a Hand

10.50 Include a hand in the illustration only when it is important to show the user exactly how to hold or manipulate a device. This practice may be particularly useful in wordless instructions, in assembly and setup instructions, and wherever the risk of personal injury exists. Using a hand may also have implications for international documents (see also chapter 2; see 10.32).

Highlighting

10.51 Shading highlights illustrations by using screens, blocks of color, or shades of gray that cover a portion of an illustration or text to attract the reader's attention.

10.52 When used carefully, a screen can

- Draw the reader's attention to a portion of an illustration;
- Keep an illustration from floating on the page; and
- Highlight featured text, such as a title or a list of promotional features.

10.53 Screen, or shade, areas by percentages of black. The lightest useful screen—least dense—would probably be about 10 percent, the darkest perhaps 40 to 60 percent (100 percent screen means that the screen is the pure color; 100 percent black literally would mean no screen). Be consistent with tonal values throughout a work.

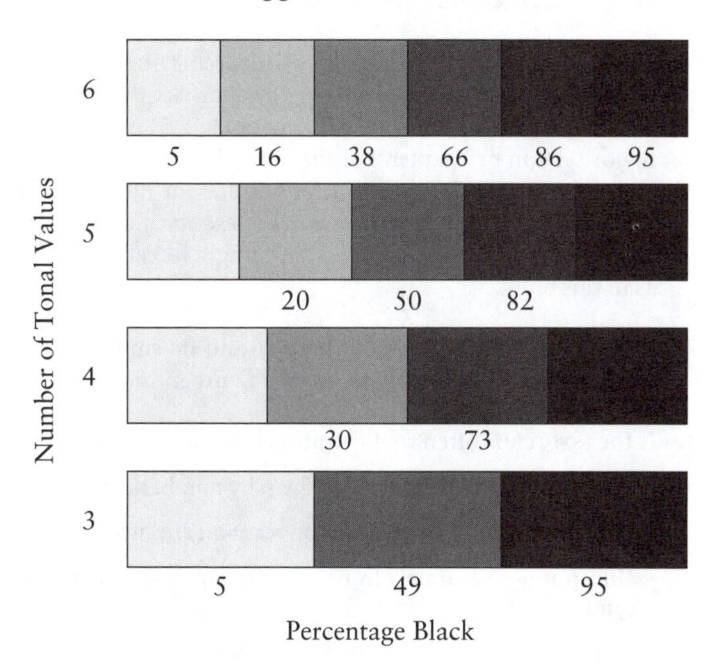

10.54 The percentage should highlight the illustration or text while allowing for maximum readability. Useful screen percentages vary with the type of printing paper and the color of the typeface. For example, since glossy (coated) paper is less absorbent than dull (uncoated) paper, a screen will appear slightly darker on glossy paper. Similarly, a 50 percent brown screen would probably be too dark over black ink, making the document difficult to read. The goal is to make the screens functional, and the screened illustration readable. To select screens for illustrations, discuss percentages with the artist.

10.55 Screens can be in color or shades of gray. While color adds appeal, it

- Varies with the qualities of the paper stock,
- Varies with the ambient lighting,
- May be used ineffectively by color-blind people,

- May have cultural implications for translated documents,
- May be reproduced by a monochromatic photocopying device,
- Is more time-consuming to prepare, and
- Costs more to print.

Numbering Illustrations

10.56 Number illustrations sequentially with arabic numerals in a manner similar to the document numbering system. Sequential numbering may continue through the entire document (1, 2, 3) or be contained within a major section or chapter. For instance, illustrations in chapter 10 would appear as Figure 10.1 (or 10-1), Figure 10.2 (or 10-2), and so forth. It may also be possible to use a single term to describe illustrations. For instance, all nontext elements in a manuscript might be called displays or exhibits, as in this book.

10.57 If the relationship between the text and its supporting illustration is obvious, the caption may not need a figure number.

10.58 If the text refers often to illustrations, number

- Every illustration and refer to them by number in the text;
- Illustrations in the order they appear and are cited;
- Illustrations separately from data displays: tables, charts, and diagrams; and
- Illustrations and data displays in a single series, citing them all with a common term: figures, displays, or exhibits.

10.59 Do not create separate numbering series for different types of illustrations:

Unacceptable	Acceptable
Drawing 7.1	Exhibit 7.1
Photo 7.1	Exhibit 7.2
Rendering 7.1	Exhibit 7.3

10.60 If appendixes contain illustrations, number them using the same method as the rest of the work. For example, if illustrations are numbered sequentially throughout the text and the last illustration in the text is Exhibit 42, then make the first illustration in the appendix Exhibit 43. If illustrations are numbered by chapter or section, the first illustration in appendix B would be Exhibit B–1.

Labeling Illustrations

10.61 Create a consistent labeling technique for all illustration elements. This section provides guidance on

- Captioning illustrations (10.62),
- Subtitling illustrations (10.68),
- Callouts (10.70), and
- Notes (10.72).

Captioning Illustrations

10.62 Give the illustration a unique caption that clearly distinguishes it from other illustrations. Feature the distinction; do not obscure it in the caption. Avoid a series of captions that all begin with the same words. Such long captions obscure the most important and useful part of the caption:

Unacceptable	Acceptable
Sodium concentrations of Mississippi	Mississippi Sodium Concentrations
Sodium concentrations of Amazon	Amazon Sodium Concentrations

10.63 The caption should not

- Repeat information found in the text,
- Provide unnecessary background information, or
- Attempt humor (Mental Illness in Illinois: Nonstandard Deviants).

10.64 Use longer and more descriptive captions for a lay audience than those for a purely technical or scientific audience. Such captions should contain complete sentences. They typically tell the reader what the illustration contains and how to interpret it:

For Technical Readers	For Lay Readers
Groundwater Contaminants in Chicago Metropolitan Area	Groundwater Contamination Increases in Chicago Area. Arsenic shows largest increase.

10.65 Write the caption as a noun or noun phrase. Avoid relative clauses; use participles instead:

Unacceptable	Acceptable
Number of Patients Who Showed Advanced Symptoms	Patients Showing Advanced Symptoms

10.66 Center one- or two-line captions horizontally over the illustration. Centering the caption helps visually associate it with the illustration and distinguish it from the text.

10.67 Left-justify captions over three lines.

Subtitling Illustrations

10.68 Use a subtitle for essential, but secondary, information that applies to the whole illustration:

- Scope of a study (Madison County, Alabama),
- Units of measurement (in thousands of dollars),
- Units of analysis (by month),
- Conditions of an experiment (T = 450 °F),
- Statistical criteria (s = 6.223), and
- Number of test subjects (n = 104).

10.69 Enclose subtitles in parentheses centered below the caption:

Depth Gauge
(metric dimensions)

Callouts

10.70 Use callouts, text that appears within the illustration itself, to explain specific illustration features. This text attracts the reader's attention to selected features and adds specific details. Leader lines or arrows often connect callouts to illustration features.

10.71 Position a callout to the left, right, above, or below its related feature. If the callout's leader line or arrow runs through a dark portion of the illustration or general text, use white space around the leader line or arrow to make it visible. Where possible, keep the leader lines or arrows straight, rather than broken at an angle.

Illustration Notes

10.72 See also 8.97.

10.73 Put specific notes at the bottom of the illustration to which they refer. If the same notes occur on multiple pages, gather all such notes at the end of the illustration.

10.74 Place specific notes in the order they occur in the illustration: starting in the upper left corner and proceeding clockwise around the illustration.

11. Creating Usable Data Displays

11.1 Tables, charts, and diagrams—data displays—present technical information in ways that text alone cannot. These data displays condense, summarize, and organize complex details to reveal relationships and provide specific facts. All data displays share some common elements; information developers need to consider:

- What are data displays? (11.2),
- Characteristics of data displays (11.4),
- Using technology to create data displays (11.22).

What Are Data Displays: Tables, Charts, and Diagrams?

11.2 Various authorities do not always agree on how to define table, chart, graph, or diagram. In this section the terms mean:

Table (11.26)

StubHead	Column 2	Column 3	Column 4
Row 1	Item 1	Item 1	Item 1
Row 2	Item 2	Item 2	Item 2
Row 3	Item 3	Item 3	Item 3
Row 4	Item 4	Item 4	Item 4
Row 5	Item 5	Item 5	Item 5
Row 6	Item 6	Item 6	Item 6
Row 7	Item 7	Item 7	Item 7
Row 8	Item 8	Item 8	Item 8
Row 9	Item 9	Item 9	Item 9
Row 10	Item 10	Item 10	Item 10

Collection of individual pieces of information arranged in rows and columns.

Chart
(11.117)

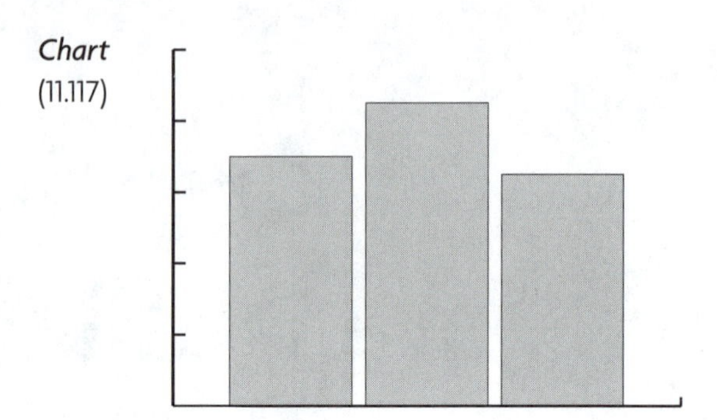

Visual display in which scaled dimensions represent numeric values. The term *chart* includes *graphs.*

Diagram
(11.268)

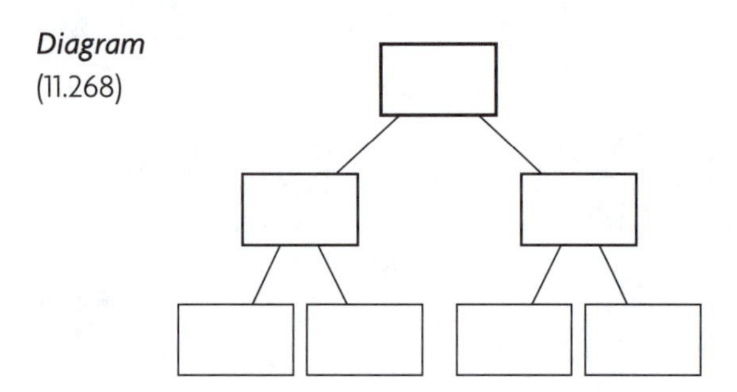

Visual representation of the relationships among the components of a system, process, organization, or procedure.

Which Should I Use?

11.3 Tables, charts, diagrams, and text all show relationships among numerical data or other forms of information. Use the right form for a particular situation:

If Relationships in Data Are ...	And ...	Then Use ...
Qualitative	Simple	Text
	Complex	Diagram
Quantitative	Exact values most important	Table
	Patterns and trends most important	Chart

Characteristics of Data Displays

Legible

11.4 Data displays should be compatible with surrounding text. Lines and lettering must be large and heavy enough so they do not fade out when reduced, printed, or photocopied (see 11.62).

11.5 It is impossible to offer type sizes and styles in a general guide like this, which must cover publications produced on typewriters, desktop publishing systems, and commercial typesetters. Individual presses and publishing organizations should set up standards for type and line weights used in data displays. Varying typographic elements, for instance, can provide emphasis:

- Bold versus normal weight,
- Italic versus roman style,
- Underlined versus not underlined, and
- Larger versus smaller type (see also 12.26).

11.6 Use cap-and-lowercase for titles and headings, and initial cap for text, in data displays. ALL CAPS IS HARDER TO READ AND TAKES UP MORE SPACE.

11.7 Use one basic typeface for all text in the data display. Vary the size and weight for emphasis. Avoid elaborate, decorative fonts. Use italic or slanted lettering sparingly.

11.8 Size the data display to present the data legibly, not to fill space on the page. Never spread out the rows and columns of a table merely to fill out a column or page.

Integrated with Text

11.9 Place data displays in the normal flow of text where possible. Do not place a data display in the middle of a paragraph. Avoid clumsy cross-references and unnecessary page flipping.

11.10 Data displays that are not essential to the text's message, present large amounts of raw data, or are referenced infrequently may better fit in an appendix.

11.11 Tables, charts, and diagrams should use the same margins and typefaces as the text. Many organizations still use serifed type for text and sans serif for data displays and illustrations. For typeset text, there seems little aesthetic reason to use a different typeface for displays than the one used for text. Body type for both text and data displays should also be the same size; using smaller type implies that the data display's information is less important (see also 12.26).

11.12 Do not use unnecessary boxes or rules to segregate text from graphics.

11.13 Leave at least the equivalent of two blank text lines above and below tables. Center or left-justify the table in the available space.

Easy-to-Locate Information

11.14 Because data displays provide quick retrieval of an individual fact or a single data value, the scanning reader must be able to recognize them on the page. When integrated with the text, data displays must be visually distinct. Normally the extra margins and rectangular layout of tables provide sufficient clues for the reader. If not, announce data displays visually by:

- Placing 10 to 25 percent–screened rules before and after the display to draw attention to the display.
- Printing the display on a 10 percent–screened background.

11.15 Avoid using a dot pattern to create a gray background, especially if the display contains numbers with decimal points.

Complete Data Displays

11.16 Design displays to be complete and self-contained. A display's meaning should be obvious, even when presented without the surrounding text. The reader should never have to search the text to discover a display's meaning.

Simple Data Displays

11.17 If a display seems too complex, consider whether it offers too much information or too many types of information. If a display requires more than two levels of headings, it is probably too complex and should be divided into two or more simpler displays.

Accessible Data Displays

11.18 Emphasize the similarities and differences of displays. Group similar items and present them in the same way. Conversely, separate dissimilar items and present them differently.

Numbering Data Displays

11.19 See 10.61.

Primary Title for Data Displays

11.20 See 10.62.

Subtitle for Data Displays

11.21 See 10.68.

Using Technology to Create Data Displays

11.22 Review sections 10.5 through 10.17 for basic information about the contemporary limitations for acquiring, processing, and printing or displaying electronically created materials.

11.23 Since a considerable variety of software can produce a specific data display, scientific and technical authors need to understand the different capabilities of these applications. For example, a variety of software can create a data display, with varied results.

Potential Applications for Creating Data Displays

Software Type	Current Examples
Illustration	Adobe Illustrator™ Adobe Freehand™ Corel Draw™
Spreadsheet Statistical	Microsoft Excel™ StatView™
Page composition	Adobe FrameMaker™ or PageMaker™ Quark Express™
Word processing	Microsoft Word™ Word Perfect™
Tag language	Standard generalized markup language (SGML) Extended markup language (XML)
Animation	Microsoft PowerPoint™ Macromedia Director™ Java™ or Javascript™ Dynamic hypertext markup language (DHTML)

11.24 Writers need to consider file compatibility and the expected quality of the resulting data displays. These decisions depend on the complexity and sophistication of production support and the document's purpose.

Production Characteristics for Software to Produce Data Displays

Software Type	Characteristics
Illustration Page composition	• Sophisticated control over grid, rule, border, and other line widths; similar control over typography, including the ability to convert type into artwork. • Screening and color selection typical of traditional printing. • Compatible with tag languages (HTML), word processors, and animation applications.
Spreadsheets Statistical	• Association with supporting data files reduces the chance of error in displays. • Limited control of screening, and color and type selection. • Often compatible with word processors, page composition and animation applications, and tag languages, though these files may require editing.
Word processing	• Control over grid, rule, border, line widths, typography, screening and color selection, though lacking as strong an association to traditional printing as illustration and page composition applications. • Compatible with tag languages (HTML), other word processors, some spreadsheets, and some page composition and animation applications.
Tag languages	• Control over grid, rule, border, line widths, typography. • Screening and color selection typical of traditional printing. • Compatible with other tag languages, some word processors, and page composition applications.
Animation	• Some control over grid, rule, border, line widths, and typography. • Screening and color selection adequate for displaying images on computer screens and projection devices. • Compatible with tag languages (HTML) and illustration applications.

11.25 Information developers should recognize that this discussion of software, like that at 10.5 through 10.17, relies on the current state of electronic support. Use all of this information to help in planning an appropriate level of electronic support for a specific information need. It may be, for instance, that word processing will perform all of the production tasks adequately; or tag languages may provide better support for a distributed information group. Assess needs based on both recommendations and discussions with writers, editors, graphic and electronic artists, and print production personnel. Do not, however, forget to consider both the expected audience and purpose.

Tables

11.26 Tables organize information into rows and columns by category and type, and simplify access to individual pieces of information. They accurately present many quantitative values at one time. This section considers

- When to use tables (11.27),
- Parts of tables, (11.28),
- Table contents (11.79),
- Special tabular problems (11.95),
- Common types of tables (11.109).

When to Use Tables

11.27 Although text can express numerical relationships, use a table to present more than a few precise numerical values. Use tables to

- Present a large amount of detailed information in a small space;
- Support detailed, item-to-item comparisons;
- Show individual data values precisely; and
- Simplify access to individual data values.

Parts of Tables

11.28 Tables conventionally contain the parts illustrated in Exhibit 11–1. This section provides additional detail about such topics as

- Column and row headings (11.29, 11.51),
- Spanners (11.39),
- Stub (11.47),
- Field (11.58),
- Rules and borders (11.62),
- Notes (11.71; see also 8.97).

Exhibit 11–1: Parts of a Table

Table 7-1: Tests of Propellant Vent Ports

Propellant Vent Port	Flow		Pressure	
	Specified (psi)	Actual (psi)	Specified (psi)	Actual (psi)
	Model 635			
External	100.0	102.7	10.2	10.4
Internal				
Aft	28.0	28.0	3.0	2.9
Mid	14.0	14.7	1.5	1.6
Stern	28.0	28.1	2.5	2.5
	Model 635-AX			
External	120.0	122.7	12.0	11.4
Internal				
Aft	30.0	28.9	3.0	2.9
Mid	15.0	14.0	1.5	1.5
Stern	32.0	29.3	2.5	2.4

NOTE: Tests performed at 72°F and 65% relative humidity.

Not all tables contain all of these parts, nor should they. This example table is more typical than ideal.

Column Headings

11.29 Give each column a short heading to describe information in that column. Column headings label dependent variables. While independent variables go in row headings, column headings label dependent variables.

Table 7-1: Tests of Propellant Vent Ports

Propellant Vent Port	Flow		Pressure	
	Specified (psi)	Actual (psi)	Specified (psi)	Actual (psi)
	Model 635			
	100.0	102.7	10.2	10.4
	28.0	28.0	3.0	2.9
	14.0	14.7	1.5	1.6
	28.0	28.1	2.5	2.5

11.30 Place information in the column heading that applies to all members of that column, and is unique to that column.

11.31 If information is too detailed to go in the heading, put it into a table note and cite the note in the column heading:

Thickness	Dielectric Strength[1]	Volume	Surface Resistance
Data	Data	Data	Data

Note: Explain the content of note 1 here because it is lengthy and will not fit in the column head. Notice the change in type size and leading.

11.32 Capitalize column headings the same as the table title, but set them in less emphatic type. Column headings may be singular or plural. Center one-line column headings; left-justify longer column headings.

11.33 All text on the table should be set horizontal:

Thickness	Dielectric Strength[1]	Volume	Surface Resistance

11.34 Turn long column headings counterclockwise so they read from bottom to top. Turn headings only if they cannot be shortened to make them fit:

Both Acceptable

Power Generation in Southeast States

Alabama Arkansas Florida Georgia Louisiana Mississippi North Carolina South Carolina Tennessee Virginia

Power Generation in Southeast States

Alabama Arkansas Florida Georgia Louisiana Mississippi North Carolina South Carolina Tennessee Virginia

11.35 If a heading is a lead-in phrase completed by a table entry, end the heading with an ellipsis:

If the weather is . . .
hot and dry
cool and wet

Column Subheadings

11.36 Include subheadings in column headings if needed. Typically, subheadings show:

- Scope of a study (Madison County, Alabama),
- Units of measurement (in thousands of dollars),

- Units of analysis (by month),
- Experimental conditions (T = 451 °F) (see also 7.42),
- Statistical criteria (s = 6.223), and
- Number of test participants (n = 104).

11.37 Enclose column subheadings in parentheses and center them. Do not put units or other information in a column subhead unless it applies to all items in the column:

Thickness (mm)	**Dielectric Strength** (Kv/mm)	**Volume** (Ω/cm)	**Surface Resistance** (Ω)

Column Numbers

11.38 To number columns for reference from the text use consecutive arabic numerals in parentheses. Such numbers appear as column subheadings. Number only the lowest levels of column headings, not column spanner headings:

Thickness (1)	**Dielectric Strength** (2)	**Volume** (3)	**Surface Resistance** (4)

Sometimes column heads contain combinations of information. In such cases, always place the column number as the last element:

Thickness (mm) (1)	**Dielectric Strength** (Kv/mm) (2)	**Volume** (Ω/cm) (3)	**Surface Resistance** (Ω) (4)
Data	Data	Data	Data

Note: Explain the content of notes here if they are lengthy and will not fit in the column head. Notice the change in type size and leading.

Column Spanners

11.39 Column spanners, also called decked heads, label two or more column heads.

Table 7-1: Tests of Propellant Vent Ports

Propellant Vent Port	Flow		Pressure	
	Specified (psi)	Actual (psi)	Specified (psi)	Actual (psi)
		Model 635		
External	100.0	102.7	10.2	10.4

11.40 Use column spanners to show the hierarchy of columns. In the example above, the column spanners include *Flow* and *Pressure*, which each span a pair of columns headed *Specified* and *Actual*. In the example below, *B* is a column spanner over columns *B.1*, *B.2*, and *B.3*. *B.3* is itself a column spanner over columns *B.3.a* and *B.3.b*:

Stub Head	A	B			
				B.3	
		B.1	B.2	B.3.a	B.3.b
Row 1	data	data	data	data	data

11.41 Include column spanners only in horizontally ruled tables. Extend the rule beneath the column spanner over all columns to which it applies. If column spanners are used, use horizontal rules to show which columns are spanned:

	Thermal Conductivity			
	Wet		Dry	
Material	Rated	Tested	Rated	Tested

In such unboxed headings, align column heads with the bottom. Place rules about a half line below the heading; otherwise, the rule may be mistaken for an underline.

11.42 Place an ellipsis (. . .) after any column spanner continued in lower column headings:

If the test solution is . . .			
red and . . .		blue and . . .	
cloudy	clear	cloudy	clear

Field Spanners

11.43 Field spanners, also called cut-in heads, cross all field columns and apply to all items below them. Field spanners identify a third variable. Repeat row headings for each field spanner.

Table 7-1: Tests of Propellant Vent Ports

Propellant Vent Port	Flow		Pressure	
	Specified (psi)	Actual (psi)	Specified (psi)	Actual (psi)
	Model 635			
External	100.0	102.7	10.2	10.4
Internal				
Aft	28.0	28.0	3.0	2.9
Mid	14.0	14.7	1.5	1.6
Stern	28.0	28.1	2.5	2.5
	Model 635-AX			
External	120.0	122.7	12.0	11.4
Internal				
Aft	30.0	28.9	3.0	2.9
Mid	15.0	14.0	1.5	1.5
Stern	32.0	29.3	2.5	2.4

NOTE: Tests performed at 72°F and 65% relative humidity.

11.44 Place the top field spanner just below the column heads. Never place the first field spanner above the column heads. Do not extend the field spanner into the stub column.

11.45 Field spanners occur only in table fields, never in the headings. They essentially start the table all over again for another data category. Hence, all of the stub headings are repeated beneath each field spanner. For this reason, it is never correct to have a single field spanner.

11.46 Using field spanners is an alternative to having separate tables, each with the same column headings and row heads.

Stub

11.47 The stub, the leftmost table column, lists the items about which the remaining columns provide information.

Stub Head

11.48 Title the stub with a column heading that describes the headings listed in the stub.

Table 7-1: Tests of Propellant Vent Ports

Propellant Vent Port	Flow		Pressure	
	Specified (psi)	Actual (psi)	Specified (psi)	Actual (psi)
	Model 635			
External	100.0	102.7	10.2	10.4

11.49 Avoid diagonally split stub heads. If information applies to all column headings put it in the title, in a column spanner (see 11.39), or in table notes (see 11.31):

Unacceptable

Resistance / Material	Wet		Dry	
	Rated	Tested	Rated	Tested

Acceptable

Material	Resistance			
	Wet		Dry	
	Rated	Tested	Rated	Tested

11.50 Omit the stub head only if the table's title clearly identifies the stub headings.

Row Headings

11.51 Row headings generally represent independent variables or categories. Row headings should

- Clearly and concisely label the table's rows,
- Uniquely identify each row, and
- Apply to all items in the row.

Table 7-1: Tests of Propellant Vent Ports

Propellant Vent Port	Flow		Pressure	
	Specified (psi)	Actual (psi)	Specified (psi)	Actual (psi)
	Model 635			
External	100.0	102.7	10.2	10.4
Internal				
Aft	28.0	28.0	3.0	2.9
Mid	14.0	14.7	1.5	1.6
Stern	28.0	28.1	2.5	2.5
	Model 635-AX			
External	120.0	122.7	12.0	11.4
Internal				
Aft	30.0	28.9	3.0	2.9
Mid	15.0	14.0	1.5	1.5
Stern	32.0	29.3	2.5	2.4

NOTE: Tests performed at 72°F and 65% relative humidity.

11.52 Keep row headings short by omitting secondary information or putting it into a table note (see 11.31).

11.53 Left-justify row headings. Indent subordinate row headings about four spaces to emphasize the row's hierarchical organization. For example, in reporting budgetary information, indent row headings for totals, subtotals, grand totals, and averages:

Power Generation
Hydroelectric
    Public
    Private
        Subtotal
Geothermal
Fossil fuel
    Coal
    Oil
    Wood
    Refuse
        Subtotal
        Total

11.54 For consistently spaced rows, indent runover lines about two spaces, but less than subordinate heads:

Characteristic
Force of
   gravity at
   surface
Distance from
   earth
   (average)

If rows are not closely spaced, left-justify runover lines.

11.55 If the row headings are uneven in length, use leaders to direct the eye to the next column:

Material

Asbestos . . . .

Asphalt . . . .

Bakelite, wood mixture . . . .

Jute, impregnated with asphalt . . . .

Pressboard . . . .

11.56 Some useful alternatives to leaders include

- Using horizontal rules to separate rows, although this can cause some visual confusion.
- Printing alternate rows on a background tone or color. Although this method creates a visual difference, it also reduces the legibility of the screened rows.
- Right-justifying the row headings. This results in a ragged left margin that is difficult to scan.

11.57 For English text capitalize the first word, proper nouns, and adjectives in row headings as sentences, not headlines. For other languages, follow the capitalization style in that language for sentences. Omit periods at the end of row headings. Do not number items in the row:

Unacceptable	**Acceptable**
1. Tensile strength (psi)	Tensile strength (psi)
2. Yield strength. (psi)	Yield strength (psi)
3. Melting Range (°C).	Melting range (°C)

Field

11.58 The field or body of the table is the array of cells below the column headings and to the right of the row headings.

Table 7-1: Tests of Propellant Vent Ports

Propellant Vent Port	Flow		Pressure	
	Specified (psi)	Actual (psi)	Specified (psi)	Actual (psi)
	Model 635			
External	100.0	102.7	10.2	10.4
Internal				
Aft	28.0	28.0	3.0	2.9
Mid	14.0	14.7	1.5	1.6
Stern	28.0	28.1	2.5	2.5
	Model 635-AX			
External	120.0	122.7	12.0	11.4
Internal				
Aft	30.0	28.9	3.0	2.9
Mid	15.0	14.0	1.5	1.5
Stern	32.0	29.3	2.5	2.4

NOTE: Tests performed at 72°F and 65% relative humidity.

11.59 Order rows and columns with a logical scheme that the reader can easily understand and use to locate the appropriate column and row headings quickly:

- Use time order for information recorded in a sequence,
- Arrange major and minor items in order from whole structure to its parts,
- Group items of the same type or size,
- Place items that will be compared close together, and
- Put numbers that will be summed or averaged in the same column.

11.60 Avoid mixing too many types of information in a single column. Information may vary by column or by row—but not both at the same time. Especially avoid unnecessary changes of units, for example, from inches to centimeters.

11.61 Align cell entries vertically with the row heading for that row. If the row heading is more than one line, align one-line cell entries with the baseline of the last line of the row heading. If both row heading and column entries are more than one line, align the first lines of both:

```
Amino Acid       Isoelectric Point    Isolated by
Alanine         ........6.0.......    Schutzenberg
Threnonine      ........6.2.......    Gartner and
                ..................      Hoffman
Aspartic        ........2.8.......    Ritthaus and
  acid          ..................      Prentis
Glutamine       ........3.2.......    Ritthaus
```

Rules

11.62 Any number of rules—horizontal and vertical lines—can be used to differentiate a table's parts. Selecting an appropriate rule design, both number of rules and their graphic characteristics, depends on the table's complexity and expected information retrieval practices. Use the same rule design for all tables in a single work.

11.63 Rule widths, or weights, depend on the number of rules. Start with the thinnest rule possible based on the expected printing process—for example, a hairline—and increase width as more discrimination is needed. For printed works, use rules in this order:

- hairline
- ½-point
- 1-point
- 2-point

Try to create a rule design that has a structure similar to that of typographic decisions. That is, begin with the least intrusive rule for fields, or the majority of a table, and increase the width for each major table component that needs visual differentiation. Do not forget, however, that data displays for nonprint display require considerably more differentiation and larger type sizes. The tables in this text, for example, use

Rule Selection

Table Part	Rule
Rows (when necessary)	Hairline with 50 percent screen
Columns	Hairline
Border (when necessary)	Hairline with 85 percent screen
Head-Stub-Field	¼-point
Side-by-side table separator	½-point

11.64 Avoid overuse of rules; they should be neither decorative nor obtrusive. Rules differentiate and categorize relationships in tabled information.

11.65 For simple, informal tables—no more than three columns and five rows—rules can be omitted altogether:

Abrasive	Knoob Hardness (Kv/mm)	Melting Point (°C)
Quartz	820	1700
Emery	2000	1900
Corundum	2000	2050
Garnet	1360	1200
Diamond	6500	1000

Abrasive	Knoob Hardness (Kv/mm)	Melting Point (°C)
Quartz	820	1700
Emery	2000	1900
Corundum	2000	2050
Garnet	1360	1200
Diamond	6500	1000

Row Rules

11.66 Row rules separate rows in the field. Although it is seldom required, use row rules only if vertical space is not sufficient to separate rows. In long tables use row rules to divide major sections of the table. Extend row rules the full width of the table, including the stub column. If used, row rules should be the lightest rules in the table.

Column Rules

11.67 Use column rules, the same weight as border rules, if white space alone is not sufficient to separate closely spaced columns. These should be lighter than head-stub-field separators but heavier than row rules.

Table Border

11.68 The table border extends around the outside of the table. If the table appears on a single-column page with margins to left and right, the left and right vertical borders may be omitted. Make the table border the same weight as column rules. These should be lighter than head-stub-field separators but heavier than row rules.

Head-Stub-Field Separators

11.69 Head-stub-field separators, heavy or double vertical and horizontal rules, separate column and row headings from the table's body. Make separator rules more emphatic than border, column, or row rules but not as emphatic as side-by-side table separators.

Side-by-Side Table Separators

11.70 If a table is too long and narrow to fit the available space, divide it into two halves presented side by side. Separate the two tables with an emphatic rule or extra space:

Abrasive	**Knoob Hardness** (Kv/mm)	**Melting Point** (°C)	**Abrasive**	**Knoob Hardness** (Kv/mm)	**Melting Point** (°C)
Quartz	820	1700	Alumina	2000	2050
Emery	2000	1900	Tripoli	820	1700
Corundum	2000	2050	Carbide	2400	2500
Garnet	1360	1200	Diatomite	830	1700
Diamond	6500	1000	Borazon	4700	2000

Table Notes

11.71 See 8.97.

Specific Notes

11.72 Specific notes give more information about a column, row, or cell. Cite specific notes only in column heads, row headings, and cells with a superscripted symbol. Do not cite specific notes in the table title or stub head; use a general note instead.

11.73 Mark specific notes with raised (superscripted) symbols above and to the right of the referenced item. Separate note references that occur together with spaces, not commas. If a note occurs in a blank cell, put the note reference in parentheses:

Single note[1]	Double note[2 3]	(4)

11.74 Choose symbols that differ from the table's data:

Data Type	Use	Avoid
Numeric	Letters	Numbers
Text	Numbers	Letters
Formula	Special	Numbers, letters
Statistics	Letters	* and **

Special symbols include the asterisk, dagger, double dagger, section, parallel, and number mark. Use the same type of symbols throughout the table. To create additional special symbols, double or triple them (**, ***, ##, ###). Do not mix special symbols (*#, #*@). In statistics * and ** are abbreviations for probabilities of .05 and .01 respectively (see also 7.49).

11.75 Put specific notes at the bottom of the table on the page on which they occur. If the same notes occur on multiple pages, gather all such notes at the end of the table.

11.76 Place specific notes in the order in which they occur in the table: starting in the upper-left corner and down the table row by row.

11.77 Extend general notes across the full width of the table. Several short notes, however, may appear on a single line:

[1] First general note could be a long one.

[2] Second note. [3] Third note.

11.78 Set table notes in smaller or less emphatic type than table entries (see also 12.31).

Table Contents

11.79 Table cells can contain text, numbers, or graphics. Align and format data within cells consistently and logically. The following sections provide further information about

- Text (11.80),
- Numbers (11.82),
- Graphics (11.92),
- Values (11.93).

Text in Tables

11.80 Table cells can contain a word, a phrase, or an entire paragraph. Punctuate all table text with the same conventions as traditional, nontabled text. Center single-line text entries in columns; left-justify multiple-line entries. Align all text in the column the same. If a column contains more than a few multiline entries or if centering the multiline entries makes them hard to read, then left-justify all entries in the column:

Short entry	Long entry that extends to multiple lines	Short entry
Another short entry	Another long entry that extends to multiple lines	Long entry that extends to multiple lines
Yet another short one		Another short entry

11.81 If space between rows does not adequately separate multiline entries, indent runover lines:

Unacceptable	Acceptable
Long entry that extends to multiple lines	Long entry that extends to multiple lines
Another long entry that extends to multiple lines	Another long entry that extends to multiple lines
	Better still, add space between rows.

Numbers as Table Content

11.82 Tables commonly organize numbers into an easily scanned collection.

11.83 Put units of measurement in a column subheading. If different units must be shown in the same column, omit the units from the column subheading and put them in each cell. Do not put units in column subheads unless they apply to all numbers in the column (see also 7.27, 7.31, 7.36, and 7.41).

Unacceptable	Acceptable	Acceptable
Thickness	Thickness (mm)	Thickness (mm)
620 mm	620	6.2×10^2
24 cm	240	2.4×10^3
2 cm	20	2.0×10^1

11.84 Use only arabic numerals, and do not spell out numbers (see also 7.16):

Unacceptable	Acceptable
IX	9
Six	6
Zero	0

11.85 Right-justify whole numbers. Align decimal numbers on the decimal point and on commas, if present (see also chapter 4). Align numbers on the actual or implied decimal point. Align plus (+) and minus (–) signs preceding numbers, and percent signs (%) following numbers:

3,468.62	+47 %
27,891.74	– 6 %
497.65	+29 %

11.86 Omit plus (+) or minus (–) signs, percent signs (%), and currency symbols ($, ¥, £) if the table title, column heading, or common usage makes the units and sign obvious.

Dates in Tables

11.87 Express all dates in the same format (see also 7.8 and chapter 4). Center dates, except military dates (2 Jun 88). For military dates, align the three components:

3/11/88	March 11, 1988	11 Mar 88
12/16/89	December 16, 1989	16 Dec 89
4/1/92	April 1, 1992	1 Apr 92

Times in Tables

11.88 Express all times in the same format. Align hours, minutes, and seconds. Include AM or PM after the time if not using 24-hour format. Include the time-zone reference (GMT, UTC, EST, etc.) in the column subheading to avoid confusion (see also chapter 4 and 7.7):

Unacceptable	Acceptable
Time	Time (UT)
10:42	10:42:00
1:24 PM	13:24:00
9:14:28	21:14:28

11.89 For duration, make the units of measurement obvious. Experimenters may know that the units report hours and minutes, but readers may interpret them as minutes and seconds:

Unacceptable	Acceptable
Duration	Duration (hours:minutes)
1:35	1:35
2:02	2:02
1:45	1:45

Ranges in Tables

11.90 Use an en-dash surrounded by spaces to link the ends of ranges. Do not use a hyphen or the word *to*:

Unacceptable	Acceptable
1972 - 82	1972–82
1929 to 1934	1929–34

Align ranges on the dash linking the ends of the range.

Formulas in Tables

11.91 Align formulas on the equal sign (=) (see also 7.67). Follow the rules for breaking and indenting multiline formulas at 7.72.

$$F = ma$$

$$PV = nRT$$

$$E = mc^2$$

Graphics in Tables

11.92 Graphics in tables can range from realistic pictures, such as the photograph of an employee, to abstract symbols (see also 10.26), such as the product rating dots used by Consumer Reports™. If symbols are not obvious, supply a key or legend. This chart uses dots to show participation by the teams listed in the column heads.

Management Teams

Project Teams	Departments							
	Engineering	Facilities	Finance	Marketing	Production	Purchasing	Research	Sales
Amethyst	●	●	●		●			●
Diamond		●	●	●	●	●	●	
Garnet	●	●		●		●		●
Quartz	●	●				●		
Topaz	●	●	●	●		●	●	●

Missing or Negligible Values in Tables

11.93 Many situations may create ambiguous table entries. To eliminate any ambiguity

- Leave any cell blank if its column head does not apply.
- Do not put "N/A" or "Not applicable" in the cell. Explain the absence of data with a table footnote in parentheses.
- Insert a leader (. . .), as wide as the widest column entry, to direct the eye to the next column if a particular cell has no data.
- Put a 0 in any cell whose value is zero; do not leave it blank:

Unacceptable	Acceptable
N/A	(7)
......	0

Repeated Values in Tables

11.94 Show repeated values. Do not use "ditto," "do," or the quotation mark ("):

Unacceptable	Acceptable
356	356
"	356

Special Tabular Problems

11.95 This section examines more complex tables and special situations including

- Continued tables (11.96),
- Referring to tables (11.101),
- Oversize tables (11.102),
- Abbreviations in tables (11.104),
- Simplifying comparisons in tables (11.105),
- Grouping and summarizing data in tables (11.106).

Continued Tables

11.96 If a table does not fit on a single page, continue it on subsequent pages. For numbered tables, repeat the table number followed by a parenthetical phrase that indicates where the current page fits in the table's sequence:

Figure 4-2 (Page 3 of 6)

For titled tables, provide a short title followed by a similar parenthetical phrase:

Resistivity of Carbon Filaments (Page 3 of 6)

Or:

Resistivity (Page 3 of 6)

11.97 Repeat column and row headings on continuation pages. On continued turned (broadside) tables repeat the column headings on the left-hand pages only; omit them from the facing right-hand page.

11.98 In horizontally ruled tables, omit the bottom rule, except on the last page of the table.

11.99 Do not number the pages of a multipage table separately, as: 3a, 3b, 3c. Do not interrupt the text with many multipage tables. Collect such tables in an appendix. If necessary, summarize these tables in the text.

11.100 In primarily nontextual works, organize and sequence tables, other data displays, or illustrations in the same manner as paragraphs and subsections.

Referring to Tables

11.101 See 8.97.

Oversize Tables

11.102 If a table is too long or tall to fit on a single page, continue it on the next or subsequent pages or divide it into multiple tables.

11.103 If a table is too wide, try remedies in this order:

1. Use a moderately condensed typeface, but avoid reducing legibility.
2. Turn the table horizontally (broadside) on the page. Omit column heads from right-hand pages.
3. Reduce the type size, but not below 8-point for printed tables, and reduce the column spacing.
4. Extend the table horizontally across two facing pages. Align the rows so the reader can move back and forth between pages without getting lost.
5. Print the table on oversize paper and bind it as a foldout.

Abbreviations in Tables

11.104 Spell out all titles, labels, and notes. Abbreviate only where necessary. Use only common abbreviations, for instance, for units of measurement (see also chapter 6):

Unacceptable	**Acceptable**
Strut Dsgn Crit	Structural Design Criteria
(pounds per square inch)	(lb/in^2)

Simplifying Comparisons in Tables

11.105 Convert numbers to common units and round off to the same accuracy. Express such numbers in the same format and with the same number of decimal points (see also 7.38):

Unacceptable	**Acceptable**
12 mm	12.000 mm
.023 mm	0.023 mm
23.4 mm	23.400 mm
1.03456 mm	1.035 mm

This rule does not apply to original observations, which should represent the actual units and accuracy measured (see also 7.42).

Grouping and Summarizing Data in Tables

11.106 In complex tables, group or summarize data. Avoid braces in the field or columns to show groupings. Use multilevel row headings or field spanners to group data.

Unacceptable

Propellant Vent Port	Pressure		
	Specified (psi)	Actual (psi)	
External	100.0	102.7	
Aft	28.0	28.0	Model 635
Mid	Internal { 14.0	14.7 }	
Stern	28.0	28.1	
External	120.0	122.7	
Aft	30.0	28.9	Model 635-AX
Mid	Internal { 15.0	14.0 }	
Stern	32.0	29.3	

Acceptable

Propellant Vent Port	Pressure	
	Specified (psi)	Actual (psi)
	Model 635	
External	100.0	102.7
Internal		
Aft	28.0	28.0
Mid	14.0	14.7
Stern	28.0	28.1
	Model 635-AX	
External	120.0	122.7
Internal		
Aft	30.0	28.9
Mid	15.0	14.0
Stern	32.0	29.3

11.107 Divide long columns of numbers by inserting a blank line every five to ten rows:

Square, Cubes, and Roots

Number	Square	Cube	Square Root	Cube Root
1	1	1	1.00000	1.00000
2	4	8	1.41421	1.25992
3	9	27	1.73205	1.44225
4	16	64	2.00000	1.58740
5	25	125	2.23607	1.70998
6	36	216	2.44949	1.81712
7	49	343	2.64575	1.91293
8	64	512	2.82843	2.00000

11.108 Other methods of grouping rows include using horizontal rules, leaders, or underprinting (see 11.62).

Common Types of Tables

11.109 The remainder of this section considers the design of specific table types:

- Look-up-a-value tables (11.110),
- Decision tables (11.111),
- Distance tables (11.114),
- Matrix charts (11.116).

Look-Up-a-Value Tables

11.110 Most tables provide easy access to specific values by combining row and column headings that describe field items. Readers locate target field items at the intersection of the row and column bearing the appropriate headings.

How We Read a Table

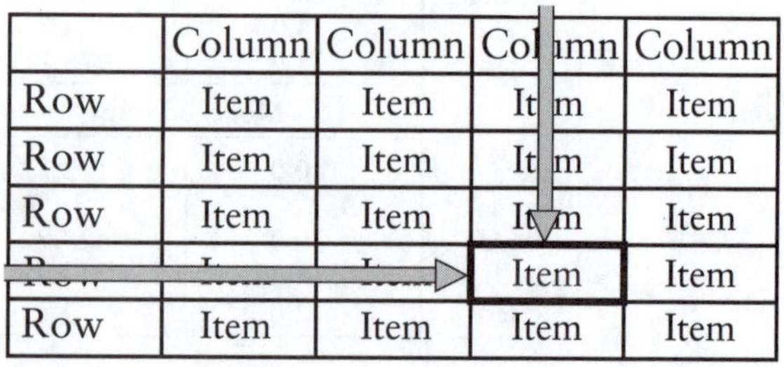

	Column	Column	Column	Column
Row	Item	Item	Item	Item
Row	Item	Item	Item	Item
Row	Item	Item	Item	Item
Row	Item	Item	Item	Item
Row	Item	Item	Item	Item

Decision Tables

11.111 Decision tables help the reader make complex decisions by simplifying alternatives and summarizing if-then conditions.

11.112 Divide the decision table horizontally into conditions on the left and actions on the right. The reader scans down the left columns to find a row that matches the current conditions and then performs the actions to the right in that row:

Typical Decision Table

Check Acceptance Policy			
Regular customer?	Check is for . . .?	ID Required	Approval Required
Yes	Purchased items	None, unless over $50	None
	Cash	Driver's license	Supervisor
No	Purchased items	Driver's license	Supervisor
	Cash	Driver's license	Manager

11.113 Conditions may be listed as yes/no questions in the column headings with the individual cells containing symbols for the answers. If the conditions and responses are few and simple, they may appear as text in the cells.

Distance Tables

11.114 Distance tables show distance or other values relating to pairs of categories. Construct distance tables by listing in a triangular grid the values related to a corresponding column and row head. One of the categories is omitted from the rows and another from the columns to prevent empty boxes.

	Tokyo	Rome	Rio	Beijing	New York	Moscow	Los Angeles	London
Bombay	4,190	3,845	8,620	2,964	7,795	3,130	8,700	4,465
London	5,940	890	5,770	5,055	3,460	1,550	5,439	
Los Angeles	5,470	6,325	6,295	6,250	2,450	6,070		
Moscow	4,650	1,475	7,180	3,600	4,660			
New York	6,735	4,275	4,830	6,825				
Beijing	1,310	5,050	10,768					
Rio	11,535	5,685						
Rome	6,124							

11.115 Distance tables, although generally used to present mileage, can illustrate any mutual relationship.

Matrix Charts

11.116 Matrix charts, a hybrid of charts and tables, present information arranged in uniformly spaced scales or categories about the combination of values represented by the cells.

Modern Periodic Table

	H 1																		He 2
	L 3	Be 4												B 5	C 6	N 7	O 8	F 9	Ne 10
	Na 11	Mg 12												Al 13	Si 14	P 15	S 16	Cl 17	Ar 18
	K 19	Ca 20		Sc 21	Ti 22	V 23	Cr 24	Mn 25	Fe 26	Co 27	Ni 28	Cu 29	Zn 30	Ca 31	Ge 32	As 33	Se 34	Br 35	Kr 36
	Rb 37	Sr 38		Y 39	Zr 40	Nb 41	Mo 42	Tc 43	Ru 44	Rh 45	Pd 46	Ag 47	Cd 48	In 49	Sn 50	Sb 51	Te 52	I 53	Xe 54
	Cs 55	Ba 56		Lu 71	Hf 72	Ta 73	W 74	Re 75	Os 76	Ir 77	Pt 78	Au 79	Hg 80	Tl 81	Pb 82	Bi 83	Po 84	At 85	Rn 86
	Fr 87	Ra 88		Lr 103	Rf 104	Db 105	Sg 106	Bh 107	Hs 108	Mt 109	Uun 110	Uuu 111	Uub 112		Uuq 114		Uuh 116		Uuo 118
lanthanides			La 57	Ce 58	Pr 59	Nd 60	Pm 61	Sm 62	Eu 63	Gd 64	Tb 65	Dy 66	Ho 67	Er 68	Tm 69	Yb 70			
actinides			Ac 89	Th 90	Pa 91	U 92	Np 93	Pu 94	Am 95	Cm 96	Bk 97	Cf 98	Es 99	Fm 100	Mv 101	No 102			

Charts

11.117 This section will enable the reader to prepare technically correct and aesthetically pleasing charts. In doing so, it will consider such topics as

- When to use charts (11.118),
- Characteristics of an effective chart (11.121),
- Parts of charts (11.133),
- Special topics and effects (11.183),
- Common types of charts (11.204).

When to Use Charts

11.118 Charts help readers understand quantitative information by illustrating changes or trends over time, patterns in numerical data, and relationships among various factors.

11.119 Use charts to

- Interest readers in the data and convince them to read the text;
- Help readers extrapolate from given values to predict future values;
- Present large amounts of complex data without overwhelming readers;
- Emphasize data trends, relationships, and patterns to make a convincing point; and
- Add credibility.

11.120 Do not use charts to

- Make boring numerical data more palatable. Find an interesting aspect of the data, put it in an appendix, or eliminate it.
- Break up text. Charts are too expensive to use as decoration. Use charts to emphasize ideas in the text.
- Show precise numbers. Use a table to allow detailed comparisons.
- Communicate to an audience unfamiliar with charts. Engineers and scientists sometimes forget that charts may confuse a lay audience.
- Interpret, explain, evaluate, and review data. Use text instead.

Characteristics of an Effective Chart

Coherent

11.121 Each chart should demonstrate coherent relationships among the data. Design charts to focus on the values and variations in the data. The chart's design must be subdued; it must not vie with the data for attention.

Simple

11.122 The chart's main idea should be simple and visually obvious to the reader. Each chart should express one main idea. Effective charts direct attention to the data's significance, not the chart's form or style.

Unacceptable

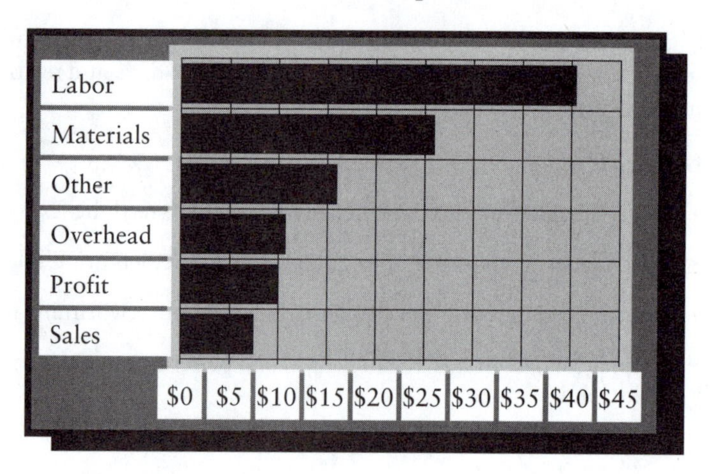

Acceptable

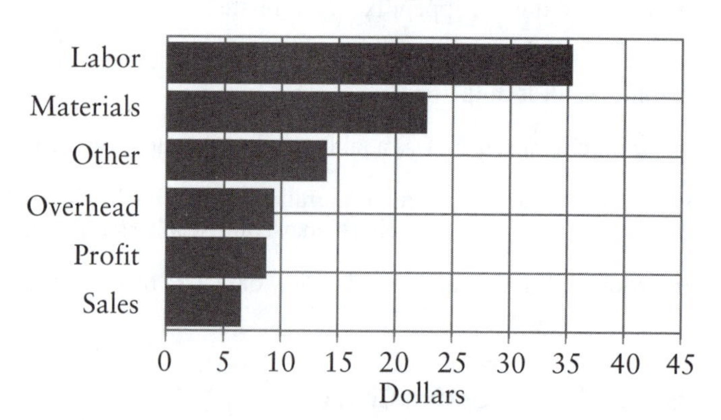

11.123 Avoid complex coding schemes of color, shading, or symbols. Likewise, avoid using chart design as decoration.

Concise

11.124 Conciseness means communicating the maximum amount of information to the greatest number of people in the shortest time with the smallest amount of print.

11.125 Allow redundancy that provides a context and organization for the data or simplifies comparison across the data, but do not repeat information without good reason.

11.126 If the complete data set cannot be charted, simplify it first by clustering, averaging, or smoothing.

Honest

11.127 Graphs should never lie, mislead, or confuse. Distortion happens when the data representation is inconsistent with its numeric values. Express all data in common units. For instance, dollars are not dollars, unless first compensated for inflation. Similar factors that can change the basis of measurement include population growth, changes in foreign exchange rates, and shifts in the value of assets.

Graphical lies frequently occur when the graphic's size is not a simple, visually obvious multiple of the number it represents. For instance, a bar chart may be composed of cubes whose edges represent a variable's value. A tripling of value results in a 27-fold increase in the cube's volume.

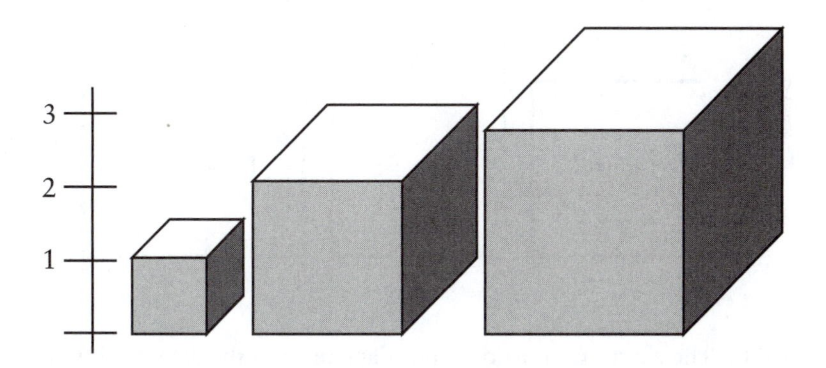

Such distortions are common in political advertisements showing the effects of inflation or growth in jobs.

Labels, annotations, and posted values can compensate for graphical distortion, but they cannot eliminate it.

Visually Attractive Charts

11.128 Each chart should have a center of attention with visual elements balanced about this center. In charts, this center will be near the bottom, above the horizontal scale, in the field.

11.129 Provide sufficient contrast to separate adjacent elements so primary information stands out from secondary details.

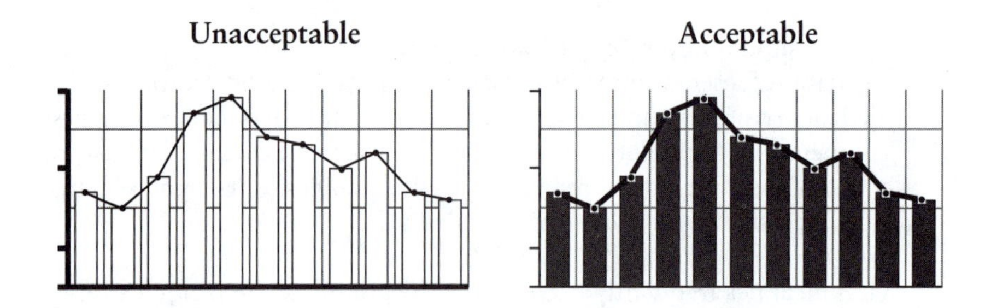

11.130 Use landscape (horizontal), as opposed to portrait (vertical), orientation for charts because they can show longer data series and text labels fit better.

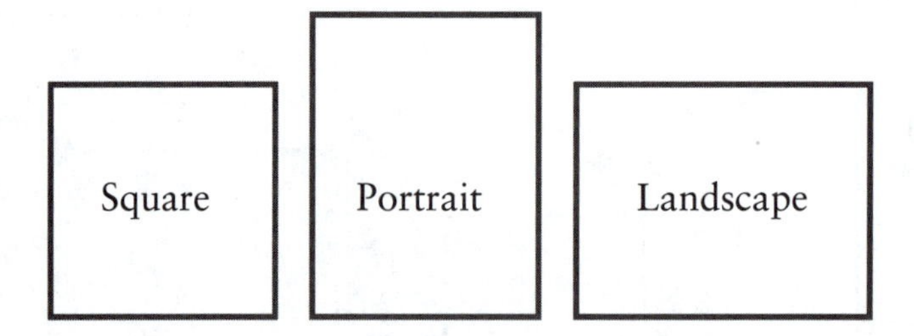

11.131 The chart's proportions and shape must fit the data. Within that restriction, recommended proportions range from 1:1.2 to 1:2.0. Right in the middle of this range is the golden rectangle, with proportions of 1:1.618.

1 1.2

1 1.618

1 2

Fit Charts to the Medium

11.132 Design charts to take advantage of the features of their expected presentation medium and to respect the limitations of that medium:

Medium	Characteristics	Special Consideration
TV	Color	Use color
	Low resolution	Only very simple graphs
	Lay audience	Only common types
	3 × 4 aspect ratio	Size accordingly; minimum text size = 1/20 chart height
Book	Read at leisure	Much detail possible
	Color is expensive	Avoid color
	High resolution	Minimum text size 6-to-8-point (10-points if reproduction is poor or lighting is dim)
Slide	Shown in dark room	Light subject on dark background
	Color common	Use color
	Paced by presenter	Keep simple, trends only
	Medium resolution	Minimum text size = 1/50 chart height
	2 × 3 aspect ratio	Size accordingly
Viewgraph	Shown in lighted room	Dark subject on light background
	Medium resolution	Minimum text size = 1/50 chart height
	Paced by presenter	Keep simple, trends only
	3 × 5 or 5 × 3 aspect ratio	Size accordingly
Computer	Color	Limited across platforms
	Low resolution	Only simple graphs, low-resolution photographs and graphics
	Limited text sizes	Simple type faces, limited across platforms
	Animated images	Limited by available memory, processor, and modem speed

Parts of a Chart

11.133 Despite the diversity of charts, most share common design elements:

Exhibit 11–2: Parts of a Chart

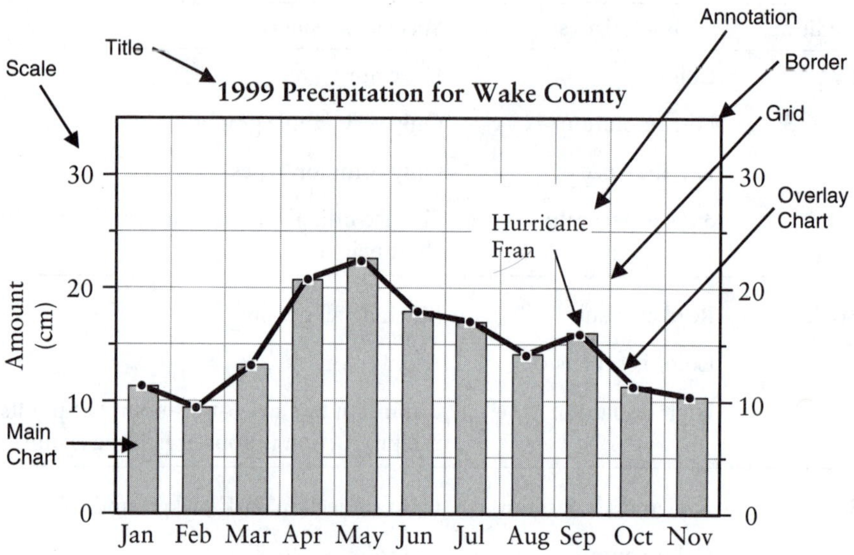

11.134 A chart should contain the following information (from most to least important):

- Data values (11.135),
- Title (see also 10.61),
- Scales (11.150),
- Grids (11.172),
- Annotations and inserts (11.175).

Data Values on Charts

11.135 The most important information on a chart is the dot, line, curve, area, or other symbol that represents the data values.

11.136 Group each item in a single data series by using the same size and symbol for data points, or the same width and tone for bars.

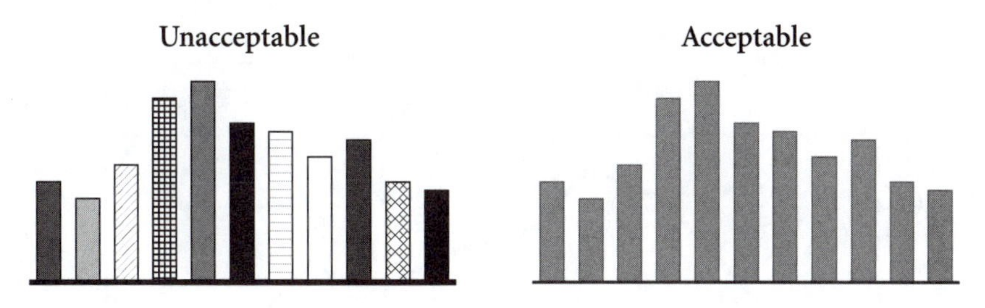

Data Point Symbols for Charts

11.137 Symbols for data points include crosses, Xs, circles, triangles, squares, and diamonds. Fill open shapes or superimpose shapes to provide more symbols. For data points, use different shapes rather than various sizes, styles, and weights of the same shape.

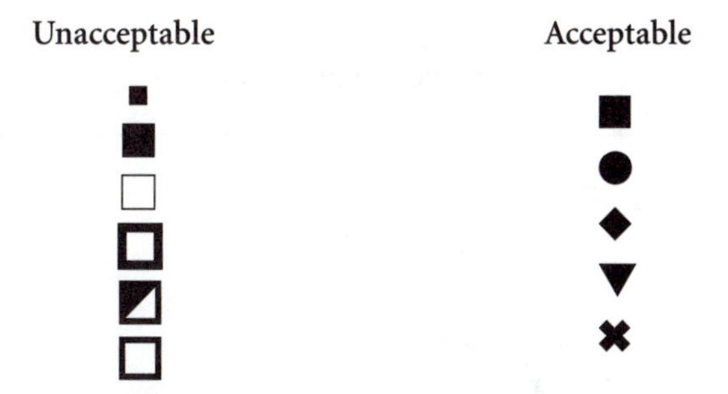

Avoid open or half-filled symbols, especially those with thin borders, for reduced or photocopied charts.

11.138 Plot data representing specific points in time (point data) on the grid line for that point in time. Plot data representing spans of time (period data) between the grid lines for the beginning and ending points of the time span.

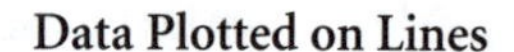

Data Plotted on Lines

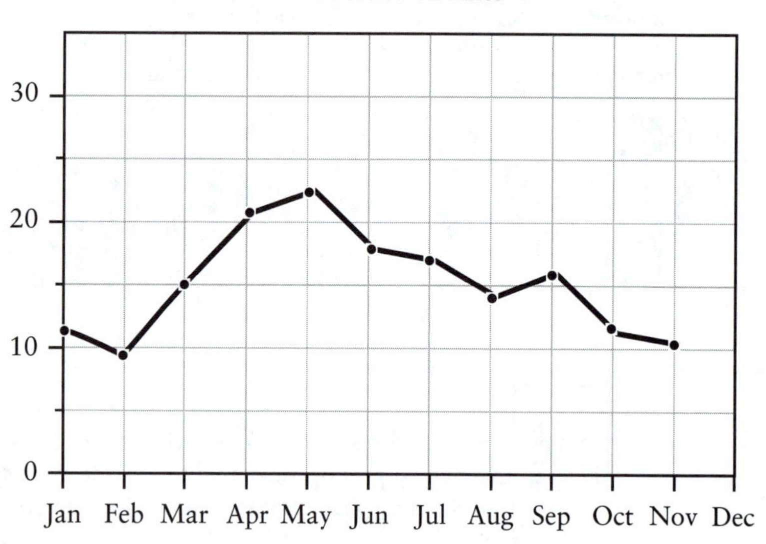

Data Plotted in Spaces

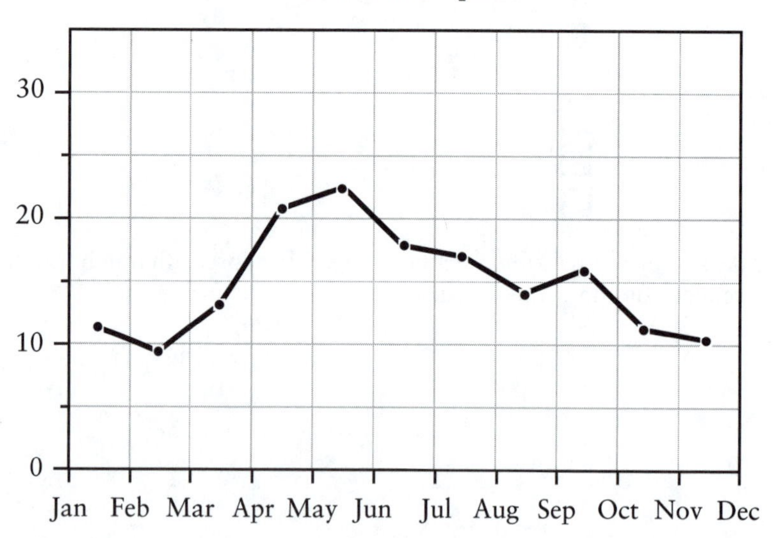

Area or Volume Designation for Charts

11.139 Since readers cannot quickly compare areas and volumes, especially of irregular shapes, avoid circles and spheres.

11.140 Since comparing overlapped areas is difficult, separate compared areas.

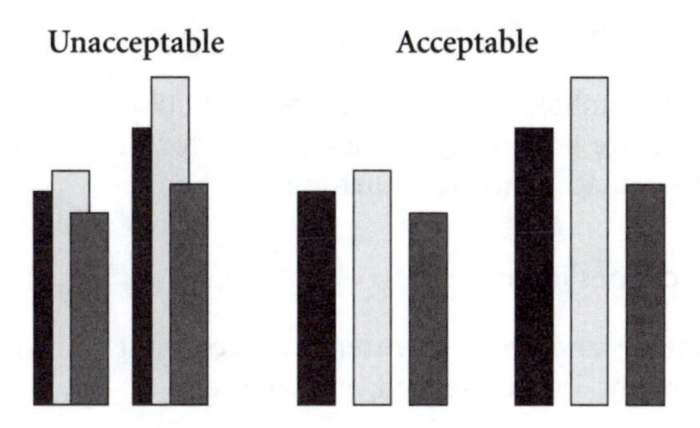

11.141 If using area to express values, show the area composed of countable blocks. The same technique works for volume as well. Make the blocks simple geometric shapes—all of the same size.

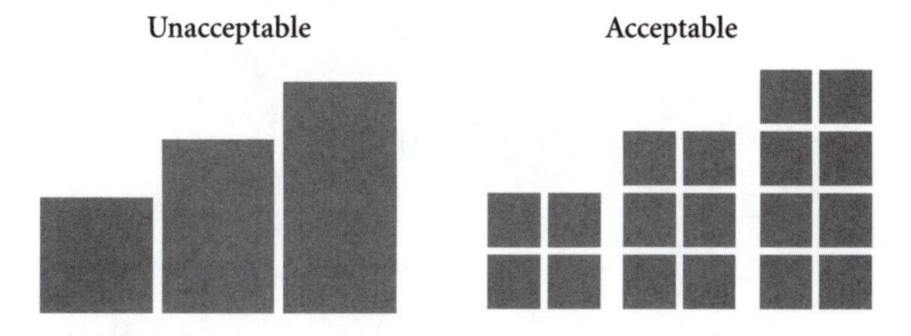

Symbols or Pictures for Charts

11.142 Pictorial symbols add little to charts for a technical audience but may add interest for a lay audience. Use pictorial symbols as counting units. Do not vary two or three dimensions of a pictorial object to show changes in quantity.

11.143 Symbols used in a series should all have the same value. Make symbols

- Self-explanatory;
- Simple and distinct in form and outline;
- Represent a definite unit of value;
- Represent a general concept, never a specific instance;
- Approximately the same size; and

- Easily divisible. Irregular shapes do not divide well. Never subdivide a human form. Round off numbers instead.

Avoid

- Dated or cultural symbols. Almost any hand sign is an obscene gesture somewhere in the world (see also chapter 2), and
- Symbols that require labels to be recognized.

11.144 Unit symbols do not require special graphic talent. Typewriter unit symbols are possible for many fields: $ for money, # for amount, % for percentages, or O for other units.

Curves and Lines in Charts

11.145 Curves and lines show trends and connect data values. Curves may be straight, jagged, stepped, or smoothed.

Straight lines show linear correlation and simple trends

Jagged lines connect actual data points.

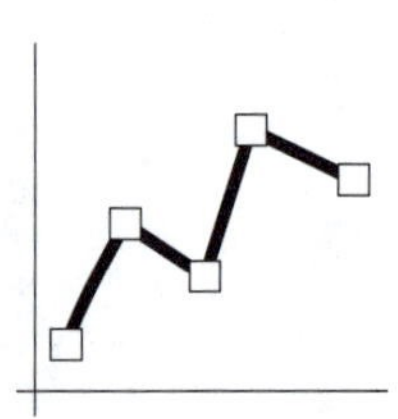

Stepped lines show changes in a variable at regular intervals.

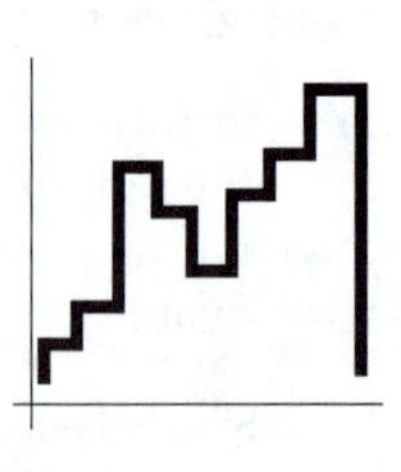

Smooth curves show more sophisticated trends and averaged data.

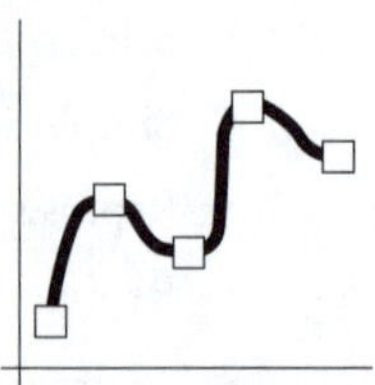

Line Weight for Chart Curves

11.146 Draw all curves the same width, about twice as thick as scales (see 11.62).

Line Style in Charts

11.147 Use different line styles (solid, dotted, or dashed) to distinguish close or crossing curves. If some variables lack data, leave this zone blank or use a dotted line to bridge zones of no data. Clearly indicate that the dotted line is not based on actual data.

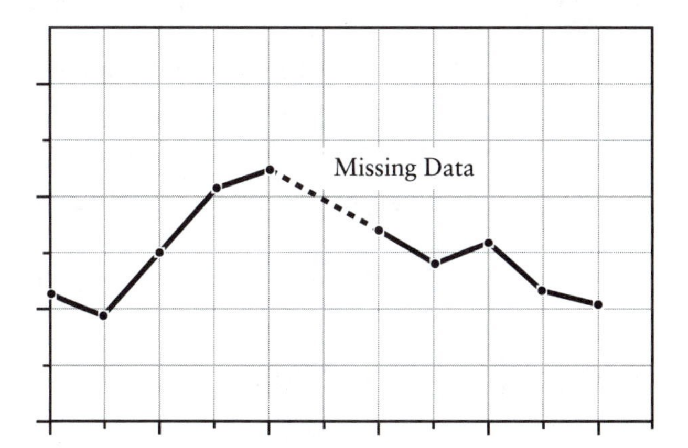

11.148 When lines cross, interrupt one of them, as if it passes under the other.

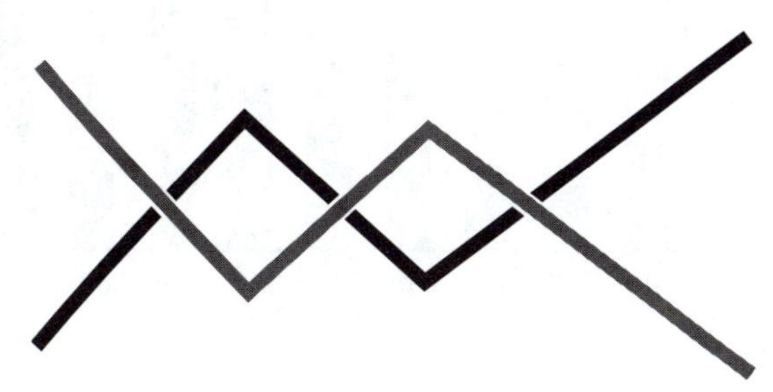

Curve Labels in Charts

11.149 Place unboxed curve labels horizontally; do not align curve labels along curves. Do not use arrows or leader lines to connect labels to curves unless absolutely necessary.

Scales for Charts

11.150 Vertical and horizontal scales indicate the value of data plotted in charts. The vertical scale is commonly called the *y-axis* or *abscissa*; the horizontal scale, the *x-axis* or *ordinate*. Place independent variables along the horizontal axis and dependent variables along the vertical axis. Show time on the horizontal axis. Values increase from bottom to top and from left to right.

11.151 Scales should start from zero. The horizontal axis corresponds to zero. However, if values fall below zero, extend the vertical axis below the minimum value and show the zero value as a horizontal reference line.

11.152 Place the vertical scale on the left side of the chart. However,

- If the data rise to the right or if the data on the right require special attention, put the vertical scale on the right side of the chart; and
- If the data vary throughout their range, put a vertical scale on both left and right.

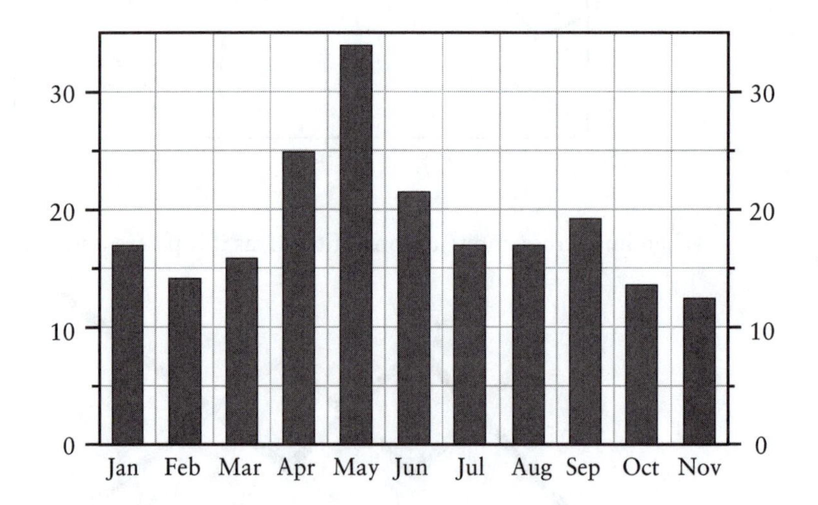

11.153 If the chart is used to look up accurate values, repeat the scale on left and right and on top and bottom. If data are plotted to two different scales, place one scale on each side and make it obvious how the two scales apply to the data. Start both scales from a common baseline of zero and use the same distance between major divisions.

Placing Scale Labels in Charts

11.154 Scale labels show the value, unit, or count of the variable plotted along that axis. The scale may be in absolute values, percentages, rates, or named categories:

- Label the vertical scale horizontally above or to the left of the scale, or
- Label the horizontal scale horizontally below the right end of the scale.

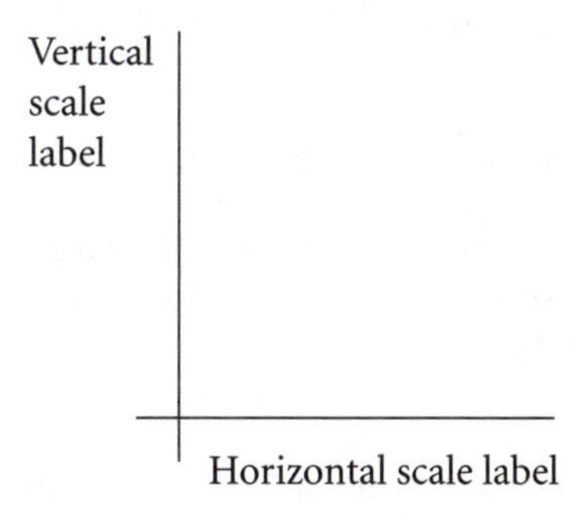

11.155 To save horizontal space the vertical scale label may be

- Turned clockwise to read from bottom to top, or
- Placed to the right of the vertical scale if this positioning does not cover the data.

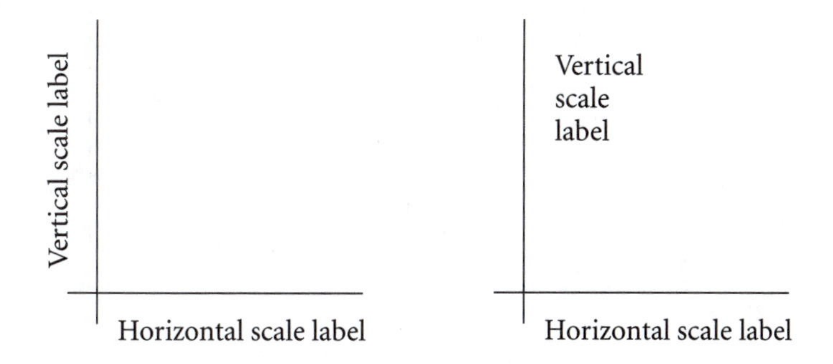

11.156 Label each scale with the variable and units of measurement. In some simple charts one or both of the scale titles may be omitted:

- If all variables are plotted against a single vertical scale and that scale is implied by the title, do not repeat the title in the scale label. Omit the vertical scale label.

- If the horizontal scale is obvious from the intervals plotted, the horizontal scale label may be omitted. For instance, if the horizontal scale is divided into intervals labeled January, February, March, and so on, the scale for months need not be labeled.

Scale Divisions

11.157 Indicate major scale divisions with tick marks and minor divisions with smaller tick marks. On a scale, place figures at every major scale division. Major scale divisions should be every second, fourth, or fifth minor division.

11.158 Tick marks can go inside, outside, or across scale lines.

11.159 Label the scale starting with 0 at the left or bottom and proceed in even steps along the scale. Label scales with multiples of 2, 5, and 10; for instance, 0, 5, 10, 15, 20; not, 1, 4, 7, 11, 14. Minimize the number of major scale divisions; 5 or 6 are acceptable but 10 or 12 are too many.

11.160 If scale figures represent units of 1000s or higher, drop the trailing zeroes and express unit size as a phrase in the scale label ("in $millions") (see also 7.36).

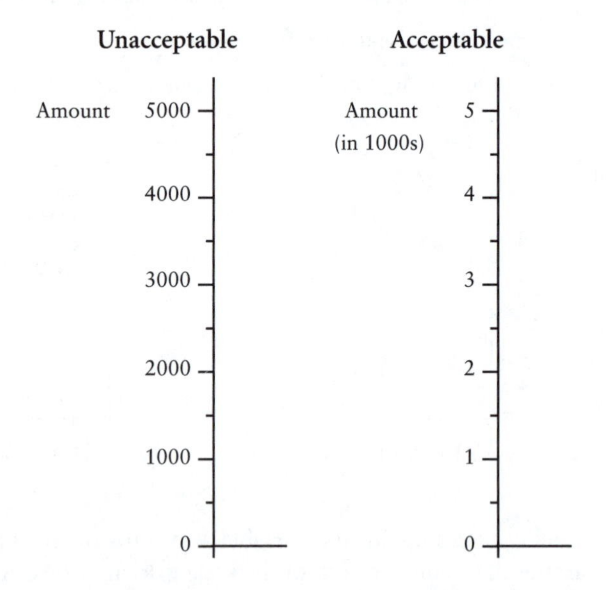

11.161 Charts for technical readers may avoid excessive zeroes by using the *Systéme Internationale* (SI) prefixes for scientific units (see also 7.41).

Scale Ranges for Data Plotting

11.162 The scale range should be slightly larger than the data range. Sometimes the plotting intervals extend beyond the end of the data to help predict future data.

11.163 Always show a zero baseline except when the data represent index numbers relative to some other reference value, for instance, 100.

11.164 When the data values do not warrant using the entire chart area, break the grid and scale to make better use of available space. Keep the zero baseline and clearly indicate the break in the grid and scales.

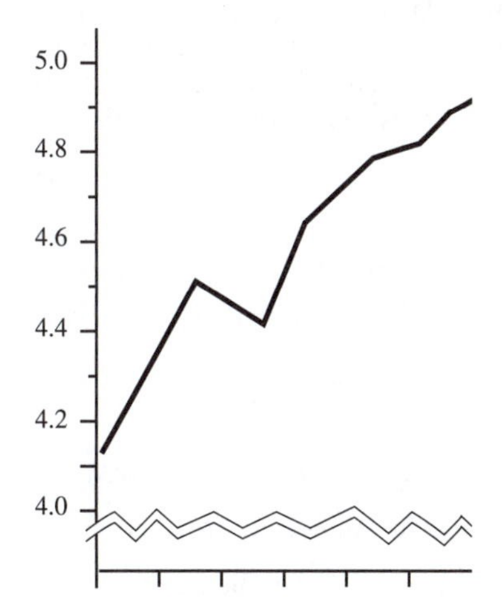

Scaling Function

11.165 Charts represent numeric values as distances on paper or on screen. The scaling function determines the relationship between the numeric value and the distance on the chart. Several scaling functions are used: linear, logarithmic, and geometric.

Linear Function

11.166 In linear scales, the chart's length is a constant multiple of the value it represents. If a 1-inch bar represents 100; then, a 2-inch bar equals 200:

$$\text{Distance} = \text{value} \times \text{scale factor}$$

Use only linear scales in charts for a lay audience.

Logarithmic Function

11.167 With a logarithmic scale, the distance or size is proportional to the logarithm of the value:

Distance = scale factor × log value

11.168 Use a logarithmic scale to

- Show proportional or percentage relationships;
- Show proportional rates of change by emphasizing change rate instead of absolute values or amounts of change;
- Compare relative changes for data series that differ greatly in absolute values (e.g., comparing a city's growth rate to that of a state). Use a logarithmic scale when a single data series includes both very small and very large values or when baseline quantities vary considerably;
- Compare rates of change measured in different units;
- Compare a large number of curves of widely varying scales. Stack separate logarithmically scaled charts atop one another on the same page. Logarithmic scales accurately display numerical values throughout a wide range of values.

11.169 Do not use logarithmic scales

- To compare exact amounts of widely varying values,
- If the change in values is less than about 50 percent, or
- For an unsophisticated audience.

Geometric Function

11.170 With a geometric scale, distances are proportional to some power of the value:

Distance = constant × value$^{\text{scale power}}$

11.171 Use geometric scales to prevent distortion when an area or volume represents a value:

Value Represented by . . .	Scale Power
Length	1
Area	1/2
Volume	1/3

Grid Lines

11.172 In curve charts used for locating specific values, include a background grid screened 70–85 percent. Make grid lines the lightest weight in the chart, about one-third the weight of data points or curves (e.g., if data lines are 2 point and axes 1, make grid lines hairlines).

Reference of Base Line

11.173 Extend reference or index lines the length of the horizontal axis. They should be lighter weight than data curves but heavier than the scales:

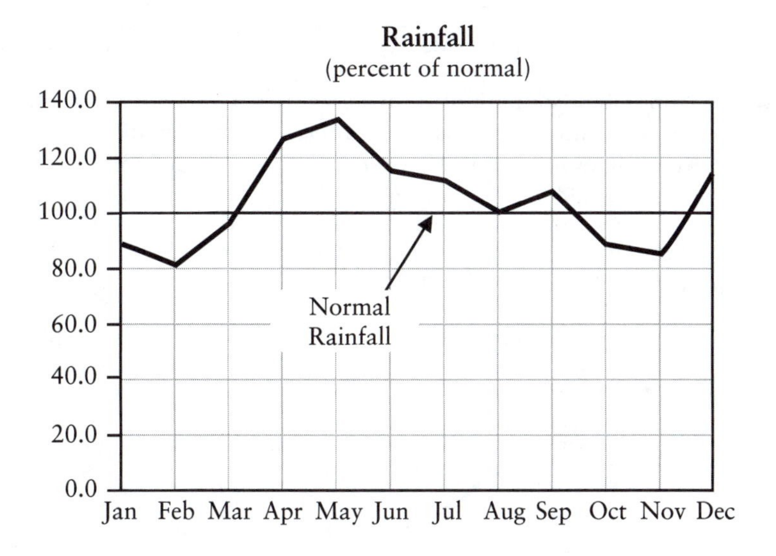

Posted Data Values

11.174 If the actual data values are important, post them on the chart next to the symbol representing the value. If the reader needs many data values to understand the chart, put these in a table or use a curve chart with a grid.

Annotations

11.175 Include annotations to explain sudden shifts in trends, errant data points, hypothetical causes for trends, and interesting data patterns.

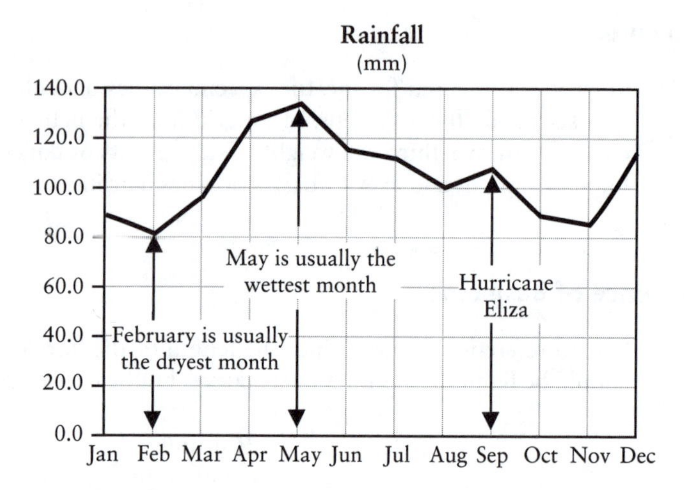

11.176 Annotation should not compete with data. If there is insufficient room for full annotations, post key numbers or letters on the chart and provide the full annotations elsewhere. Follow the guidelines for table notes (see also 8.97).

11.177 If the chart has a background grid, blank out the grid behind annotations and labels.

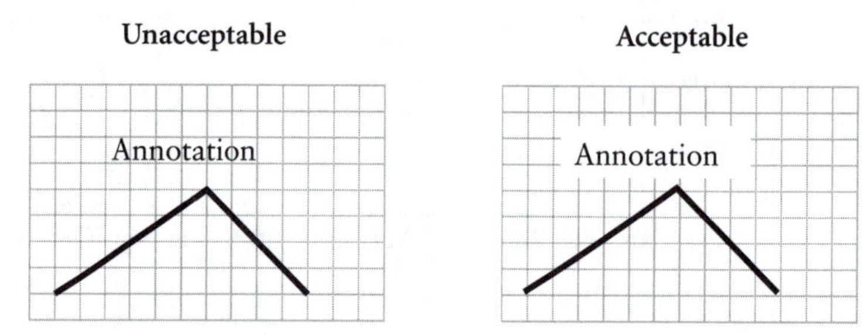

Inserts

11.178 Inserts are areas of additional information within a chart to supplement or extend its basic elements. Typical uses for inserts include

- Explanatory text,
- Source of data,
- Related chart,
- Table of source data,

- Formulas,
- Key or legend of symbols,
- More recent data,
- Enlargement of a region of dense data, and
- Totals.

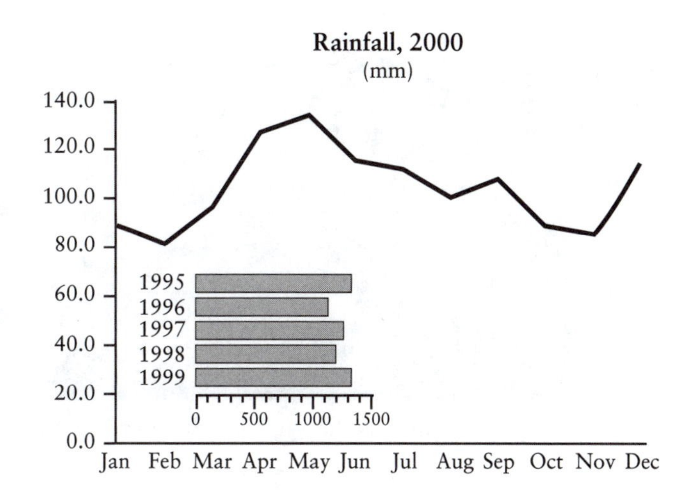

11.179 Box the insert only if it might otherwise be confused with the main chart. For example, box a line chart inserted on another line chart.

Notes for Charts

11.180 See 8.97.

Border or Frame for Charts

11.181 Although borders are seldom necessary to separate charts from text, they are useful to

- Reinforce page margins,
- Repeat reference lines,
- Establish visual horizontal and vertical directions, and
- Contain grids or scale lines.

Keys and Legends for Charts

11.182 Label chart parts unambiguously. If necessary, connect labels to their referent with arrows. If shading is used, place a key or legend, which contains small swatches of the shades, in a vacant area of the chart or to its right.

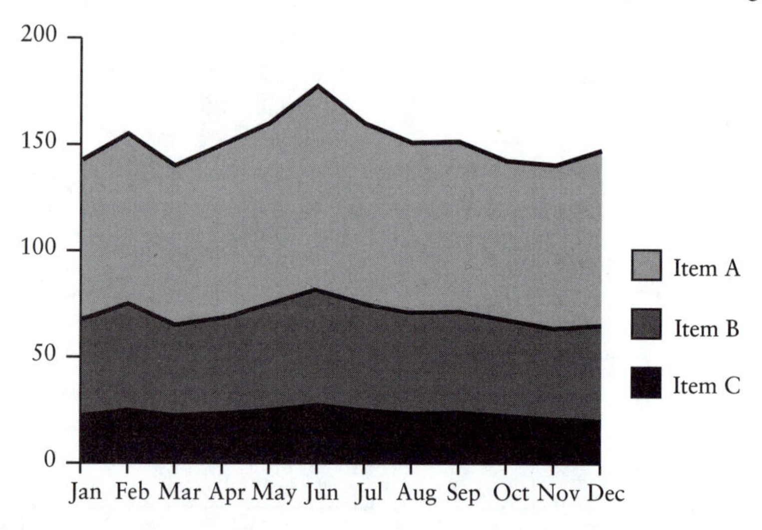

Special Topics and Effects

Size of Chart Typography

11.183 Type size for charts depends on how they are displayed and where they appear. For printed charts in books, technical reports, and other published works, the following are typical sizes.

Part	Typical Size (pts)	Notes
Title	12 bold	Same size as third-order heading
Scale or axis labels	12	Same size and style as body text
Scale figures	10	
Notes	10	Same size as text footnotes and endnotes
Data values	10	
Annotations	8	Smallest legible size

Weights of Chart Lines

11.184 Charts should use line weights that allow data and trends to stand out from the background of reference marks:

Component	Weight (× 1-width*)	Weight (points)
Curves	2–3	2–3
Scale axes	1	1
Grid, tick marks	0.5	hairline

*Units of l-width are the width of the letter *l* in the text letter size.

Abbreviations

11.185 Spell out all titles, labels, and notes. Abbreviate only where necessary. Use only common abbreviations, for instance, for measurement units:

Unacceptable	Acceptable
Strut Ld	Structural Load
(pounds per square inch)	(lb/in^2)

Patterns, Textures, and Shading

11.186 Emphasize the differences among areas by applying a different texture, tone, or color to each area. In using tones and textures, follow these rules:

- Put the darkest tone (or brightest color) in the smallest area. When using different shadings within a bar, put the darker shading at the left. In a column, put it at the bottom.
- Alternate tones or colors to maintain adequate contrast between adjacent areas.
- Avoid stripes and cross hatches that make comparing areas difficult and create an unpleasant visual effect.

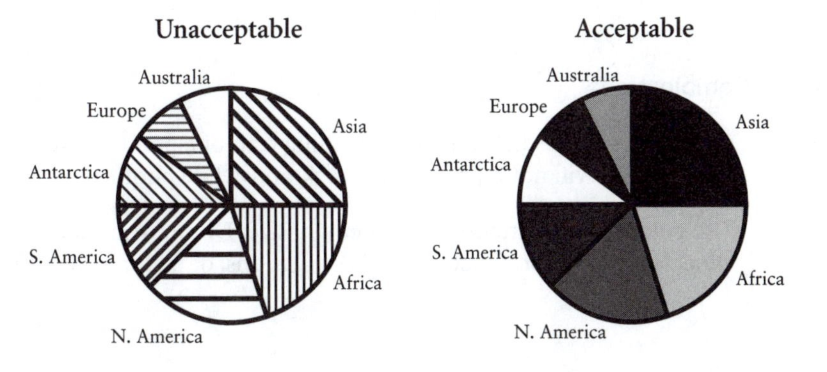

- When using stripes and patterns, establish a hierarchy similar to tonal values that makes discriminating them easier. Use the same tone or pattern for all areas of a data series.

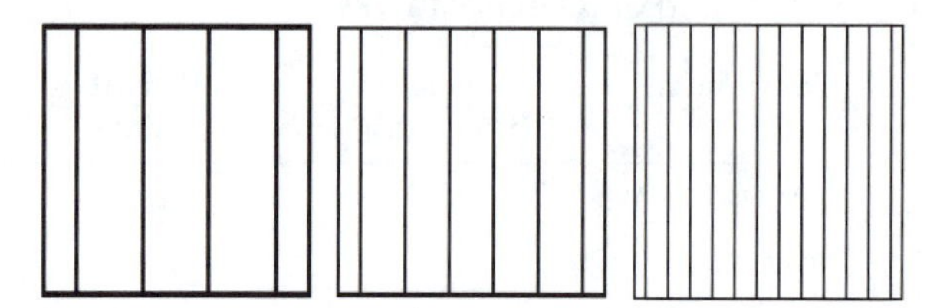

- Do not use black to shade large areas of any chart. Large areas of black can overwhelm the rest of the chart.

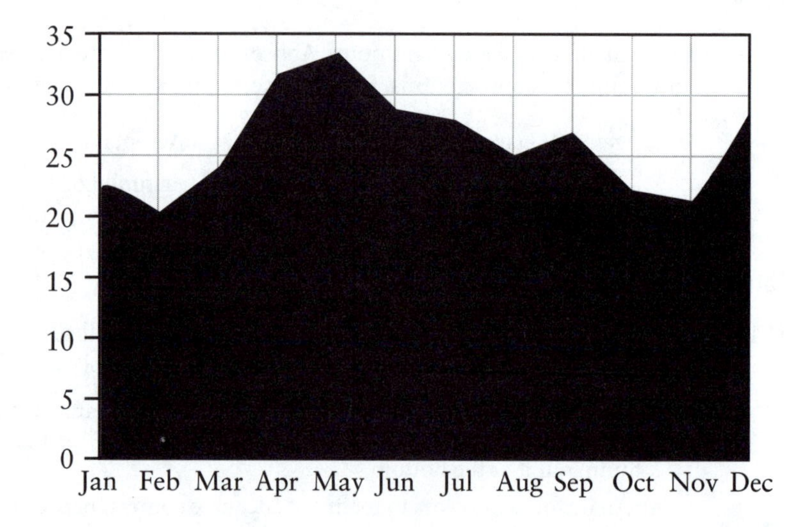

- When using shading, avoid a key or legend by labeling areas directly. If you cannot label areas, then a key should be provided.

Color Combinations

11.187 Never use color as the sole distinguishing trait of a point, curve, or area. Color codes will not reproduce in black-and-white photocopies.

11.188 Use color to draw attention or to emphasize. Do not use color to make objects recognizable, make critical distinctions, or add information.

11.189 Rather than a multicolor scale, use gradations in lightness for a single color or gradations between two compatible colors. Use bright warm colors to draw attention to important data and muted, cool colors to subordinate data.

11.190 Adjacent colors should have at least 30 percent contrast in lightness. Separate colors with a thin black line.

11.191 Maintain contrast between an element and its background. The most important element in the graphic should have the brightest color and the greatest contrast with its background. Warm colors, such as yellow-green, yellow-orange, orange, and red-orange, attract attention, especially on a darker color.

11.192 Do not place complementary colors of similar value or colors from opposite ends of the spectrum (red and blue, for example) next to each other. They will appear to pulse.

11.193 Avoid color choices that will make the chart incomprehensible to color-deficient or color-blind persons. Red is especially troublesome; blue can be distinguished from other colors by most people (see also 10.51):

Some Legible Color Combinations

Subject	*Background*
Black	Yellow
Yellow	Black
Black	White
Dark green	White
Dark blue	White
Dark white	Blue

Three-Dimensional Effects

11.194 Use three-dimensional symbols to make the chart appear to jump, slide, or float off the page. Such effects, while adding interest, can distort data representation and make accurate comparison difficult. Use them where only approximate comparisons are important.

Oblique Views

11.195 To create an illusion of depth, draw the chart as if viewed from an oblique angle. To maintain legibility, however, do not tilt text.

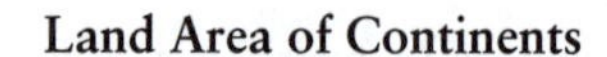
Land Area of Continents

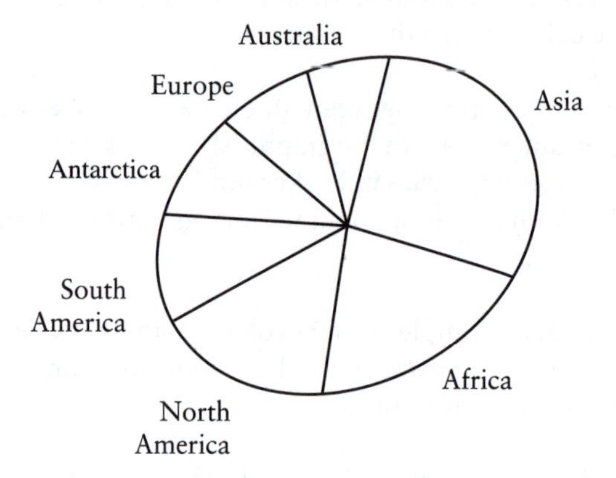

Caution: On an oval pie chart the size of a slice depends on its position around the circle. Slices on the sides appear smaller than the same slices on front or back.

Thickness

11.196 Draw a pie chart that looks as if it were cut from a squat cylinder.

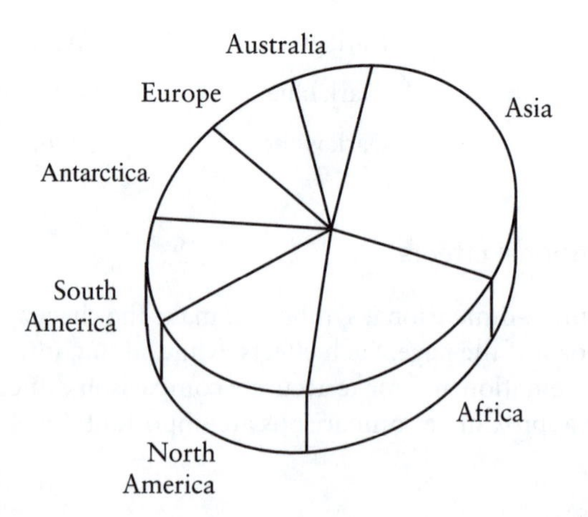

Drop Shadows

11.197 Draw a drop shadow behind the chart or its border to make the chart appear to float above the surface of the paper.

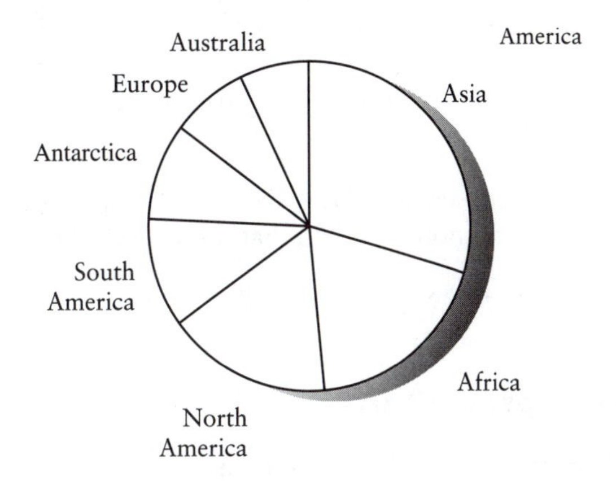

Pictorial Elements (see also 10.26)

11.198 To add interest for a lay audience use such pictorial elements as

- A picture as an insert or background,
- A simple picture as a border, and
- Symbols used as text labels on charts.

Overlays and Combinations

11.199 To compare dissimilar data series or to show two aspects of the same data, combine two or more types of charts into one. For example, to show cyclical data, superimpose data for corresponding time periods (days or years).

Referring to Charts

11.200 Number every chart in sequence with other figures and illustrations and refer to them directly in the text by number. Position charts in the text in the order in which they are cited:

> Our study, as shown in figure 6.4, confirms Ripley's results.

Or parenthetically:

> Our study confirms Ripley's results. (See figure 6.4.)

11.201 If the parenthetical reference occurs within a sentence, capitalize and punctuate it as a phrase. If it occurs between sentences, treat it as a sentence:

> Our study confirmed Ripley's results. (See figure 6.4.) Later studies added further corroboration.

11.202 Do not refer to a chart by its position, for example, "the above chart" or "the following chart."

11.203 Avoid unnecessary cross-references by repeating short, simple charts. In the general notes, explain that this chart is a repeat:

> Note: This chart is the same as figure 6.2.

Common Types of Charts

11.204 This section examines the special requirements of common types of charts including:

- Scale (11.205),
- Scatter (11.208),
- Triangular (11.211),
- Circular (11.215),
- Curve (11.218),
- Column and bar (11.231),
- Column and bar combinations (11.248),
- Pie (11.258).

Scale Charts

11.205 Scale charts show the ranking or value of a single variable along a scale.

11.206 Do not use scale charts for ordered data values, such as a time series; use a bar or column chart instead.

11.207 Arrange data items along the scale in positions corresponding to their values. Place labels next to their scale positions. If the values are too close together, use arrows to connect labels to scale values.

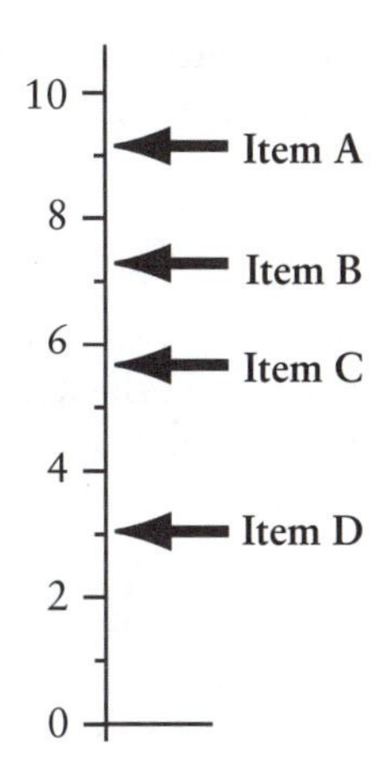

Scatter Charts

11.208 Scatter charts, also known as dot charts or scattergrams, show the correlation between two variables.

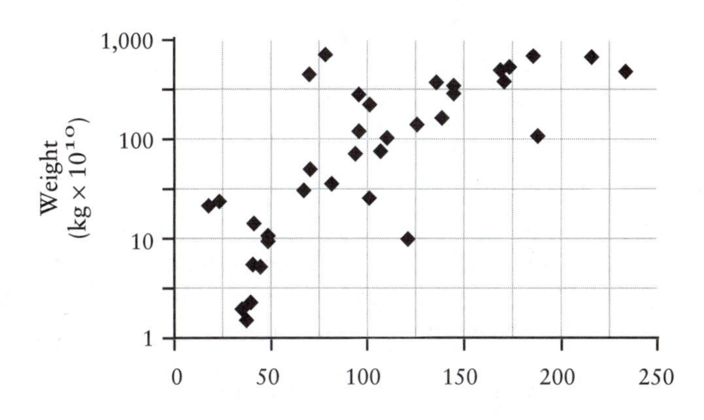

11.209 Use scatter charts to

- Present many data values (hundreds or thousands),
- Show correlations between two variables, and
- Draw conclusions about relationships in the data.

Do not use scatter charts for time series data or a few data points.

11.210 Use the same symbol for each point in a single data series; use different symbols for different data series.

Triangular Charts

11.211 The triangular chart, also known as a trilinear chart, shows the makeup of three-component items. Each edge of the triangle provides a scale of the percentage or proportion of each component present.

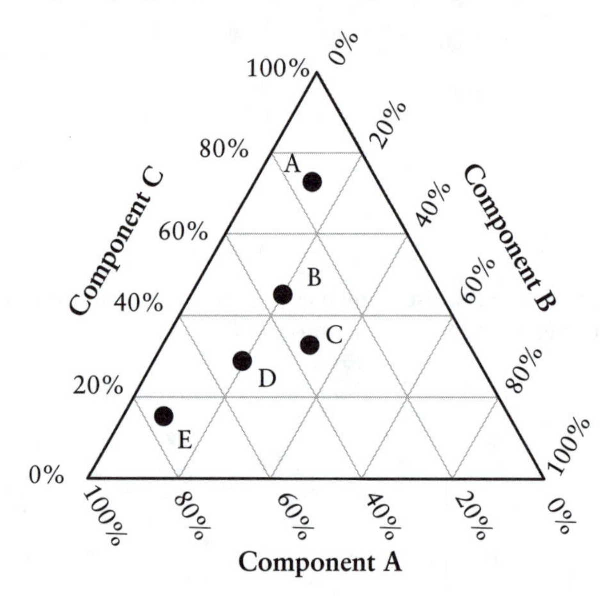

11.212 Do not use triangular charts for subjects made up of more or fewer than three components or for which the three components do not add up to 100 percent. Use triangular charts to show the chemical composition of compounds and alloys with three components.

11.213 Use distinctive symbols for data points. Label each data point and the scales outside the triangle.

11.214 Include a light grid, parallel to each edge, inside the triangle (see 11.172, 11.184).

Circular Charts

11.215 Circular charts, also known as rose diagrams and polar plots, use angle and distance from a center point to show how values vary over cyclical ranges.

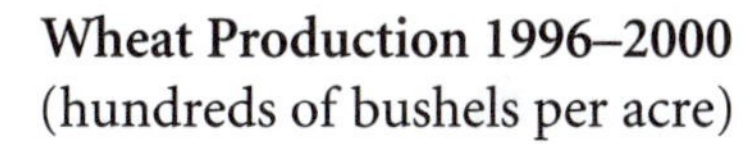

Wheat Production 1996–2000
(hundreds of bushels per acre)

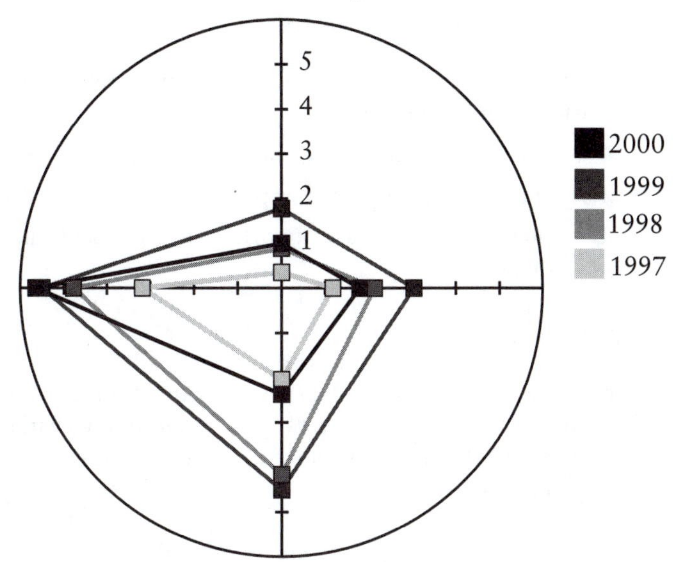

11.216 Use circular charts to show how a small number of variables vary over a circular range, such as the azimuth, hours of the day, or months of the year. Do not use circular charts if the variable is not cyclical. Circular charts may prove too technical for lay audiences.

11.217 Use the angle, which must be an actual angle or some other variable that varies cyclically and returns to its starting value, to represent the independent variable. Represent the dependent variable by a distance from the circle's center (starting at zero). Include no more than one or two variables on a circular chart.

Curve Charts

11.218 Curve charts, also known as line graphs, line charts, and curve graphs, use lines to show the relationship between two variables measured along the horizontal and vertical scales. The term *curve* refers to trend or data lines, regardless of their shape: straight, jagged, smooth, or stepped. Use curve charts to

- Show relationships among variables,
- Present large amounts of data in a limited space,
- Summarize data that must be understood quickly,

- Show a trend over time,
- Compare several data series,
- Interpolate known data values.

11.219 Do not use curve charts when the amount or rate of change is more important than trends; instead

- Use column charts for lay audiences,
- Use column or bar charts to compare relative size,
- Use a log scale if ratios are important or if the data values vary over a wide range.

Frequency Distribution Charts

11.220 Use frequency distribution charts to plot the number of occurrences of individual values within some range of a continuous variable. Annotate the curve to explain deviations from a normal (bell shaped) or other expected distribution.

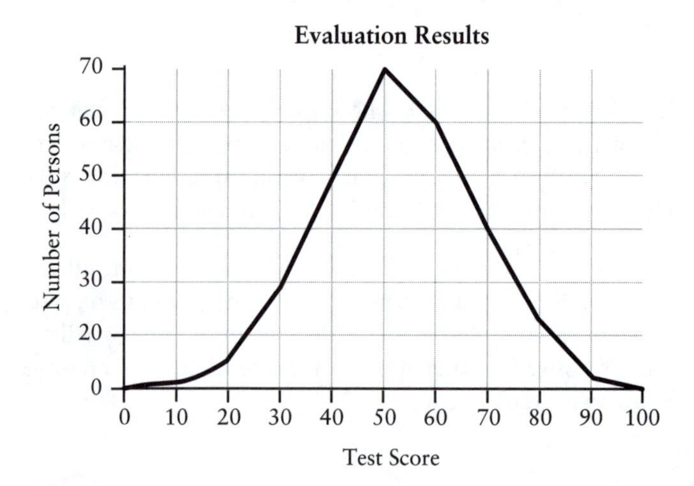

Cumulative Curve Charts

11.221 For each horizontal unit, cumulative curve charts plot the total of its amount and all prior units.

11.222 Label the curves to the right of the corresponding lines. Always label cumulative charts as such.

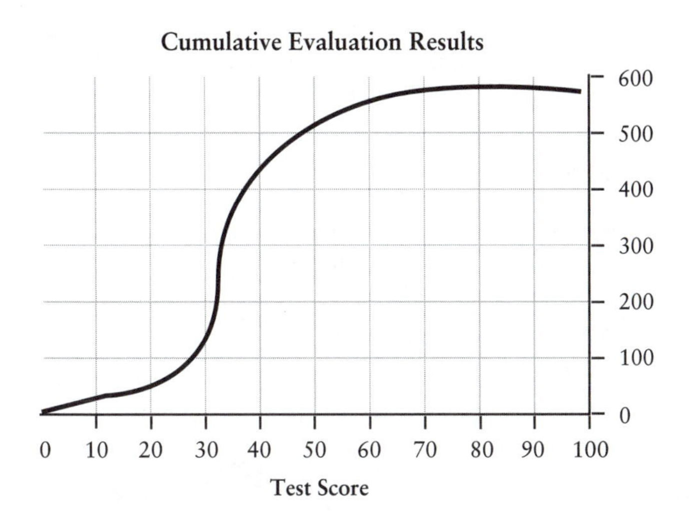

Surface Charts

11.223 Surface charts, curve charts with a shaded area between the curve and the baseline, emphasize the difference between two series.

11.224 Label the curves to their right and use arrows, if necessary, to connect labels with curves. Avoid using a key or legend. Always title cumulative charts as such.

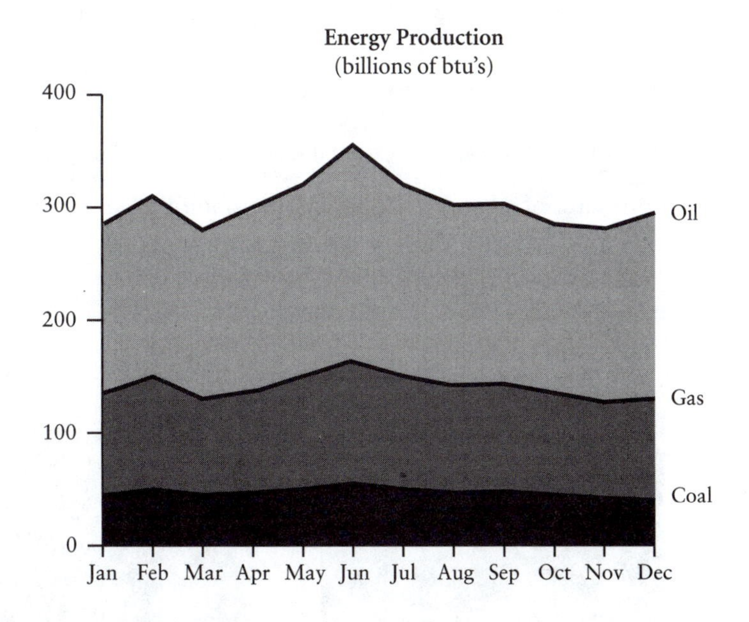

11.225 Do not confuse surface charts with band surface charts. Since surface charts merely shade the area beneath the curve, they cannot be used for overlapping curves. The curve's height still represents its value.

Band Surface Charts

11.226 In surface band charts, the width of each band represents the values of data series. Only the tops of the first and top (total) band plot absolute values.

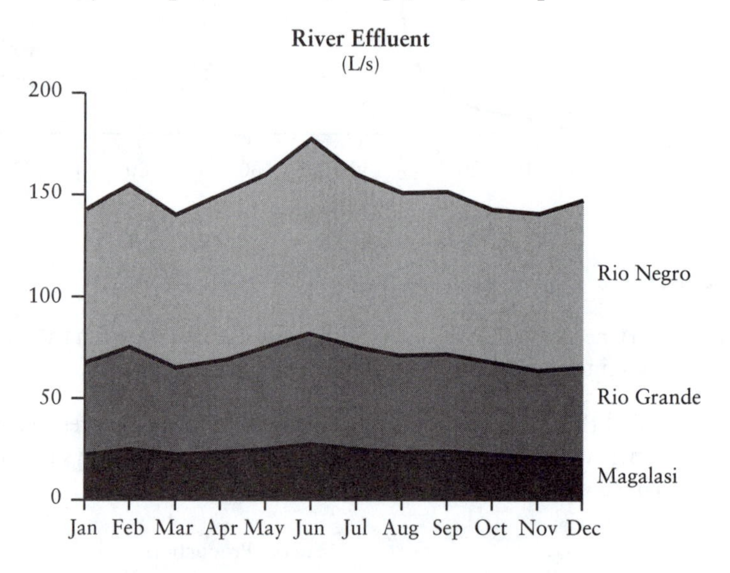

11.227 Use surface band charts to compare trends of components and a whole. Do not use a band chart if several layers vary considerably or if strong upward trends create an optical illusion of squeezed layers.

11.228 If one layer of a band chart has more pronounced variations, place it at the top.

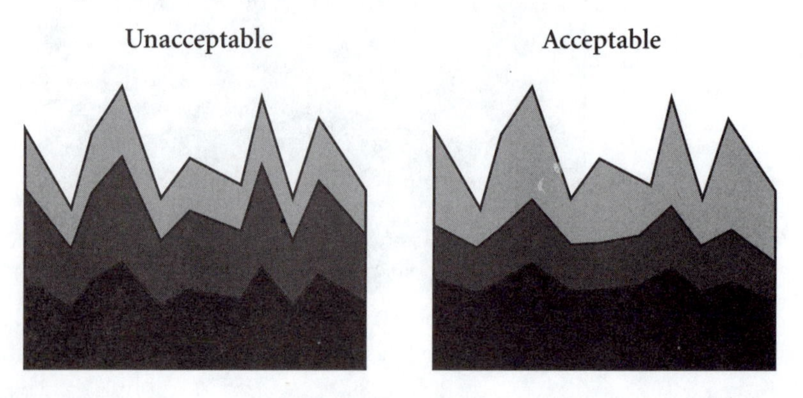

100 Percent Surface Charts

11.229 One hundred percent surface charts show trends among components of a whole. The values of the bands always total 100 percent.

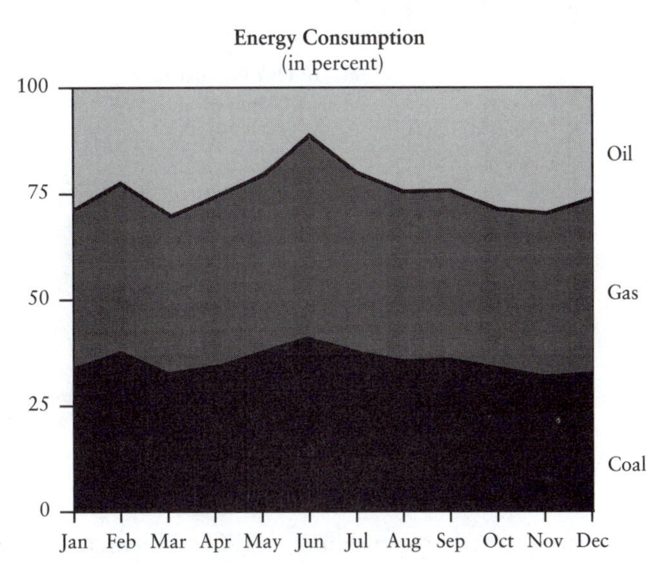

Net-Difference Surface Charts

11.230 Net-difference surface charts show surpluses and deficits or profits and losses. Shade the difference between two series with a different tone or color for pluses and minuses.

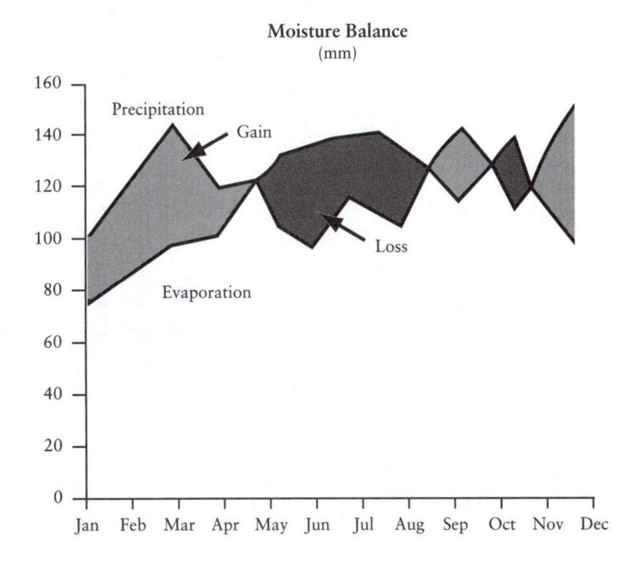

Column and Bar Charts

11.231 To represent quantities for a series of categories, column charts use vertical bars, and bar charts use horizontal bars. The categories may be specific values of an independent variable or named categories:

Sales of Our Product Line
(in $millions)

$25
$20
$15
$10
$5
$0
1998 1999 2000

11.232 Use column/bar charts to

- Compare the size or magnitude of different items or one item at different times,
- Emphasize the individual amounts or differences in a simple, regular data series.

11.233 Do not use column/bar charts to show a trend; use a curve chart instead. Do not use a column chart simply for comparing values; use a bar chart instead. Do not use column charts if quantities do not vary significantly.

11.234 Shade columns/bars to emphasize the data and sharpen the contrast between items being compared.

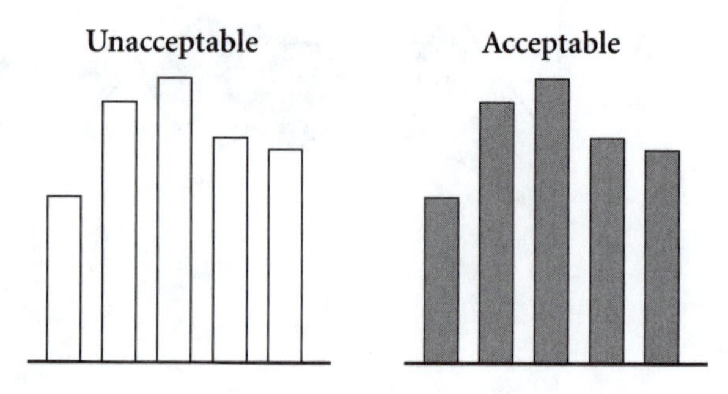

11.235 Arrange columns and bars in a consistent, logical order. Keep the same order for all columns and bars that represent data in a related series.

11.236 Columns and bars should be of a uniform width and evenly spaced. Their width and spacing depends on the number of columns/bars and the space available. The best spacing ranges from 25 to 100 percent the width of a single column/bar, with 50 percent the ideal. Begin the first column/bar about half a width from the scale. Columns/bars should be taller than wide.

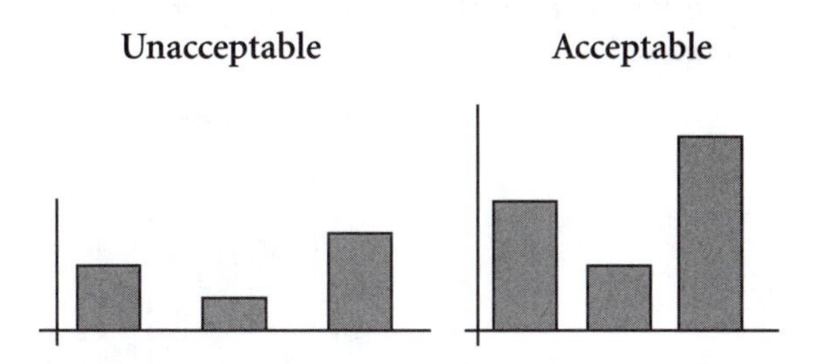

11.237 Label columns along the horizontal axis. Right-justify labels for bar charts to the left of the vertical axis. Place no labels within the columns/bars.

11.238 Place data values based on these rules:

- Post values beyond the end of shaded columns or bars, not inside them. Use a small, but legible, type size and add sufficient leading to ensure that these values do not obscure column/bar length.

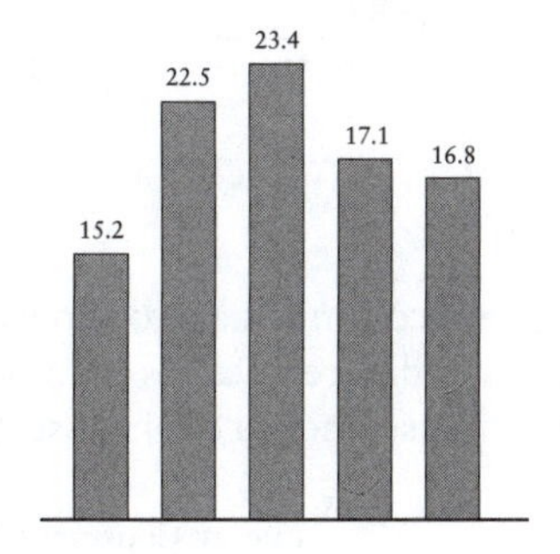

- For unshaded columns/bars, post the value within the columns/bars, if it will not obscure their length; otherwise, post the value at their base.

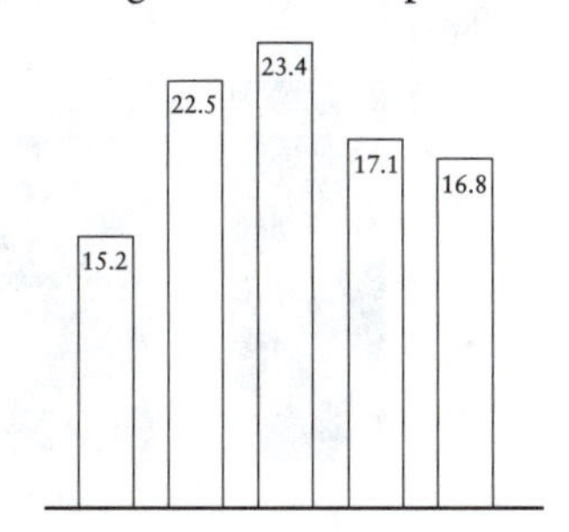

11.239 In column/bar charts, include grid rules to help readers interpret the values of specific columns.

High-Low Charts

11.240 Use high-low charts to show daily, weekly, or yearly fluctuations in prices, temperature, or similar variables.

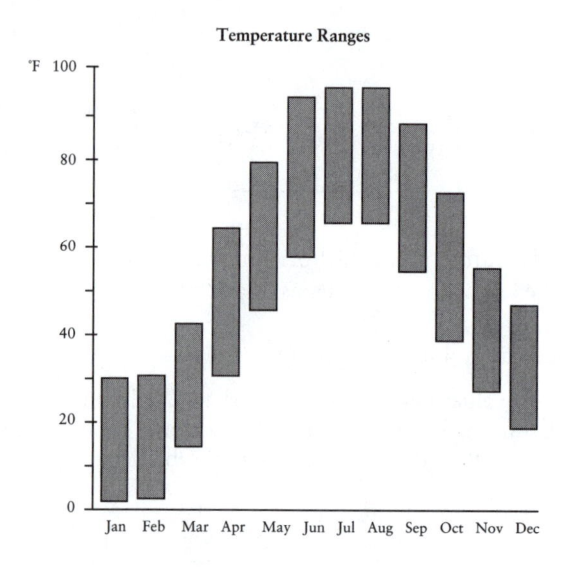

100 Percent Column Chart

11.241 Use the 100 percent column chart to show the proportional composition at different times or different values of some other variables. In this chart, space the bars and use lines to connect sections.

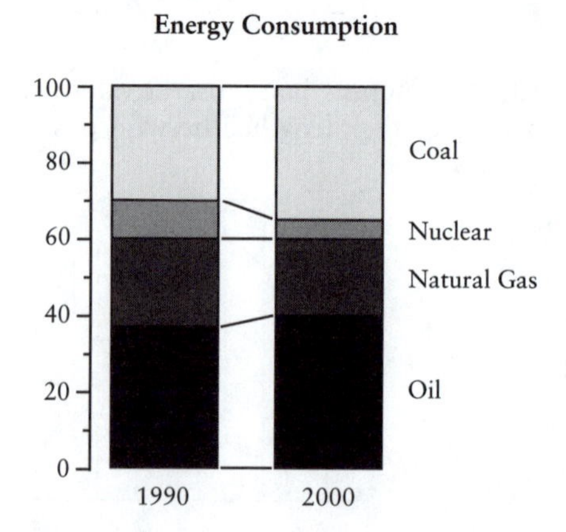

Histogram

11.242 Use a histogram to show the relative frequency of various values within ranges. Normally columns should all be the same width. In histograms with unequal intervals, the column width should equal the interval's width and its height reduced so that the column's area represents the range's frequency.

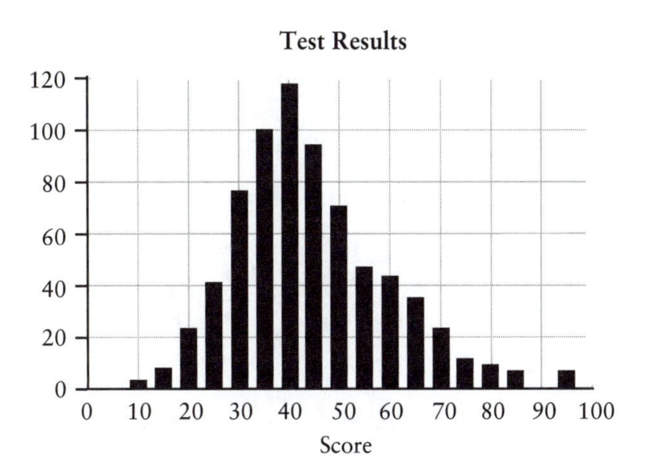

Time Lines

11.243 Use time lines to summarize a complex schedule or procedure by cataloging its timetable. Time lines list activities on a grid to show the period during which each activity occurs.

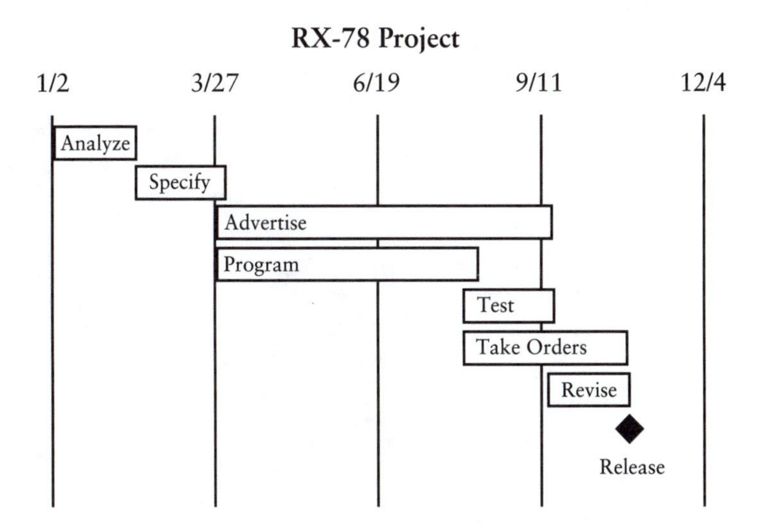

11.244 Use vertical grid lines to mark intervals: seconds, minutes, hours, days, weeks, months, quarters, years, or decades. List activities in chronological order. Place labels at the left edge or within bars, not beside them.

11.245 Use small symbols—circles, triangles, diamonds—to indicate project milestones.

Paired-Bar Charts

11.246 Paired-bar charts compare two members of a category moving in opposite directions from a common baseline. Use paired-bar charts to compare two different types of data, perhaps with different units or scales.

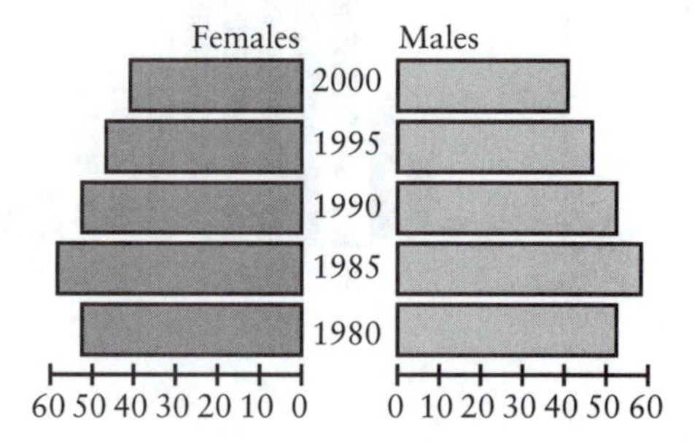

Sliding-Bar Charts

11.247 Sliding-bar charts represent ranges with bars whose horizontal ends mark the beginning and ending values of the ranges.

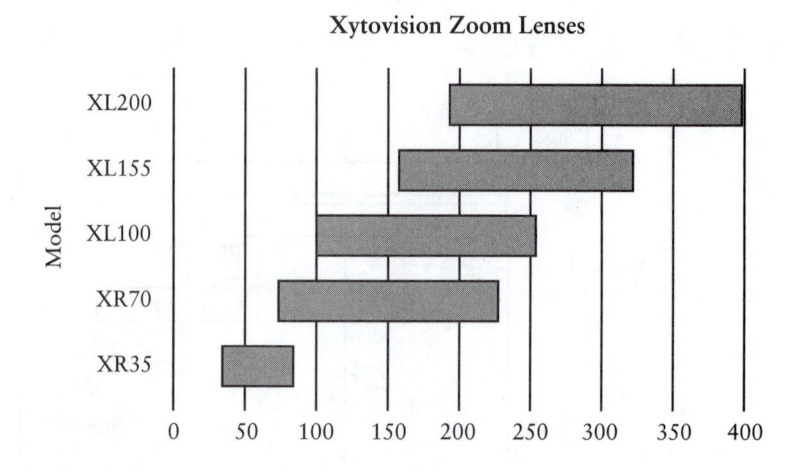

Column/Bar Chart Combinations

Grouped Column/Bar Charts

11.248 Grouped column/bar charts present the values of variables at a different time or under different conditions. Such charts help contrast changes among variables over time.

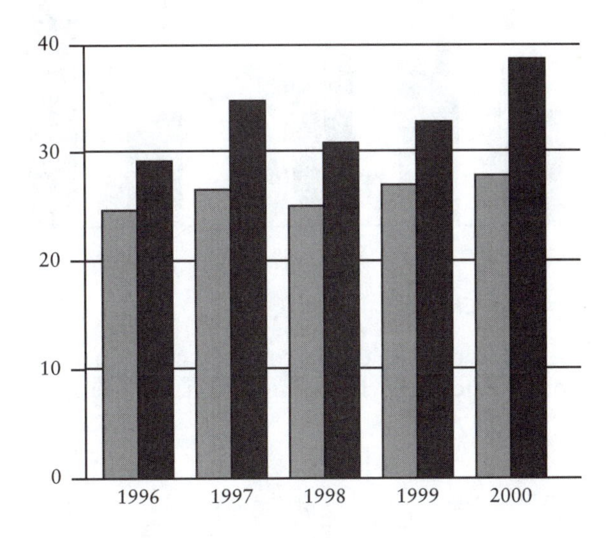

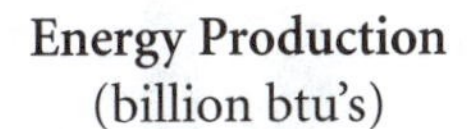

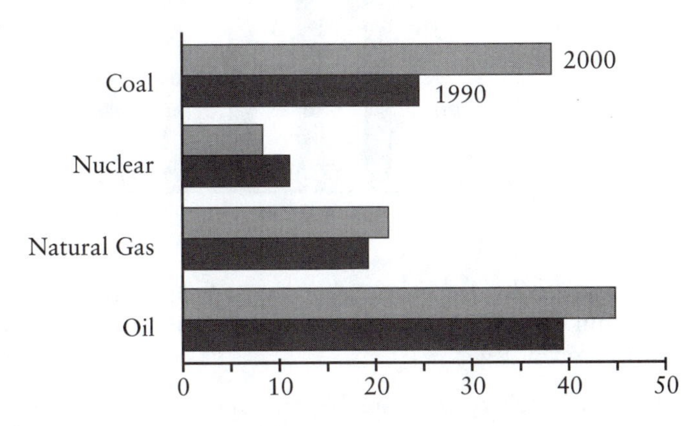

11.249 Do not use grouped column/bar charts to compare the composition of various items at one point in time; use a 100 percent column chart instead.

11.250 Separate groups of columns/bars by at least one-quarter the width of the group or the width of a single column/bar.

11.251 Grouped columns/bars may be closely spaced or touching, but not overlapping. Use a different color or tone for each column/bar in the group.

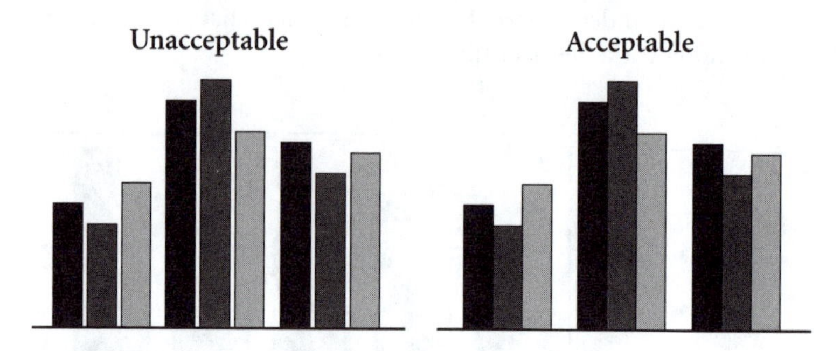

Divided Column/Bar Charts

11.252 Divided column/bar charts compare components and wholes over a period of time or for changes in similar variables.

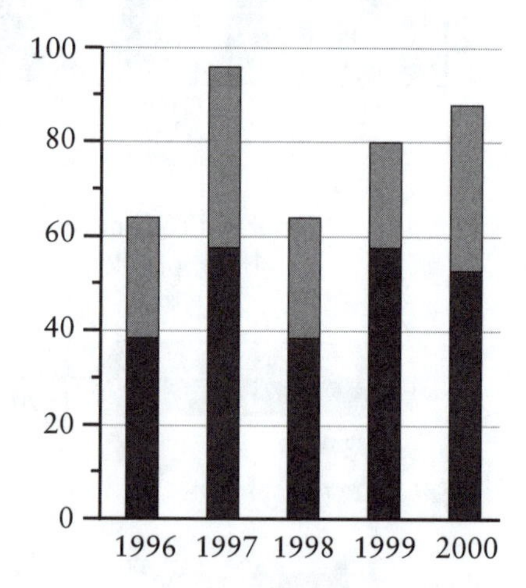

11.253 Do not use divided column/bar charts for precise comparisons. Label column segments to the right of the last column, not within columns or with a key or legend.

11.254 Divided bar charts compare the component parts of a series of items, and components and totals for various items at one time. Arrange the bar segments in a consistent, logical order. Keep the same order for all bars of the series.

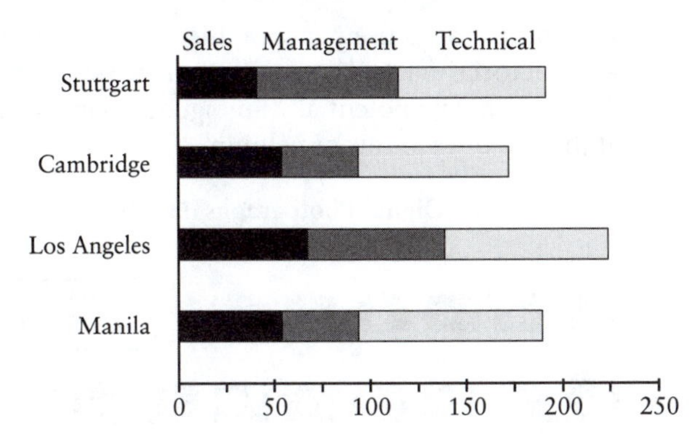

Deviation Column/Bar Charts

11.255 Deviation column/bar charts compare positive and negative values that extend beyond a zero baseline.

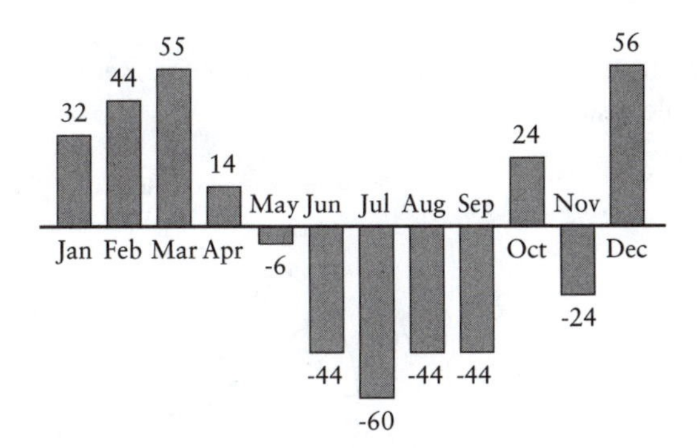

11.256 In designing deviation bar charts show positive and negative values starting at a central zero baseline. Place positive values to the right, negative to the left.

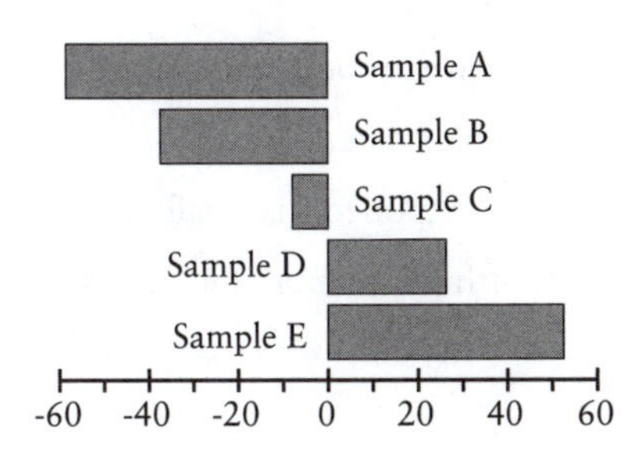

Pictographic Column/Bar Charts

11.257 For simple comparisons use pictographic column/bar charts when collections of pictorial symbols represent quantities. Omit the scale but include a key to define potentially ambiguous symbols. Also, post total values at the end of each row or column.

Pie Charts

11.258 Pie charts use slices to represent either percentages or fractions of the whole, usually depicted as a circle.

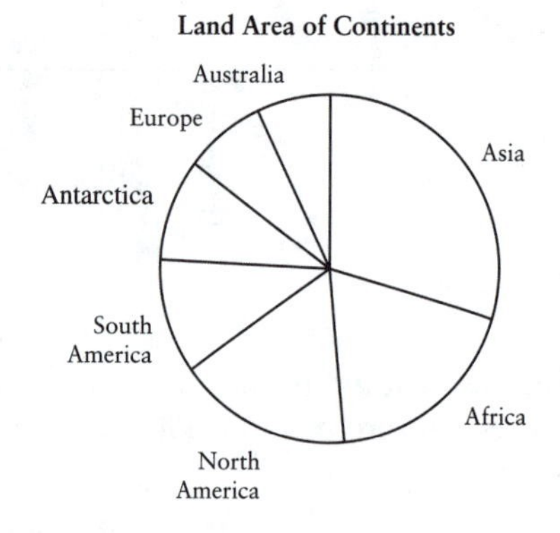

11.259 Use a pie chart when

- Showing the relative proportions of component parts,
- Simplifying complex data,
- The number of components is small, and
- General relationships are more important than precise quantities.

11.260 Do not use a pie chart when

- Comparing parts of two or more wholes,
- The number of compared components is large,
- Some of the components are small, or
- The exact percentages are important.

11.261 Arrange segments by size by placing the largest slice at the top and moving clockwise from vertical. Continue placing progressively smaller slices clockwise. If there is an "other" or "miscellaneous" segment, place it last, even though it may be larger than other slices.

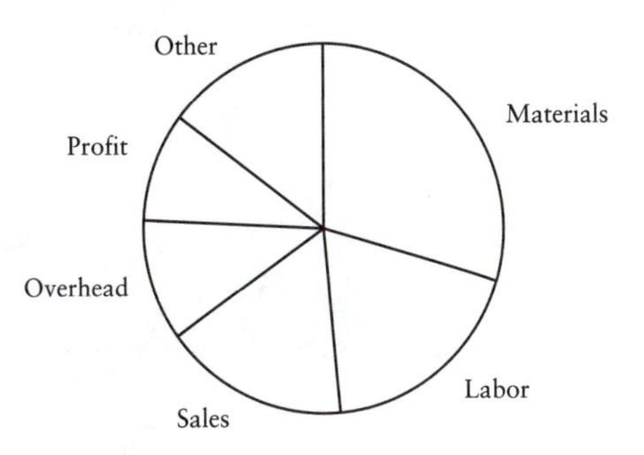

11.262 Limit the number of slices to an upper limit range from five to eight. Include no slices smaller than about five percent (18 degrees). If you have more than a few categories of data, consolidate them into a smaller number of main categories:

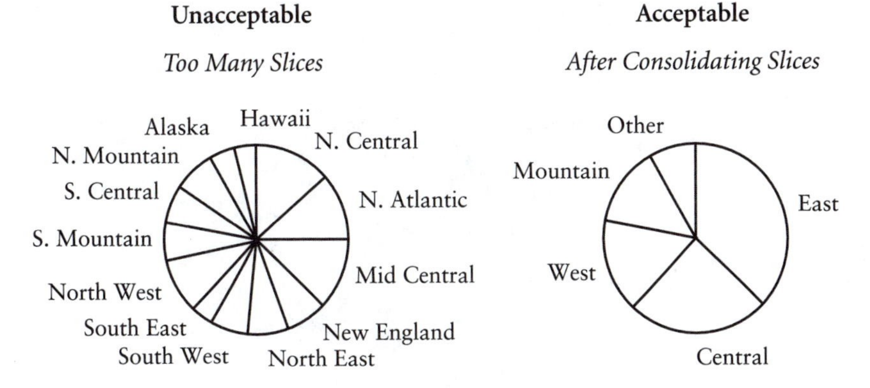

11.263 Place short, descriptive labels for each slice outside the circle. Never apply labels across segment boundaries. Avoid arrows, leader lines, or legends if possible.

11.264 If space permits, include the absolute quantity, the percentage, or both in the slice label in parentheses: France (28 percent).

Pictorial Pie

11.265 To add graphical interest to a pie chart, cut slices from a picture of some circular, cylindrical, or spherical object (see also 10.26).

State Cheese Sales
(by major counties)

11.266 Choose an object that symbolizes the type of data shown, for example:

- Coin for money,
- Clock face for time, or
- Wheel for automotive sales.

Allocating Teachers' Time

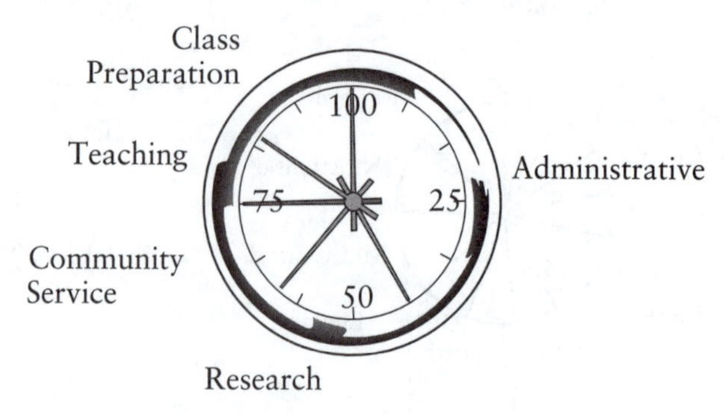

Separate Segment

11.267 To emphasize a single slice, detach it from the rest of the pie, sliding it outward about 10 to 25 percent of its length.

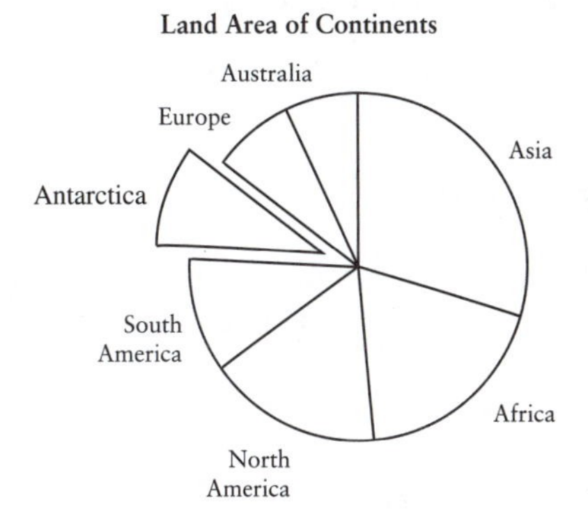

Diagrams

11.268 Use diagrams to display the makeup of a system and the interrelationships among its components. Diagrams use visual means to show qualitative differences and relationships:

Relationship	Diagram
Spatial	Map
Compositional	System organization
Responsibility	Personnel organization
Sequence	Flowchart
Dependency	Project tracking chart
Connection	Topology
Influence	Causal diagrams

11.269 If a technical field (computer science, architecture, etc.) has a formal diagramming technique, follow its principles and rationale.

11.270 Diagrams are useful for

- Understanding how something operates,
- Recording interrelationships among components,
- Stimulating thinking about complex systems,

- Communicating detailed technical information concisely among technical experts,
- Documenting a system,
- Enforcing rigorous and logical organization, and
- Automating analysis by computer tools.

Components of Diagrams

11.271 Diagrams consist of *symbols*, which represent the objects of the system, and *links*, which represent the relationships among them.

Symbols

11.272 Use symbols, special labeled visual images, to represent the system components. Symbols can be

- Solid or outlined geometric shapes: circles, squares, diamonds, or triangles;
- Silhouettes or icons of familiar objects;
- Detailed pictures of objects; or
- Words alone.

11.273 Use larger symbols for important objects in a diagram. To avoid visual confusion, do not use more than five or six symbols.

11.274 Label all symbols. If a label will be illegible inside a symbol, put it next to the image.

11.275 If space does not permit a complete label, place a short label in or near the shape and use a footnote to provide more information.

Links

11.276 A link should connect each symbol to at least one other symbol. The linkage pattern among symbols is the diagram's message. Each link shows a relationship among system items. Links can show relationships in one direction, in both directions, or in no particular direction.

11.277 Lines, symbols, and labels define links. Various types of connections can be shown by using different types of lines:

- Line weight or width:
  light, medium, or heavy

- Number of grouped lines: single or double

- Line style: solid, dotted, dashed

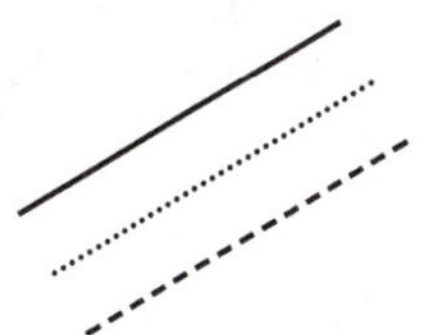

- Line direction: vertical, horizontal, diagonal

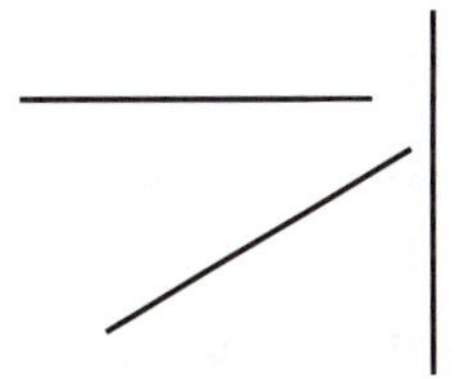

- Line curvature: straight, jagged, right-angle, or smooth

11.278 Use conventional graphic shapes, such as arrowheads, to show the direction of a relationship and the kind of link symbolized by the line.

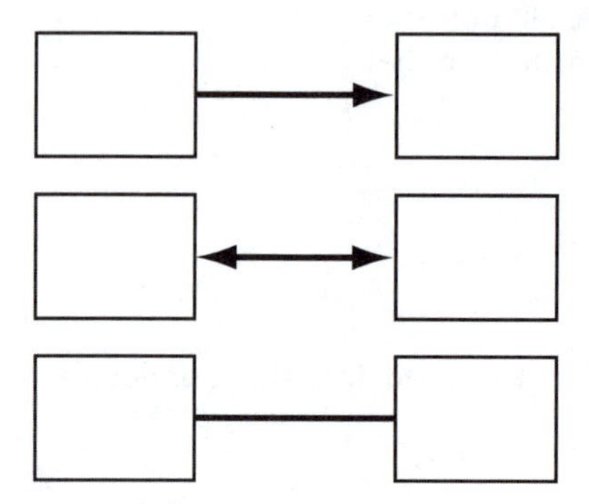

11.279 Links may be labeled. Do not use arrows to connect labels to links. If possible, place the label horizontally next to the link. If it is necessary to label an angled link, rotate the entire label.

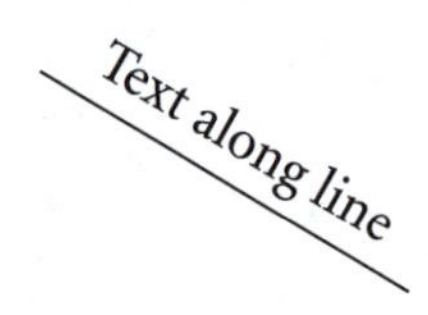

Patterns of Linkages

11.280 Linkages commonly occur as follows:

- Sequence linkages depict two choices: forward or backward.

- Grid linkage organizes and presents information along two logical dimensions. For instance, the reader can read down the columns or skim along the row to compare command syntax. Because tables have an inherent grid structure, most systems with a grid structure can be presented in a table as well as a diagram.

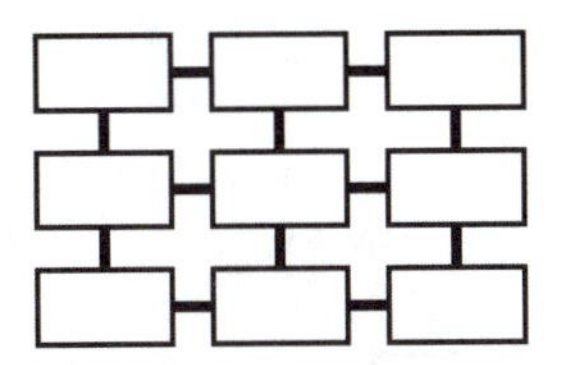

- Tree structure, a common technique for organization charts, presents simple hierarchy.

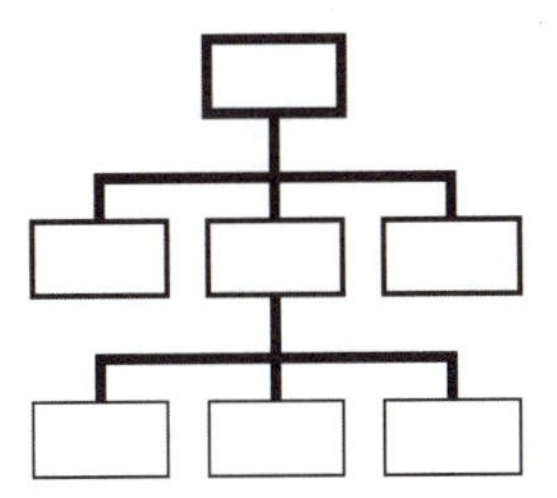

- Web linkage illustrates complex relationships.

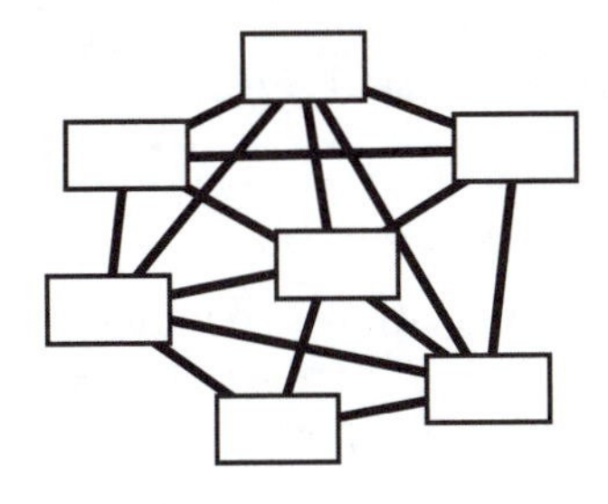

Common Types of Diagrams

Organization Charts

11.281 Use organization charts to display the interrelationships within a hierarchical organization. They originate from one symbol and divide into more and more symbols.

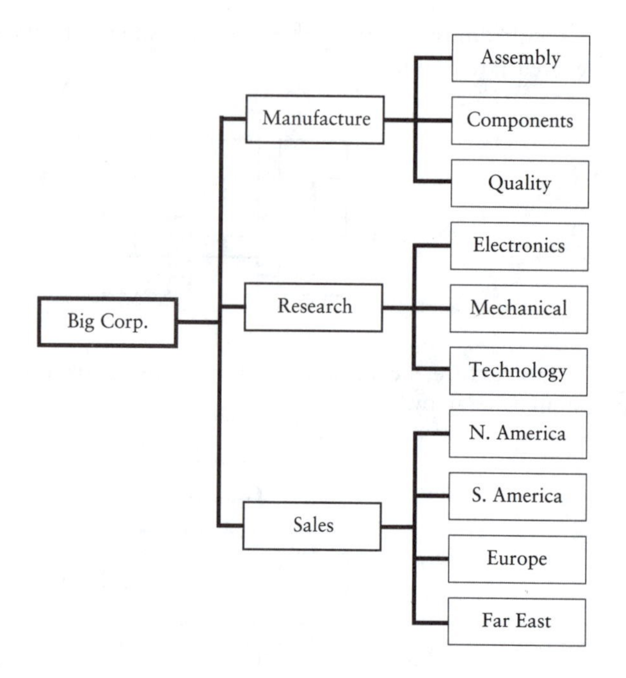

11.282 Do not mix different types of units in a single organization chart.

11.283 Organization charts can be drawn vertically (as a pyramid or tree), horizontally, or circularly. Multilevel organization charts often require too much space for boxes. Solutions include starting in the center and branching outward or setting the chart landscape.

11.284 In management organization charts, place the more important position on the top. Box size or border weight can also establish importance.

11.285 In an organization chart, use solid lines to show official command and control relationships. Use dashed, or screened, lines to show coordination, communication, and temporary relationships.

Flowcharts

11.286 Flowcharts

- Show the sequence of activities, operations, events, or ideas in a complex procedure or interrelated system of components;
- Summarize long, involved, detailed descriptions and instructions;
- Describe the specific series of tasks, activities, and operations necessary to perform a procedure; and
- Illustrate the movement through a process.

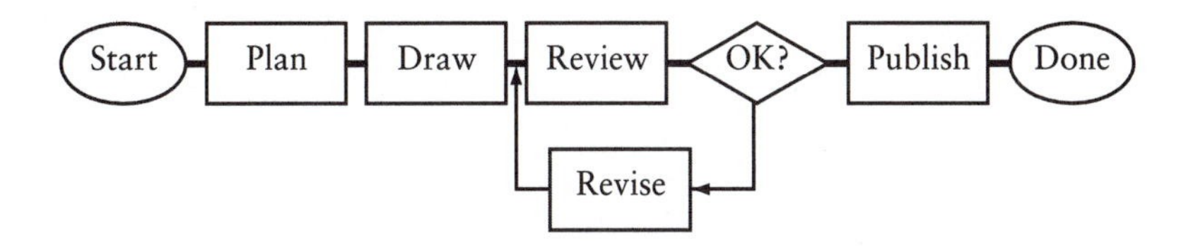

11.287 The assumed direction of flow is from top to bottom and left to right. Links that move in these directions may omit arrowheads. Links that move counter to these directions must include arrowheads.

11.288 Subdivide complex procedures into multiple simple flowcharts.

Connection Maps

11.289 Use connection maps to show permitted or existing connections. A map of a subway system is an example of a connection map.

Dependency Diagrams

11.290 Use dependency diagrams to show causal effects and influences among interrelated components. Dependency diagrams consist of labels, sometimes in circles, connected by arrows showing the direction of influence.

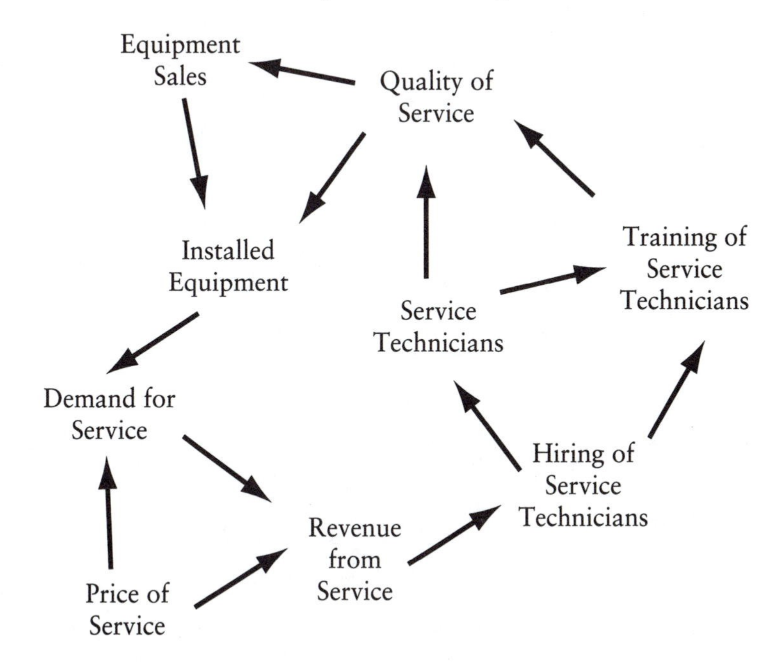

12. Designing Useful Documents

Designing a Document's Format

12.1 Planning a document's graphic design requires an understanding of audience, purpose, and scope. Before making specific design decisions, define these needs by asking

- How will the document be used? Good design can promote a document's intended usage pattern. For example, if a technical manual will be read in a crowded workspace, a small size will be more effective.
- How often will the document be used and updated? If a document must survive constant use, design it to minimize wear and tear. If it requires frequent updates, the binding must permit pages to be added easily for paper documents; electronic documents must allow quick and accurate changes to links, graphics, and other supporting techniques.
- How will the document be distributed? Distribution requires knowledge of storage, packaging, and shipping costs and methods for paper documents, storage, formats, programming, and accessibility for electronic documents.
- Will the document compete with other information for readers' attention? If so, it should have interesting packaging. Whether a document exists as paper or electronic information, it still conveys an image.

12.2 Good information design

- Invites people to read the document,
- Helps readers find information easily in the document,

- Offers a good impression of the content and authoring agency, and
- Provides a sense of corporate or organizational identity.

12.3 A document's graphic design, whether in printed or displayed form, controls its physical appearance by considering such issues as

- Using technology to support document design (12.5),
- Designing the page and screen (12.9),
- Page and screen elements (12.38),
- Page and screen grids (12.48),
- Designing specific information types (12.51),
- Controlling large document sets (12.107), and
- Graphic production considerations (12.119).

12.4 These sections present general design rules. Although these rules are usually good ones to follow, remember that any rule can be wrong in a particular instance. Good design is not simply a matter of aesthetics; it is also a matter of function and improved communication.

Using Technology to Support Document Design

12.5 Since a considerable variety of software can support document design and production, scientific and technical authors need to understand the different capabilities of these applications (see also 11.22).

12.6 Information developers should recognize that this discussion of software relies on the current state of electronic support (see also 10.5–10.13). Use all of this information to plan an appropriate level of electronic support for a specific information need. It may be, for instance, that word processing will perform all of the production tasks adequately; or tag languages may provide better support for a distributed information group.

12.7 Assess needs on the basis of both recommendations by and discussions with writers, editors, graphic and electronic artists, and print production personnel, as well as an analysis of the document's expected audience and purpose. Begin by cataloging specific document types that can satisfy the needs of specific information classes, or *genres* (see the table on the next page).

12.8 Use an appropriate technology that offers the ability to apply the level of control needed to produce the intended document. For example, an illustration program, such as Adobe Illustrator™, could be used to produce brochures, booklets, and newsletters, especially those that use color or incorporate images. A basic word processor might easily produce some of these same texts (see 11.22 and following).

Document Types and Information Classes

Document Type	Information Class			
	Tutorial	Marketing	Procedural	Reference
Brochure		X		X (quick)
Booklet	X	X	X (getting started)	X (quick)
Newsletter		X		
Article				
Journal				X
Magazine	X	X		
Manual	X		X	X
Electronic				
On-screen	X	X	X	X
Disk	X		X	X

Designing the Page and Screen (see also 12.48)

12.9 People read most technical documents for those paragraphs or sections that describe a specific procedure or concept, rather than cover to cover. Use obvious and simple formatting principles to help readers move directly to specific information without distractions. While printed documents have their own methods for annoying readers, on-screen information that combines too many design elements—unusual color combinations, odd grid structures (if any), multiple frames, and the like—and techniques (animations, sound, etc.) can frustrate readers.

12.10 After developing a format, use that format throughout the document. The single most important factor in developing a page format is consistency. Presenting information in too many ways confuses readers. When designing the page, consider all the formal format elements:

- Page and screen size (12.11);
- White space (12.18);
- Margins (12.23);
- Typefaces (12.26);
- Type sizes (12.31);
- Leading or line spacing (12.33);
- Line width (12.35);

- Interactions of type, leading, and line width (12.36); and
- Highlighting techniques (12.37).

Page and Screen Size

12.11 When selecting an appropriate page or viewing area size, keep in mind the document's expected usage pattern. For documents that will be read on a crowded desk, select a smaller page size; however, for lengthy text, a larger page size may prevent the final document from looking too imposing. Give the same consideration to probable viewing characteristics for displays.

12.12 Fashion can affect page size. For example, small-format manuals, typically 5 by 8 inches, have become customary for office documents. Larger equipment documentation usually employs the traditional 8½-by-11-inch page size. Remember, too, that documents destined for foreign markets are often printed on other size pages (see 12.50, 12.86).

Typical Page Sizes and Characteristics

Document Type	Size (inches)	Orientation	Advantages and Disadvantages
Brochure, one-fold	7 × 10	Portrait	
Booklet	5 × 7	Portrait	Uses little space in a work area; but too narrow for two columns
Newsletter	8 × 10	Portrait	
Professional journals	8 × 10 5 × 8	Portrait	
Commercial magazine	8 × 10	Portrait	
Tutorial manual	7 × 10	Portrait	Uses little space in a work area; can accommodate two columns
Procedure or reference manual	8½ × 10	Portrait	Offers a large printable area; can accommodate two columns
Office automation products	7 × 10 5 × 8	Portrait Landscape	Offers a large printable area; can accommodate two columns

12.13 Do not neglect to consider the differences between displayed information destined for computer screens (see 12.14, 12.92). Typically, computer screens vary considerably in resolution (pixels per inch), refresh rate (Hertz), and dot pitch (usually stated in millimeters). While it would be difficult to catalog all the potential differences here, at least a representative example of contemporary computer monitors will help emphasize this variety:

Sample Display Screen Sizes

Viewing Size (inches)	Dot Pitch (mm)	Pixels per Inch (ppi)
13.9	.28	1280 × 1024
14	.27	1024 × 768
15.9	.23	1024 × 768
19.8	.25	1800 × 1440
20	.22	1600 × 1200
	.26	2048 × 1536

Note: 1024 × 768 ppi is a common resolution.

12.14 Electronic information designers suggest that, since the typical workplace monitor has a 14- to 15-inch viewing area, any displayed information should be "set" or "imposed" to respect those limitations. While this problem may disappear with changes in technology, the installed base of such monitors makes this practice important for reaching the widest possible audience.

Basic Screen Sizes (ppi)

Viewing Area	Printable	Maximum
640 × 480	535 × 295	595 × 295

Note: Printable indicates that the screen will be printed as a paper document.

12.15 For printed documents, consider that paper is cut from stock-size sheets. For instance, 3-by-6-inch pages are often printed on $6^{1}/_{4}$-by-$12^{1}/_{2}$-inch sheets that have been cut from 25-by-38-inch standard size sheets (see 12.50).

Some Standard Page Sizes and Printer Sheets
(inches)

Page Size	Paper Size	Press Sheet Size
3 × 6	25 × 38	6 1/4 × 12 1/2
5 × 7	32 × 44	10 1/4 × 14 1/2
5 × 8	35 × 45	10 1/4 × 14 1/2
7 × 10	32 × 44	16 × 22
8 × 10	35 × 45	17 1/2 × 22 1/2
8 1/2 × 11	35 × 45	17 1/2 × 22 1/2

12.16 The available press size may control page size. The table above offers only a limited selection of press sheet sizes. Typically, these sheets range from 6 1/4 by 12 1/2 inches to 38 by 50 inches.

12.17 Using a specific page size may not be possible because the document may be part of an existing document series for which the page size has already been determined.

White Space

12.18 Use white space, empty area around text or illustrations, to isolate and separate information. Make pages or screens readable, visually appealing, and easily accessible by using enough white space to emphasize information.

12.19 Place white space between sections, headings, and major topics to indicate a physical break; add white space to bring attention to illustrations.

12.20 To emphasize division and associate a header, illustration, or data display title visually with its referent, establish a white space ratio above and below or around page or screen elements.

Typical White Space Ratios

Level Head	White Space (points)	
	Leading	Trailing
1	18	12
2	15	10
3	12	8

12.21 In creating this ratio consider the use of a similar ratio for type size (see 12.31).

12.22 Currently, both leading and white space for on-screen information are created by scaling transparent artwork or type. Like other limitations of on-screen typography, this is likely to change as new techniques emerge. But information developers should be aware of the current practice.

Margins

12.23 Use margins, the white space around a page's perimeter, to improve legibility.

12.24 Use ragged-right margins, uneven text in the right margin, to give a document a relaxed appearance. Readability studies indicate that people read ragged-right text faster than justified text.

12.25 Use justified margins, when all text share a common right margin, to give a technical document a more formal appearance. However, justified text may create "rivers," accidental interline spacing, that cause the reader's eye to move vertically rather than horizontally:

This text has been justified left and has a ragged-right margin. Some readability studies indicate that it is easier to read than fully justified text.

This text has been fully justified. Some readability studies indicate that it is more difficult to read than left-justified text. It also has "rivers of white."

Typefaces

12.26 Typeface is the style of lettering used in a document. Typewriters and letter-quality word processing printers usually offer only a few typefaces. Laserographic printers and typesetting equipment usually offer a considerable variety of typefaces. All typefaces fall into two categories: serif and sans serif. A serif is a short line that stems from the end of a letter. Serifs give a visual illusion of connecting the letters in a word. Readability studies show that for long documents, a serif typeface is easier to read than a sans serif typeface.

12.27 Use 10- or 11-point serifed type for most technical documents. Avoid too many typefaces or styles in a single document. Typically, a single serifed typeface (such as Times) is selected for the body text, a sans serif typeface (such as Univers or Helvetica) is used for degree heads, and one of these two typefaces is also used for illustrations.

Selecting Compatible Typefaces

Serif		Sans Serif						
		Franklin Gothic	Futura	Gil Sans	Goudy Sans	Helvetica	Optima	Univers
Old Style	Bembo	■		■				
	Caslon		■					
	Garamond	■	■	■	■			
	Palatino	■	■		■		■	
	Sabon		■				■	
Transitionals	Baskerville		■					
	New Baskerville		■	■				
Moderns	Bodoni		■					
Slab Serif	Clarendon	■			■			■
	Glypha	■						

Try these general guidelines for typeface selection:

- Consider the medium for which the original typeface was designed. For instance, Frutiger designed a number of sans serif faces for signage that needed to be readable at a distance. Many of these same faces remain in common use and can serve for viewgraph presentations as well as display faces in text.
- Check the weight and conformity of the face to its original throughout the proofing process. Often what appears to be a good typographic choice does not reproduce well. Stroke widths, for example, may be too fine for newsprint.

- Select faces that suit both the task and the subject matter. An italic face, like Goudy Sans, that has an exaggerated lean, for instance, might be perfect for a popular book on cycling.
- If type choices are severely limited, select only one and use it to its best advantage.
- Be sure that the selected face provides all of the characters needed to discuss the subject and meet the needs of the intended audience:

Selecting Typefaces for Documents

Typeface	Brochure	Booklet	Newsletter	Article	Manual
Old Style					
Bembo		■		■	■
Caslon		■		■	■
Garamond	■	■	■		■
Palatino	■		■		■
Sabon	■		■		
Transitionals					
Baskerville		■			
New Baskerville		■		■	■
Moderns					
Bodoni	■			■	■
Slab Serif					
Clarendon	■			■	■
Glypha	■		■		

Style	Example Typefaces
Mathematics and symbols	Helvetica Fractions New Century Schoolbook Lucida Math Mathematical Pi Universal Greek and Math Pi
Miscellaneous styles	Tekton (script) Studz (unical) Kette Fraktur (black letter)
Miscellaneous Pi	Bundesbahn Pi Carta European Pi Warning Pi Zapf Dingbats
Foreign languages	Edo Euro Hira Katakana Khmer Korean

12.28 Recent studies of displayed information indicate that sans serif typefaces are more legible on-screen. In selecting useful typefaces, the major difficulty centers on the possible interactions among platform, display, and web browser. While this difficulty will likely be a short-term problem, the safest course for ensuring that displayed text can be read by any user is to select default typefaces or those designed specifically for display. Two serif faces, Times New Roman and Georgia, and one sans serif face, Verdana, are representative examples. Several other typefaces can also be used:

"Safe" Typefaces for Displayed Information

Serif	Georgia Minion® Web Times New Roman
Sans serif	Arial Caflisch Script® Web Mezz® Web Bol Myriad® Web Verdana Trebuchet

Note: Caflisch Script is a cursive typeface.

12.29 At present, Cascading Style Sheets (CSS) and Server Side Includes (SSI) have emerged to alleviate typographic display problems. For those organizations creating a considerable volume of information intended to be read only on-screen, employing these, and future, techniques and monitoring their potential usefulness should be an integral part of the document design process.

12.30 Use as few different typefaces as possible in any one document or document library. Typically, very useful documents use no more than two different typefaces—one for the text and a second for any nontextual (graphic) information or for headers.

Type Sizes

12.31 Type sizes are measured in points (72 points = one inch). When compared with typewriter letter sizes, 12-pitch typewriter letters are about equivalent to 10-point type; 11- or 12-point type is about equal to 10-pitch typewriter letters.

12.32 For typeface and type size considerations, documents are divided into headings and body text:

- For body text, 10- or 11-point serif type is the most comfortable typeface and size for the average reader. On-screen text should be sans serif.
- Captions and callouts for illustrations are preferably the same size as the body text, though, in practice, they tend to be smaller. Using the same typeface for both body text and illustration can reduce production costs. If effective discrimination methods—white space, rules, boxes, and the like—have been used, this practice should not make the text more difficult to read.
- Use larger type sizes for headings. In addition, establish a ratio among the various heading sizes to provide useful clues for the reader seeking specific heading levels. Select type sizes at least 3 points different in size to distinguish headings, since the typical user cannot discriminate type size differences of less than 3 points. In creating this ratio, also consider the use of a similar ratio for white space (see 12.18).

Typical Header Algorithm

Level Head	Header Type Size	Leading White Space	Trailing White Space
1	18	24	16
2	15	20	14
3	12	16	12

Leading (Line Spacing)

12.33 The space from the bottom of the descenders, such as the tail of *p* or *g*, on one line to the top of the ascenders—usually capitals but sometimes lowercase *h*, *l*, or the like—on the next line is referred to as line leading. The recommended amount of leading for 10- or 11-point type is 1 or 2 points; more or less leading tends to slow reading. Note, however, that typeface as well as line width also influences leading decisions.

12.34 For documents prepared with typewriters or character printers, use single-spaced text rather than double-spaced or one-and-a-half-spaced text; it is usually easier to read. The double-spaced manuscripts required by many publications are convenient for reviewers because this provides space for annotations; the final document is usually produced with closer line spacing for readability.

Line Width

12.35 Format text with no more than 10 to 12 words per line. Most text set in 10-point type should have a line width of no more than 4 to 5 inches (24 to 30 picas), with 1 point of leading for shorter lines and 2 points for longer lines.

Interactions of Type Size, Line Width, and Leading

12.36 Type size, line width, and leading interact in a variety of ways. The best source for information on these interactions is a graphic designer. However, some general observations can be offered based on contemporary research. These observations should be incorporated into design decisions:

Interaction of Type Size, Line Width, and Leading

Type Size (points)	Line Width (picas)	Leading (points)
8	14–36	2–4
9	14–30	1–4
10	14–31	1–4
11	16–34	0–2
12	17–33	1–4

Highlighting Techniques

12.37 Use highlighting techniques such as boldface type, italic type, reverse type, underlining, and different type sizes to help readers find important text. Use the following guidelines for highlighting techniques:

- Keep highlighting simple; do not use all available type styles, type sizes, and paragraph formats.
- Avoid decorative uses for and combinations of type styles. Italics, boldface, underlining, and other typographic emphasis techniques work well when only about 1 word in every 200 is emphasized. Readers begin to ignore specialized typography when about 1 word in every 30 is emphasized.
- Use all-capital type only for text elements that deserve unusual attention, since this type slows reading considerably. Thus, all caps may be ideal for warnings when danger to a human operator is involved.
- Boldface type, set in mixed case, is best for cueing, especially in headings. However, boldface inside paragraphs disrupts the balance of the page, creating a spotty and disorganized appearance.
- Color can also be used for highlighting. Color adds interest and improves reading proficiency when it does not overpower the text. Like any highlighting technique, color should be used sparingly and for special and specific purposes. Useful color combinations are shown.

Useful Color Combinations

Paper Color	Ink Color				
	white	black	red	blue	yellow
white					
red					
green					
blue					
cyan					
magenta					
yellow					
black					

Best Combination

Second Best Combination

Not a Good Combination

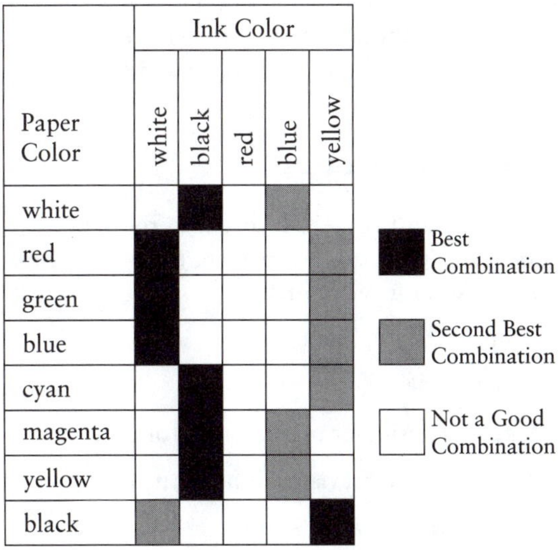

Page and Screen Elements

12.38 Once general formatting and page design decisions have been made, additional decisions about specific features that recur throughout a document can be considered:

- Page numbering (12.39),
- Running headers and footers (12.41),
- Source notes and references (12.45),
- Lists (12.46),
- Revision notices (12.47).

Page Numbering

12.39 Number pages in a visible position:

- Centered in the top or bottom margin of every page,
- Top left corner margin of left-hand (verso) pages, or
- Top right corner margin of right-hand (recto) pages.

12.40 Long manuals frequently use modular page numbers (1-1, 1-2, . . . , 5-1, 5-2, or 1.1, 1.2, . . . , 5.1, 5.2) rather than sequential page numbers (1, 2, 3, . . . , 19). Such a numbering system allows these manuals to be updated easier; if information changes in only a few chapters, those chapters can be reprinted without changing the entire manual. Only the pages in the revised chapters need to be repaginated.

Running Headers and Footers

12.41 Use headers and footers to provide reference information at the top (header) and/or bottom (footer) of every page. Besides the page number, headers and footers can include

- Company logo or name,
- Manual's title,
- Manual's revision or printing date (or chapter or page),
- Chapter title (abbreviated, if necessary),
- Current heading, and
- Inventory control information.

In addition, on-screen headers and footers might include

- Copyright and legal notices,
- Statement of ownership,
- Address for authoring agent or corporation,
- Date of last change, and
- Text-only menu of possible links.

12.42 Use typeface, size, placement, and (perhaps) color in designing the header and/or footer to help the reader understand where they are in the text. Use type for running headers and footers that is either the same size as or slightly smaller than body type. It is unnecessary to use a different type style for these page elements; their placement is a good enough separation and distinguishing feature.

Specialized Notices: Warnings, Cautions, and Notes

12.43 Use warnings, cautions, and notes to offer three levels of emphasis for special situations. Definitions of these components vary among organizations, and are usually listed in the style standards. Some organizations use these guidelines:

- Warnings inform the reader of possible bodily injury if procedures are not followed exactly,
- Cautions alert the reader to possible equipment damage if procedures are not followed correctly, and
- Notes inform the reader either of a general rule for a procedure or of exceptions to such a rule.

12.44 Designers use not only typeface, type size, and color to make these distinctions, but also line width, placement, and ruled boxes. Specialized notices should be designed so that readers can

- See at a glance, without any ambiguity, that these notices have a different kind of importance than either the body text or the headings.
- Distinguish among their levels of urgency. In most documents, warnings (physical danger to humans) are most urgent, then cautions (danger to the equipment), followed by notes (noncritical information of general interest).

Source Notes and References

12.45 Use a footnote or endnote when citing the authority for a statement. Because footnotes and endnotes are subordinate to a manual's text, use smaller type sizes (see also 8.101, 8.107).

Lists

12.46 Set lists off from major text sections and use the same type conventions found in the rest of the body text. Use numbers for lists in which the reader is expected to perform an activity in a specific order. Use bullets if there is no precedence.

Revision Notices

12.47 Use revision notices for material that is updated or revised frequently. Such notices usually appear in two forms: text or revision bar. Textual revision notices often appear as part of the running footer and include some corporate coding scheme that accounts for the state of each page. Revision bars are vertical rules placed in the margin that indicate which sections of a text have been changed recently.

Page and Screen Grids

12.48 Page grids are based on a geometrical process that accounts for reading and viewing eye motion. Several bibliographic sources for this section illustrate this geometry (see 13.12). Although this process should be rigorously applied, the resulting grids can be adjusted in small increments of perhaps 2 points or less without destroying their value.

12.49 Create page grids to establish where specific page elements will appear on a sheet of paper. Page grids provide guidance across product lines, among writers working on the same project, or across many corporate sites. They provide consistency and continuity that help readers use documents successfully.

12.50 The following sections offer sample grid designs for five major size documents—5 by 7 inches, 5 by 8 inches, 7 by 10 inches, $8^{1}/_{2}$ by 11 inches—A4; and a typical (640 × 480 pixels) computer monitor (see 12.12–12.15). It should be understood that these sample grids do not exhaust the variations a graphic designer could propose for a document. Instead, they are intended as guidelines to help create similar designs for both paper and electronic composition systems (such as the many electronic paste-up programs available for computers).

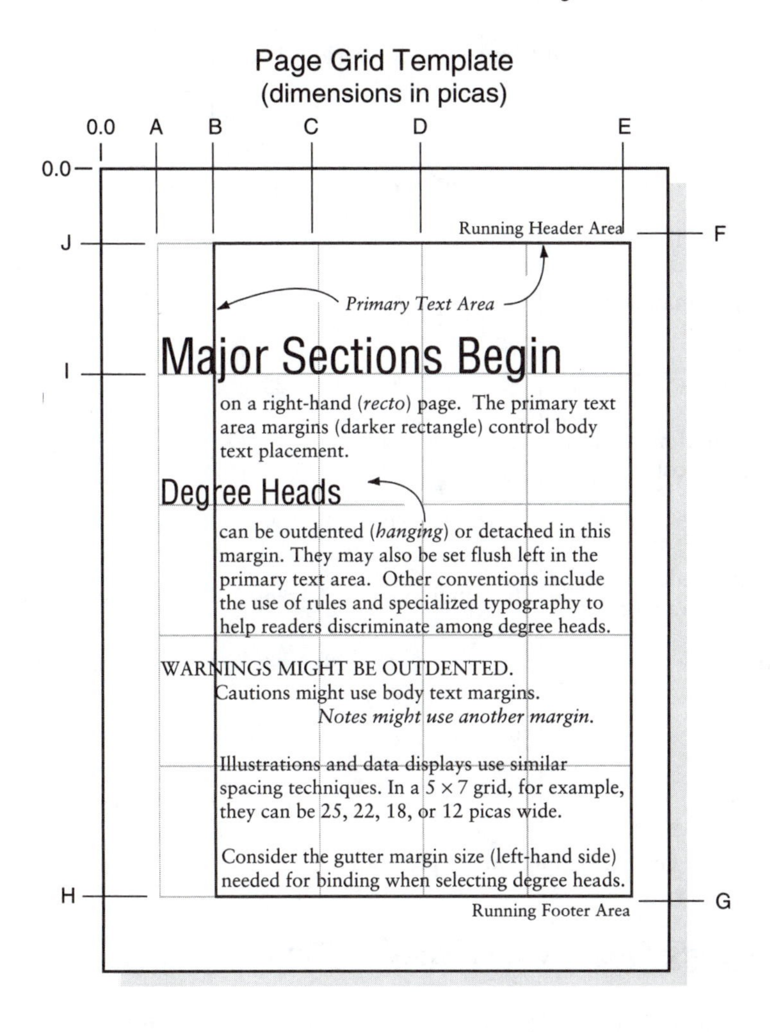

A	Left margin of the page	F	Baseline for running header
B	Left margin of primary text area	G	Capline for running footer
C	Right edge of left-hand column	H	Bottom margin of primary text area
D	Left edge of right-hand column	I	Baseline for chapter or major section head
E	Right margin of primary text area	J	Top margin of primary text area

(See "Grid Dimensions for Various Page Sizes" on the following page.)

Grid Dimensions for Various Page Sizes

Page Size (inches)	Dimension Points (picas)									
	A	B	C	D	E	F	G	H	I	J
5 × 7 booklet portrait, one col.	3	6			28	3	39	38	11	4
5 × 8 copier manual landscape, one col.	3	13.5			45	3	28	27	13.2	4
two col.		3	23	25	45	3	28	27	13.2	4
7 × 10 tutorial manual portrait, one col.	3	6			39	3	58	57	12.6	4
two col.		3	20	22	39	3	58	57	12.6	4
8½ × 11 procedure manual, portrait, one col.	6	16.5			48	3	64	63	15.8	4
two col.		6	26	28	48	3	64	63	15.8	4
A4 one col.	6	16.1			46.5	3	67.5	66.5	16.5	4
two col.		6	30.3	32.3	46.5	3	67.5	66.5	16.5	4

(See "Page Grid Template" on the previous page.)

Designing Specific Information Types

12.51 The following sections discuss design issues related to five major information types:

- Brochures and newsletters (12.52),
- Papers or articles for publication (12.54),
- Technical manuals (12.63),
- A4 documents (12.86), and
- On-screen information (12.92).

Designing Brochures and Newsletters

12.52 Well-designed brochures and newsletters encourage people to invest the time and effort to read them. When preparing a brochure or newsletter, consider the audience profile, objectives, and needs:

- What purpose will the document serve? Is it a marketing tool? Will it educate customers or users?
- What is the personality or character of the product (or subject) and its market (or audience)?
- How much information must be presented? How many illustrations are needed, and what kind?
- What is the most important part of the document? What is the least important part?
- Will the document be used in-house (entirely by people within an organization), outside, or both?
- Does the organization have an established corporate identity? Are there corporate colors, logos, and styles?

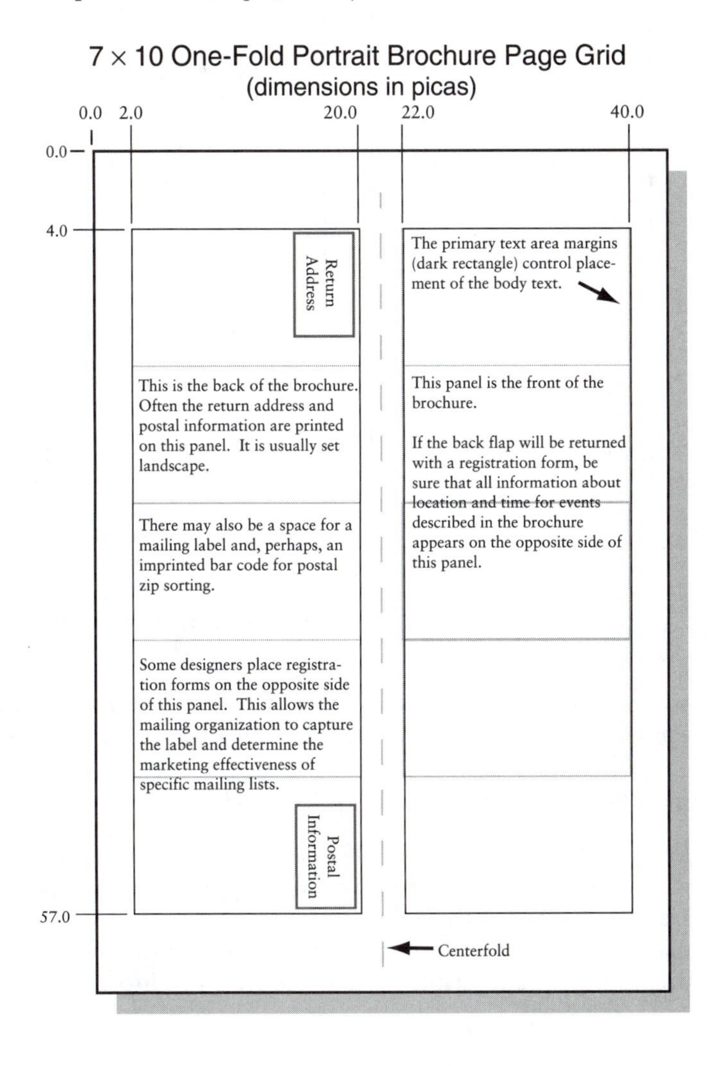
7 × 10 One-Fold Portrait Brochure Page Grid
(dimensions in picas)
0.0
2.0
20.0
22.0
40.0
0.0
4.0
57.0
Return Address
The primary text area margins (dark rectangle) control placement of the body text.
This is the back of the brochure. Often the return address and postal information are printed on this panel. It is usually set landscape.
This panel is the front of the brochure.
If the back flap will be returned with a registration form, be sure that all information about location and time for events described in the brochure appears on the opposite side of this panel.
There may also be a space for a mailing label and, perhaps, an imprinted bar code for postal zip sorting.
Some designers place registration forms on the opposite side of this panel. This allows the mailing organization to capture the label and determine the marketing effectiveness of specific mailing lists.
Postal Information
Centerfold

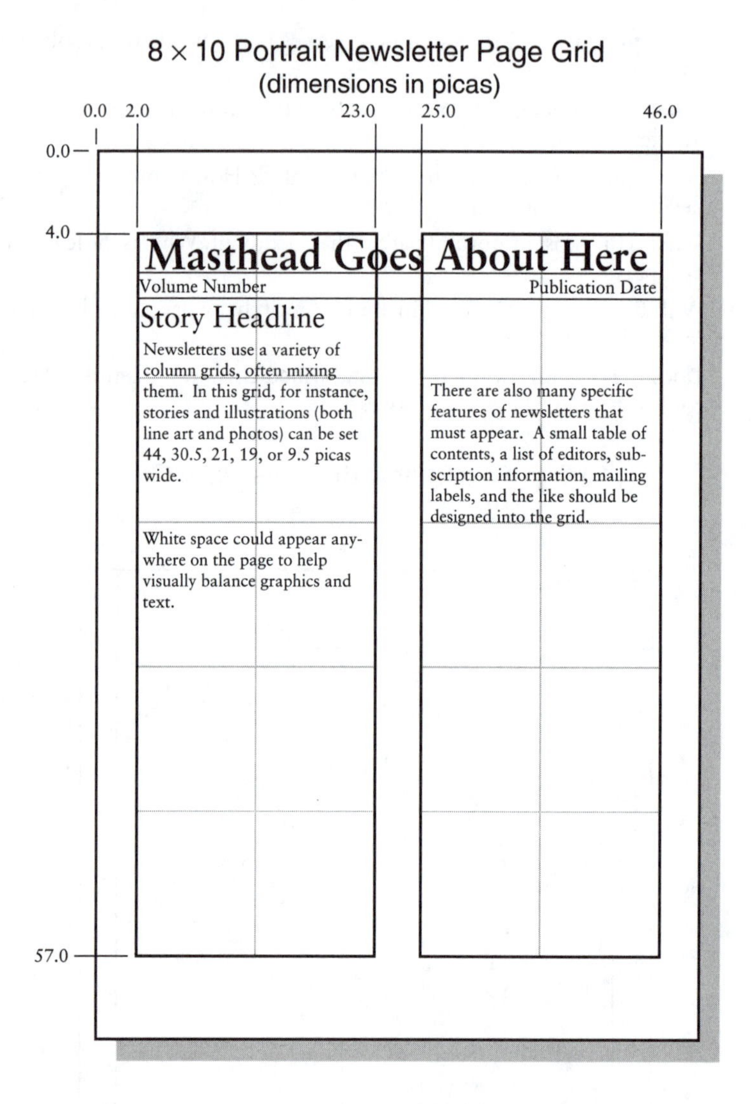

8 × 10 Portrait Newsletter Page Grid
(dimensions in picas)
0.0
2.0
23.0
25.0
46.0
0.0
4.0
57.0
Masthead Goes About Here
Volume Number
Publication Date
Story Headline
Newsletters use a variety of column grids, often mixing them. In this grid, for instance, stories and illustrations (both line art and photos) can be set 44, 30.5, 21, 19, or 9.5 picas wide.
White space could appear anywhere on the page to help visually balance graphics and text.
There are also many specific features of newsletters that must appear. A small table of contents, a list of editors, subscription information, mailing labels, and the like should be designed into the grid.

12.53 Use brochures, short-shelf-life documents, to advertise or summarize a product or service. Use newsletters to convey news about an organization.

Preparing a Paper or Article for Publication

12.54 Before writing an article, select a potential journal or magazine and prepare the article according to that publication's requirements. Because there are so many journals and magazines that cover such a variety of subject matter, it is impossible to provide a single set of rules to follow for preparing a manuscript. Always refer to the editorial policy and/or manuscript guidelines provided by the publishing agency. In the absence of such a policy, either refer to earlier issues of the same publication or use the advice offered in this section for standard parts of most papers and articles (see also chapter 8).

Title

12.55 Create a specific title that describes the paper's content and emphasis. There is a fine balance between brevity and descriptive accuracy; a 3-word title may be too cryptic, but a 14-word title is probably too long:

Applying Television Research to Online Information:
A Case Study

12.56 Place the byline beneath the title. Some journals require the name of the institution and city in which the author conducted the research:

Carbon-Dating Evidence Based on Mesoamerican Cooking Utensils

Margaret Wooldridge
Bian University, England

Abstract

12.57 Provide an abstract for academic journals that summarizes the work reported in the article; most trade journals do not require an abstract. Unless the journal must approve an abstract before considering the article, write the abstract last to ensure that it reflects the article's content.

12.58 Since most journals generally accept a 200-word abstract, keep the abstract succinct. A typical abstract might be:

> This article examines the genealogy of recent innovations in online information and hypermedia. In doing so it provides an overview of the more important developments and applications of these advances. While the essay celebrates those successes that are clearly exceptional, it also assesses the more typical, and sometimes equally useful, techniques found in less dramatic and innovative environments. Finally, the essay asks the reader to consider some of the more optimistic scenarios for the future of online information and hypermedia in relation to the concept of literacy.

Introduction

12.59 In the introduction, clearly state the problem or theory and indicate the work's significance in the context of what is already known about the subject. If appropriate, provide a brief review of relevant literature.

Body

12.60 Whatever the journal type, use headings and subheadings to help readers understand the information's structure and content. The body's organization and length depend on the subject matter and the journal. Many academic journals, for instance, have strict organizational rules. When appropriate, include data displays or illustrations to improve clarity (see also chapters 10 and 11).

Conclusion

12.61 Most academic journals require a conclusion, summary, discussion, or suggestions for additional research. Make sure such conclusions are interpretive and not repetitious. Most trade journals do not require formal conclusions; however, such summaries can be used to emphasize an article's main points.

References

12.62 Methods for citing references in journals differ widely; check recent issues of a journal for preferences (see also chapter 8).

Designing a Technical Manual

12.63 In planning a technical manual, maintain consistency and make decisions that create a simple and straightforward format.

12.64 Consult existing style standards before designing a technical manual. Many organizations have documentation style standards for technical manuals that specify the format and design for most document components. However, most standards leave some format decisions to the author; when possible, make such decisions to coordinate with your organization's standards.

12.65 In designing the manual, consider these format issues:

- Reader access,
- Front cover,
- Front matter,
- Chapter or section divisions,
- Appendixes,
- Glossary,
- Indexes,
- Bibliography, and
- Back cover.

Reader Access

12.66 Provide reader access to book sections. This access is based on such questions as

- Will the manual have separator tabs?
- Will such tabs be physical tabbed pages, or will they be blind tabs printed on the edges of the text pages?
- Will every chapter and appendix have a tab, or do the chapters and appendixes fall into groups, with a tab per group?
- What text will appear on each tab?
- Are the chapter titles short enough to fit the tabs? If not, how can they be abbreviated?

12.67 Print a table of contents for each chapter on the tab page that begins a new section.

Front Cover

12.68 Design the front cover to emphasize the company's identity, the document's content, and the design features found in the document. For instance, if the document is set 5-by-8-inch landscape on a two-column grid, use those same grid dimensions to place cover elements. This technique establishes a sense of consistency and continuity from the outset.

Front Matter

12.69 Almost all organizations specify the content and format of front matter, pages that appear before the first chapter or section. Front matter can include

- Title page,
- Copyright page,
- Acknowledgments,
- Abstract or preface,
- Table of contents,
- List of figures, and/or
- About this manual.

12.70 Make format decisions for the chapter pages and the cover first, then plan the front matter to be as consistent as possible with these components. Number front matter pages with Roman numerals, beginning on the back of the title page with the number "ii."

Title Page

12.71 Documents often have a covering title page, called a half-title, that offers abbreviated publication information: short title, author's or authoring agency's name.

12.72 Use a title page to list the name, date, and publication number of the manual. (The publication number sometimes appears on the copyright/trademark page.) The title page of some technical documents and scientific reports sometimes includes approval signatures granted by various administrators in charge of projects.

Copyright Page

12.73 Provide a copyright/trademark page to display the manual's copyright notice and list of the registered trademarks used in the manual. Some manuals, especially those that support consumer electronic products, may also have an ISBN number and library catalog information on this page.

Acknowledgments

12.74 Use an acknowledgments page to list the contributors to the manual. Alternatively, they may appear on the copyright/trademark page or as part of the preface.

Abstract or Preface

12.75 Include a preface to define the manual's purpose, intended audience, organization, and prerequisites (if any) for using the manual.

Table of Contents

12.76 Use a table of contents to present the contents of the manual, according to chapters, sections, subsections, appendixes, glossary, and index. Lists of figures and/or tables are sometimes presented separately.

List of Figures

12.77 Use a list of figures and/or tables to catalog the manual's visual contents, according to chapters, sections, subsections, and appendixes.

About This Manual

12.78 Use this introduction, often a manual's first chapter, to

- Present an overview of the document's contents;

- Offer specialized instructions on how to use the manual; or
- Summarize manual conventions (such as warning, caution, and note practices).

Chapter or Section Divisions

12.79 Chapters can be considered navigation aids in a very broad sense, because they help readers locate common information.

Appendixes

12.80 Place multiple appendixes in the same order in which the manual's text refers to them. Appendixes contain details that might disrupt the flow of information in a chapter or section. Lengthy tables, error messages, and algorithms usually appear in appendixes.

12.81 Design decisions for chapters and appendixes include the layout for the opening page of each chapter or appendix, the layout for continuing pages, and the numbering method. Design the opening page of a chapter or appendix to emphasize its title.

12.82 Number chapters and identify appendixes with letters. Make specific design decisions for typeface, type size, and capitalization of the numbers and letters.

Glossary

12.83 Design a glossary to be consistent with the rest of the text. A glossary, a special kind of chapter or appendix, defines terms used in the text and provides the same alphabetical access as an index.

Indexes

12.84 Design decisions for indexes should reflect the information-access nature of an index; locating index entries quickly should be the goal. Many indexes use multiple columns and smaller type than the body text. For simplicity, maintain as much as possible of the chapter page design (see also chapter 9).

Back Cover

12.85 Continue any design principles on the back cover if there is anything printed on that surface.

Preparing A4 Bilingual and Multilingual Documents

12.86 Prepare documents that include more than one language in one of the following formats:

- Separate publications,
- Separate joined publications,
- Two-column side-by-side publications, and
- Multicolumn side-by-side publications.

12.87 In deciding which technique to use for your needs, be sure to consider the difference in length for texts written in various languages. Typically, most foreign languages require more pages than English for the same text (see also chapter 2).

Separate Publications

12.88 Use separate publications, two separate documents, when many copies will be printed in the different languages. Use this technique only when frequent revisions are not expected.

Separate Joined Publications

12.89 Use this technique when two languages must appear in a single publication. Typically, these documents are printed in normal orientation (front-to-back) or tumbled. A tumbled document has two front covers: one in the first language, the other in the second. Where the text in one language ends, the text for the second language appears upside down. To read this section of the document, the book must be tumbled or turned. Tumbled orientation is often impractical for multilanguage documents, although good examples of its use in short documents for consumer products can be found.

Two-Column Side-by-Side Publications

12.90 Use this method when the content of concurrent versions must be seen by all readers.

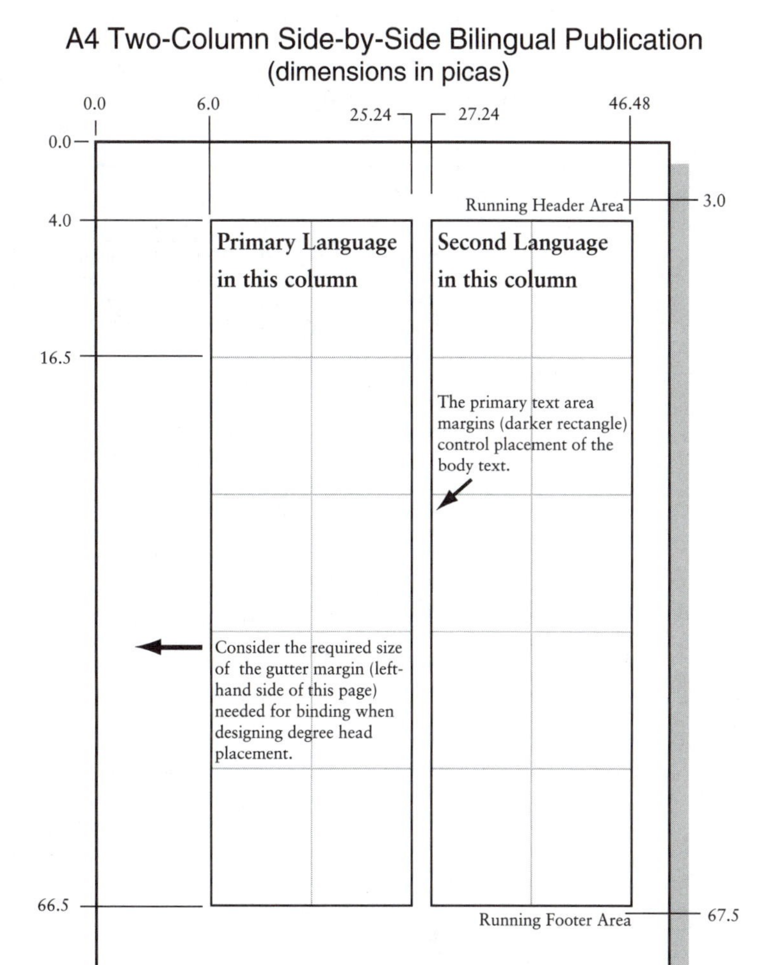

Multicolumn Side-by-Side Publications

12.91 Use this method when many languages must be present in a document. This technique is often seen in directions for small consumer products.

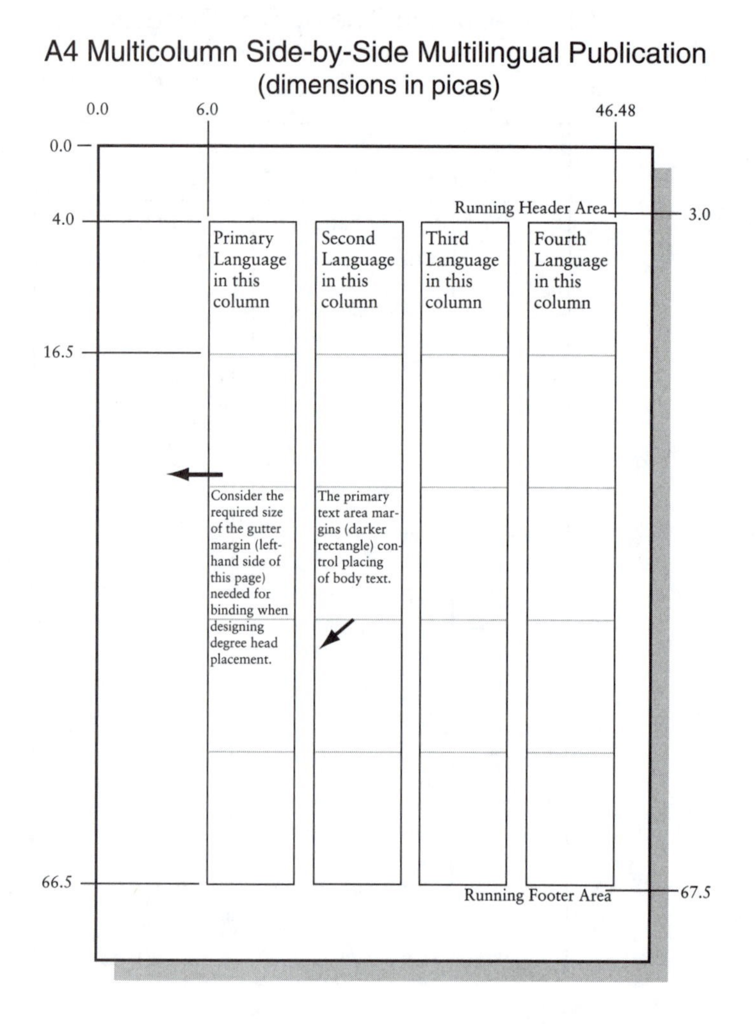

Designing On-Screen Information

12.92 On-screen information has come to mean pages displayed on the World Wide Web. It would be impossible to capture either the dynamics of on-screen information development or the constant evolution of new techniques that influence its creation. As a consequence, this section provides the same kind of guidance found for other document types; design information, specific to web pages, can be found in works cited in the bibliography (see 13.12).

12.93 Although on-screen information differs dramatically from printed texts, it would be difficult to imagine readers suddenly changing their information-gathering behaviors as radically. Thus, authoring agents still need to consider many of the qualities and methods for producing paper information, and their implications for on-screen communication.

Document Techniques and Information Class

12.94 Like paper documents, on-screen information uses specific textual and graphic techniques to support the needs of an audience and the purposes for which the website has been developed. This suggested interaction between information classes and development techniques should be viewed as a method with which any authoring agency can create its own matrix. One could argue, for example, that data capture can be used by a site offering procedural information as means of verifying that the site can provide useful information (see 1.17 and following).

Document Techniques and Information Classes

Document Techniques	Information Class			
	Tutorial	Marketing	Procedural	Reference
Static text	X	X	X	X
Static text and graphics	X	X	X	X
Static text and dynamic graphics	X	X	X	X
Data capture (forms)	X	X		
Data capture and manipulation (sales)	X	X		
Purvey archived materials	X	X	X	X

12.95 Another way to consider these interactions is to compare site access characteristics to the kind of information one wishes to convey. These interactions can, obviously, be characterized in other ways. For example, a group in an intranet may "sell" services to other groups in the same network, even though they are part of the same overall group. Similarly, one might argue that a vendor in an extranet would be ineligible for training. The basic idea is to use the matrix as a method for characterizing a specific authoring agency's profile (see 1.17 and following).

Cite Access Characteristics	Information Class			
	Tutorial	Marketing	Procedural	Reference
Internet—universally available	X	X	X	X
Intranet—used by a single group or its subunits	X		X	X
Extranet—used among a limited number of groups	X		X	X

Splash Page

12.96 Like a book cover, this page sets the tone for the remainder of the site. If it employs static text, a range of colors, and a navigation technique, these elements create reader expectations that must be satisfied in the remainder of the site.

Home Page

12.97 This page organizes and sets the tone for the remainder of the site. Unlike the splash page, users will return to this page often. Some key elements of this page include

- Navigational and organizational structure,
- Description of unusual practices,
- Copyright and legal notices,
- Contact information,
- Update history, and
- Links (either text-only or graphic) to access the site.

Section Pages

12.98 Section pages are any single page or collection of pages accessed from the home page. Their relationship to that page depends on the link structure controlling the site (see also 11.280):

- Sequential—forward and backward movement only (booklike),
- Tree—branch from major to subordinate levels (organizational chart),
- Web—complex, almost random.

In reality, websites tend to be mixtures of these linkage styles without serious consequences. For example, one might offer a "calendar" that is likely to be read sequentially; however, that source might use a tree structure to link to specific subordinate information such as personnel, deadlines, and expected outcomes.

Linked and Archived Sources

12.99 Like the *back matter*—appendixes, glossary, and bibliography—linked and archive sources can be viewed as methods for amplifying a document. For example, appendixes could include archived and downloadable information; a bibliography and glossary could provide a collection of links to external sources that provide additional information to support the site.

Text: Static Information

12.100 Like a printed text, text on-screen is simply a collection of displayed, typographic characters. If it includes no dynamic techniques (see 12.103), it should retain many of the best characteristics of printed information.

12.101 One major problem with text on-screen is that it offers some challenges for reading:

- On-screen typographic manipulation is limited (see 12.29),
- On-screen text provides few clues about the document's size, and
- Readers often report feeling lost in long documents.

While some of these problems may be resolved with time, it is difficult to ignore the fact that readers have difficulty using this information source.

12.102 Edit printed documents destined for on-screen display to reduce their length. Some of the techniques for achieving this goal include

- Using lists and tables to catalog similar information,
- Reducing sentence length and complexity, and
- Applying sentence-combining techniques to reduce the total number of sentences.

Multimedia: Dynamic Information

12.103 Multimedia include all those on-screen techniques that provide both movement and sound:

- Rollovers—text or graphics that change their state when a cursor moves over them,
- Balloons or bubbles—small, pop-up windows tied to specific on-screen elements that provide additional information about that feature,
- Windows—a secondary window that provides secondary information,
- Slide shows—either self-running or user-operated presentations,
- Animations—two or more elements change state to create an animation,
- Video—live or prerecorded videographic presentation,
- Sound—any sound source, and
- Forms—any interactive method for capturing and manipulating information.

12.104 Since each of these multimedia techniques may limit access by requiring specialized hardware and software, use multimedia appropriately. For example, a prerecorded video of a presentation requires significantly more time to download and more memory to display than a voice-over slide show of the same material. Similarly, an animation of a physics principle might be more memorable, and learned better, than text accompanied by a drawing.

Graphic Specification

12.105 Like its typographic limitations (see 12.28), on-screen information has some restrictions on the use of color and appropriate formats for graphic files:

- Colors are currently limited to a palette of 216 colors, rather than the 256 provided by the RGB palette,
- Graphic files must be saved in two formats: GIF (Graphics Interchange Format) for simple illustrations and JPEG (Joint Photographic Experts Group) for photographs and natural color.

12.106 Aside from those limitations (and any others that might be imposed by the use of multimedia) (see 12.103), on-screen design should use a grid system that emulates paper documents. Given the typical computer monitor currently in use, this means that designers have an area 535 × 295 pixels as a minimal viewing area size (see 12.13).

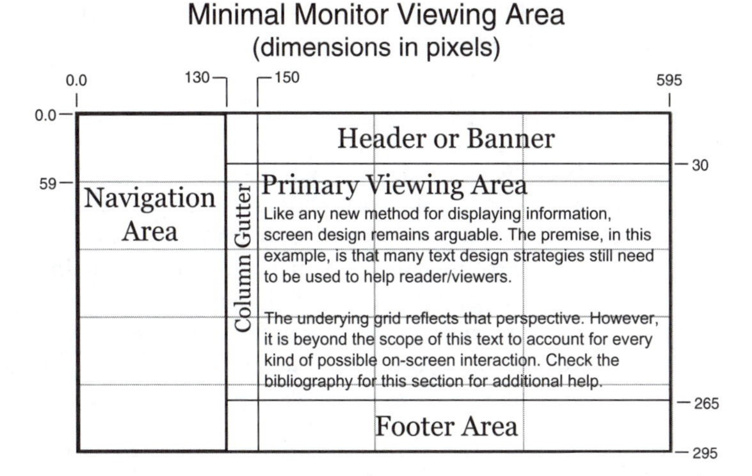
Minimal Monitor Viewing Area
(dimensions in pixels)
0.0
130
150
595
0.0
59
30
265
295
Header or Banner
Navigation Area
Column Gutter
Primary Viewing Area
Like any new method for displaying information, screen design remains arguable. The premise, in this example, is that many text design strategies still need to be used to help reader/viewers.
The underying grid reflects that perspective. However, it is beyond the scope of this text to account for every kind of possible on-screen interaction. Check the bibliography for this section for additional help.
Footer Area

Controlling Large Document Sets

12.107 Organize a document and design its format to accommodate all the readers and all the information. This effort may sometimes mean creating document sets (or suites) rather than a single document. Few readers, for instance, need to understand every detail of the installation, use, operation, preventive maintenance, troubleshooting, and repair of a large product. Consider the documentation needs of all the people who play a role in flying a large jetliner. The pilots who fly the aircraft, the ground crew who check out the system before flight, the personnel who perform routine maintenance and repairs, the flight attendants who ensure passenger comfort and safety—all need different types of information, which should be in separate documents. To address this issue, consider

- Categorizing information (12.108),
- Making information accessible (12.113).

Categorizing Information

12.108 Categorize information for a document set based on audience, frequency of use, task, and system configuration.

Audience Analysis to Classify Documents

12.109 Use audience analysis to identify documents based on who needs what information; audience identification is usually combined with one of the other three approaches. For example, an aircraft document set might include a system overview for all readers and then one or more documents each for pilots only, for ground crew only, and so on.

Frequency of Use to Classify Documents

12.110 Use frequency of use to assign information to documents on the basis of how often readers need that information. Readers generally need installation information only once, and troubleshooting information only occasionally. On the other hand, they may refer every day to information about use, operation, and preventive maintenance. For example, extensive training prepares pilots to operate their aircraft without referring to a flight manual, but they do complete a documented preflight checklist before every takeoff.

Task Orientation to Classify Documents

12.111 Use task orientation to focus on procedural information that different groups of readers need to do their jobs. Answering the following questions can help identify the essential information to include in various task-oriented documents:

- Who performs the task?
- What action begins the task?
- What steps are required to perform the task?
- What action ends the task?
- Can product variations affect the task or its results? If so, what are the variations and their effects?
- How many ways can the task be performed? What are the benefits of each method? Should each method be documented?

System Configuration to Classify Documents

12.112 Use system configuration to place information in documents according to the structure or features of the product. For example, the document set for the jetliner's maintenance personnel could include separate documents about the engines, landing gear, and so on.

Making Information Accessible

12.113 To help readers find the information they need, include overviews, master indexes, and master glossaries.

12.114 An overview describes the paths readers can take through the information. For example, a novice audience might be required to read certain documents, while an expert might not need that same information. Certain documents may contain the task information most readers want most of the time, while other documents are better for first-time or reference use.

12.115 Overview design may be as simple or as complicated as needed for a specific audience and document. Simple approaches include using tables or flowcharts to list types of information and/or tasks, along with the names of the documents that contain them. A more sophisticated overview might use a graphic presentation to show how the document set is organized.

12.116 Include an index for every document in a set, unless the index has very few entries or is unlikely to be used more than once. Also include a master index to the document set to help readers find out which documents contain which information. Readers can then use the individual document indexes to look up a particular type of information in detail. Some document sets include the master index in every document, but that approach is usually not practical for large document sets, especially in technical material that is updated or revised frequently.

12.117 Similarly, include a master glossary to define terms for a topic, product, or system that uses a technical or specialized vocabulary. Individual documents could also provide glossaries to define only the terms introduced in that document.

12.118 Master indexes and glossaries may duplicate some information. The convenience to readers justifies this duplication and outweighs the additional cost.

Graphic Production Considerations

12.119 Producing documentation includes

- Preparing camera-ready pages (12.120),
- Choosing paper (12.122), and
- Choosing the binding method (12.125).

Preparing Camera-Ready Pages

12.120 When a document is in its final form, it can be prepared for production. Follow all the graphic design principles discussed earlier and create the final, camera-ready pages. While laserographic printing and direct plate mastering have become electronic processes, the final pages before printing are still referred to as camera-ready.

12.121 Camera-ready pages (also called mechanicals) use the page size, typefaces, type styles, type sizes, line spacing, line width, and illustration specifications selected during the design process. Depending on the production system available, text and illustrations may be combined on these pages for the first time.

Choosing Paper

12.122 Printing paper varies in several ways:

- Durability—the ability of paper to withstand wear based on the ways in which readers will use the document;
- Color and/or brightness—some documents may be read under varying lighting conditions that may make text difficult to read when printed on color paper;
- Finish—special textures and finishes may be needed for specific purposes (glossy paper can print accurate photographs; textured paper can present a professional tone for an annual report);
- Weight—a proper paper weight will help a document survive repeat use (bending, folding, and the like);
- Opacity—less opaque papers are generally inappropriate for technical information because images can be seen through the page;
- Cost—since most technical and scientific information is not sold as a separate product, the cost of its production must be justified.

12.123 Designers, printers, and paper companies are all good sources of information about paper selection. Often a printer maintains a supply of floor stock paper. Using floor stock can save money if the paper is suitable for the printing job (see 12.15).

12.124 Among the most popular paper types are bond paper, coated paper, uncoated paper, and textured paper. Except for the bond paper, these papers come in book and cover weight:

- Bond paper, used primarily for letterhead and correspondence, is very strong. There are two types of bond paper: plain bond and bond with a rag or cotton content. Bond with a rag or cotton content is a higher-quality paper than plain bond, generally has a more attractive finish, and is more expensive.

- Coated paper, which has either a glossy or dull enamel finish, is a good choice when printing photographs. However, the glare created by its coating makes it unacceptable for documents containing large amounts of text.
- Uncoated paper, which is less expensive than either coated or textured paper, has either antique or smooth finishes and is available in various weights. Offset paper, typically used in quick-printing applications, is a type of uncoated paper.
- Textured paper—a high-quality, uncoated paper—is available in a variety of colors and finishes.

Binding the Document

12.125 Binding refers to how the pages of a document are held together. The type of binding is a particularly important decision for science and technical texts because many of these documents must be used under working conditions, and they must be updated often. There are three basic binding methods:

- Wire stitching,
- Mechanical binding, and
- Perfect binding.

Wire Stitching

12.126 Wire stitching is the most common binding method for short documents such as pamphlets, booklets, and newsletters; it is also the simplest and least expensive. In wire (or saddle) stitching, the document is stapled along the centerfold. For example, to make an 8½-by-11-inch saddle-stitched document, a set of 11-by-17-inch sheets of paper are folded in half and stapled at the center.

12.127 A less common wire stitching method—side stitching—punches staples through all of the pages along the spine from front to back cover. While this is a relatively inexpensive binding method, the staple ends damage any surrounding documents on a shelf.

Mechanical Binding

12.128 Mechanical binding—loose-leaf ring or coils—allows documents to lie flat when open. The primary difference is that the footprint, the total area that the open document occupies in the work area, is usually larger for loose-leaf, ring-bound books than coil-bound (which fold back on themselves).

12.129 Loose-leaf ring binding, in which the printer punches or drills holes into the margins of the pages so that the document can be inserted into a loose-leaf binder, is one form of mechanical binding. Loose-leaf binding works better than all other binding methods for large documents because it allows for easy updating.

12.130 The other methods of mechanical binding are wire-o, spiral, and plastic comb. In wire-o and spiral binding, holes are drilled through the covers and pages, which are then bound together by a wire or plastic coil. These documents can be folded back on themselves and still lie flat, creating a small document footprint. Wire-o and spiral binding are permanent binding methods, most effective for documents less than 3/4-inch thick that will not need updating.

12.131 In comb binding, holes are punched in the covers and pages, and a plastic binding comb is inserted. Low-cost equipment for comb binding is available for offices and copy centers. Comb-bound documents lie flat when open, but do not fold back on themselves easily to create a single-page footprint. Updating pages in comb-bound documents is possible but rarely done in practice, because removing and replacing the combs is difficult to do without damaging the document.

Perfect Binding

12.132 In perfect binding, assembled pages are trimmed along the back edge and bound with hot glue. A preprinted cover is then glued to the spine. Paperback books are perfect-bound; in large quantities, perfect binding is relatively inexpensive.

12.133 The disadvantages of perfect binding are the increased binding time and the fact that perfect-bound books do not lie flat when opened. Also, most printing companies do not have perfect-binding capabilities and must subcontract with a specialty firm, which requires additional processing time.

13. Bibliography

13.1 Audience Analysis and Document Planning

Caernarven-Smith, P. *Audience Analysis and Response.* Pembroke, MA: Firman Technical Publications, 1983.

Campbell, Kim Sydow. *Coherence, Continuity, and Cohesion: Theoretical Foundations for Document Design.* Hillsdale, NJ: Lawrence Erlebaum Associates, 1995.

Carroll, J. M., ed. *Minimalism beyond the Nurnberg Funnel.* Cambridge: The Massachusetts Institute of Technology Press, 1998.

Council of Biology Editors. *Scientific Style and Format: The CBE Manual for Authors, Editors, and Publishers.* 6th ed. New York: Press Syndicate of the University of Cambridge, 1994.

Lay, Mary M., et al. *Technical Communication.* 2nd ed. Boston: Irwin McGraw-Hill, 2000.

Rosenbaum, S., and R. D. Walters. "Audience Diversity: A Major Challenge in Computer Documentation." *IEEE Transactions on Professional Communications* PC29.4 (1986).

Van Buren, Robert, and Mary Fran Buehler. *The Levels of Edit.* 2nd ed. Arlington, VA: Society for Technical Communication, 1991.

13.2 Writing for Non-native Audiences

Adams, Ann H., Gail W. Austin, and Melissa Taylor. "Developing a Resource for Multinational Writing at Xerox Corporation." *Technical Communication* 46.2 (1999): 249–54.

Association for Machine Translation in the Americas. *Association for Machine Translation in the Americas.* Webpage. Available: *http://www.isi.edu/natural-language/organizations/AMTA.html.* December 20, 1999.

Barthe, Kathy, et al. *AECMA Simplified English.* November 25, 1999. Webpage. AECMA (European Association of Aerospace Industries). Available: *http://www.aecma.org/senglish.htm.* December 20, 1999.

Benson, Morton, Evelyn Benson, and Robert Ilson, eds. *Bbi Dictionary of English Word Combinations.* Philadelphia, PA: John Benjamins Pub. Co., 1997.

Boiarsky, Carolyn. "The Relationship between Cultural and Theoretical Conventions: Engaging in International Communication." *Technical Communication Quarterly* 4.3 (1995): 245–59.

Celce-Murcia, Marianne, and Diane Larsen-Freeman. *The Grammar Book: An ESL/EFL Teacher's Course.* Boston: Heinle & Heinle, 1999.

De Vries, Mary A. *Internationally Yours.* Boston: Houghton Mifflin, 1994.

DGXII of the European Commission. *EURODICAUTOM.* Webpage. Available: *http://eurodic.echo.lu/cgi-bin/edicbin/EuroDicWWW.pl.* December 20, 1999.

European Association for Machine Translation. *European Association for Machine Translation.* Webpage. Available: *http://www.lim.nl/eamt/.* December 20, 1999.

Greenbaugh, Sidney, and Randolph Quirk. *A Student's Grammar of the English Language.* Reading, MA: Addison-Wesley Pub. Co., 1991.

Hodges, John C., and Suzanne Webb, eds. *Harbrace College Handbook.* 12th ed. New York: HBJ College and School Division, 1995.

Hoft, Nancy. *International Technical Communication.* New York: John Wiley & Sons, Inc., 1995.

———. "Global Issues and Local Concerns." *Technical Communication* 46.2 (1999).

Hornby, A. S., A. P. Cowie, Jonathan Crowther, and John R. Crowther, eds. *Oxford Advanced Learners Dictionary of Current English.* Oxford: Oxford University Press, 1995.

Horton, William. "The Almost Universal Language: Graphics for International Documents." *Technical Communication* 40 (1993): 682–93.

Jones, Scott, et al. *Developing International User Information.* Boston: Digital Press, 1992.

Kohl, John R. "Improving Translatability and Readability with Syntactic Cues." *Technical Communication* 46.2 (1999): 149–60.

Kostelnick, Charles. "Cultural Adaptation and Information Design: Two Contrasting Views." *IEEE Transactions on Professional Communication* 38.4 (1995): 182–96.

Leininger, Carol, and Rue Yuan. "Aligning International Editing Efforts with Global Business Strategies." *IEEE Transactions on Professional Communication* 41.1 (1998): 16–23.

Markel, Mike. *A Step-by-Step Guide for Engineers, Scientists, and Technicians.* New York: IEEE, 1994.

Quirk, Randolph, Jan Svartvik, and Geoffrey Leech. *A Comprehensive Grammar of the English Language.* Reading, MA: Addison-Wesley Pub. Co., 1985.

Sauer, Beverly A. "Communicating Risk in a Cross-Cultural Context: A Cross-Cultural Comparison of Rhetorical and Social Understandings in U.S. and British Mine Safety Training Programs." *Journal of Business and Technical Communication* 10.3 (1996): 306–29.

Schoff, Gretchen H., and Particia A. Robinson. *Writing and Designing Manuals.* Chelsea, MI: Lewis Publishers, 1991.

Scollon, Ron, and Suzanne Wong Scollon. *Intercultural Communication.* Oxford: Blackwell Pub., 1994.

Séguinot, Candace. "Technical Writing and Translation: Changing with the Times." *Journal of Technical Writing and Communication* 24.3 (1994): 285–92.

Sinclair, John, and Henry H. Collins Jr., eds. *Collins Cobuild English Dictionary.* New York: HarperCollins, 1995.

Weiss, Timothy. "The Gods Must Be Crazy: The Challenge of the Intercultural." *Journal of Business and Technical Communication* 7.2 (1993): 196–217.

———. "'Ourselves among Others': A New Metaphor for Business and Technical Writing." *Technical Communication Quarterly* 1.3 (1992): 23–36.

———. "Reading Culture: Professional Communication as Translation." *Journal of Business and Technical Communication* 11.3 (1997): 321–38.

13.3 Grammar, Usage, and Revising for Publication

American Psychological Association. *Publication Manual of the American Psychological Association.* 4th ed. Washington, DC: American Psychological Association, 1994.

The Chicago Manual of Style. 14th ed. Chicago: University of Chicago Press, 1993.

Cook, Claire Kehrwald. *Line by Line: How to Improve Your Own Writing.* Boston: Houghton Mifflin, 1986.

Council of Biology Editors. *Scientific Style and Format: The CBE Manual for Authors, Editors, and Publishers.* 6th ed. New York: Press Syndicate of the University of Cambridge, 1994.

Dodd, Janet S., ed. *The ACS Style Guide: A Manual for Authors and Editors.* 2nd ed. Washington, DC: American Chemical Society, 1997.

Fowler, H. Ramsey, and Jane E. Aaron. *The Little, Brown Handbook.* 7th ed. New York: Longman, 1998.

Garner, Bryan A. *A Dictionary of Modern American Usage.* New York: Oxford University Press, 1998.

International Standards Organization. *International Standards Organization.* October 2, 1999. Webpage. Available: *http://www.iso.ch/.* December 20, 1999.

Rude, Carolyn D. *Technical Editing.* 2nd ed. Boston: Allyn and Bacon, 1998.

United States Department of Labor. *Occupational Safety and Health Administration.* June 29, 1999. Webpage. Available: *http://www.osha.gov/about.html.* December 20, 1999.

Van Buren, Robert, and Mary Fran Buehler. *The Levels of Edit.* 2nd ed. Arlington, VA: Society for Technical Communication, 1991.

Van Leunen, Mary-Claire. *A Handbook for Scholars.* New York: Oxford University Press, 1992.

Williams, Joseph M. *Style: Ten Lessons in Clarity and Grace.* 5th ed. Englewood Cliffs, NJ: Prentice-Hall, 1991.

13.4 Punctuating Scientific and Technical Prose

American Institute of Physics. *AIP Style Manual.* 4th ed. New York: American Institute of Physics, 1990.

Government Printing Office Style Manual. Washington, DC: United States Government Printing Office, 1984.

Gordon, K. E. *The New Well-Tempered Sentence: A Punctuation Handbook for the Innocent, the Eager, and the Doomed.* Boston: Houghton Mifflin, 1993.

McCaskill, M. K. *A Handbook for Technical Writers and Editors.* Webpage. NASA. Available: *http://sti.larc.nasa.gov/html/Chapt3/Chapt3-TOC.html.*

McMurrey, D. A. *Online Technical Writing: Common Grammar, Usage, and Spelling Problems.* 1999. Webpage. Austin Community College. Available: *http://www.io.com/~hcexres/tcm1603/acchtml/gramov.html.* December 20, 1999.

Olson, G. A. *The Writing Group: Punctuation Made Simple.* 1999. Webpage. JAC Online. Available: *http://www.cas.usf.edu/JAC/pms.* December 20, 1999.

Purdue University. *On-line Writing Lab.* 1999. Webpage. Purdue University Writing Lab. Available: *http://owl.english.purdue.edu/Files/Punctuation.html.* December 20, 1999.

Williams, Joseph M. *Style: Ten Lessons in Clarity and Grace.* 5th ed. Englewood Cliffs, NJ: Prentice-Hall, 1991.

13.5 Using Acceptable Spelling

Berg, Donna Lee. *A Guide to the Oxford English Dictionary.* New York: Oxford University Press, 1993.

Cambridge University Press. *Cambridge International Dictionaries Online.* September

28, 1999. Webpage. Cambridge University Press. Available: *http://www.cup.cam.ac.uk/elt/dictionary*. December 20, 1999.

Glazier, Stephen. *Word Menu*. New York: Random House, 1992.

Indiana University. *Biotech Life Science Dictionary*. 1998. Webpage. Indiana University. Available: *http://biotech.icmb.utexas.edu/search/dict-search.html*. October 2, 1999.

Kerstein, Bob. *Encyberpedia The Living Encyclopedia*. September 27, 1999. Webpage. Encyberpedia. Available: *http://encyberpedia.com/glossary.htm*. December 20, 1999.

Lindberg, Christine A., ed. *The Oxford American Thesaurus*. New York: Oxford University Press, 1998.

Maggio, Rosalie. *The Dictionary of Bias-Free Usage: A Guide to Non-discriminatory Language*. Phoenix, AZ: Oryx Press, 1991.

Pyle, I. C., and Valerie Illingsworth, eds. *Dictionary of Computing*. 4th ed. New York: Oxford University Press.

Thompson, Della, ed. *The Concise Oxford Dictionary of Current English*. 9th ed. New York: Oxford University Press, 1995.

13.6 Incorporating Specialized Terminology

De Sola, Ralph, Dean Stahl, and Karen Kerchelich. *Abbreviations Dictionary*. 9th ed. Boca Raton, FL: CRC Press, 1995.

Hodges, M. Susan. *Computers: Systems, Terms, and Acronyms*. 10th ed. Winter Park, FL: SemCo Enterprises, 1999.

Institute of Electrical and Electronics Engineers. *IEEE-USA*. Webpage. Institute of Electrical and Electronics Engineers. Available: *http://www.ieeeusa.org/*. December 20, 1999.

International Organization for Standardization. *International Organization for Standardization*. 1999. Webpage. International Organization for Standardization. Available: *http://www.iso.ch/*. December 20, 1999.

International Telecommunication Union. *International Telecommunication Union*. 1999. Webpage. International Telecommunication Union. Available: *http://www.itu.int/*. December 20, 1999.

Jane's. *Jane's Defence Terms*. November 26, 1999. Webpage. Jane's. Available: *http://defence.janes.com/defset.html*. December 20, 1999.

Market House Book Editors, eds. *Oxford Dictionary of Abbreviations*. Oxford: Oxford University Press, 1992.

Schneider, G. L., N. G. Roman, and N. Kuin. *Selected Astronomical Catalogs, Vol. 2, #1*. 1999. Webpage bibliography. University of Toronto. Available: *http://astro.utoronto.ca/adc2list.html*. December 20, 1999.

Scientific and Technical Information Network. *Tips for Searching DODISS Online.* 1999. Webpage. Defense Technical Information Center. Available: *http://www.dtic.mil/stinet/htgi/dodiss/dtips.html.* December 20, 1999.

Society for Technical Communication. *Society for Technical Communication.* 1999. Webpage. Society for Technical Communication. Available: *http://www.stc-va.org/.* December 20, 1999.

13.7 Using Numbers and Symbols

American Institute of Physics. *AIP Style Manual.* 4th ed. New York: American Institute of Physics, 1990.

Bureau Internationale des Poids et Mesure. *Bureau Internationale des Poids et Mesure.* 1999. Bureau Internationale des Poids et Mesure. Available: *http://www.bipm.fr/.* December 20, 1999.

———. *The International System of Units (SI).* Trans. available in English and French. 7th ed. Paris: Bureau Internationale des Poids et Mesure, 1998.

CRC Handbook of Chemistry and Physics. Cleveland, OH: CRC Press, published annually.

Gillman, Leonard. *Writing Mathematics Well.* Washington, DC: Mathematical Association of America, 1987.

National Institute for Standards and Technology. *Guide for the Use of the International System of Units (SI) [NIST Special Publication 811].* Washington, DC: National Institute for Standards and Technology, 1995.

———. *International Standards of Units from NIST.* October 1999. Webpage. National Institute of Standards and Technology. Available: *http://physics.nist.gov/cuu/Units/index.html.* December 20, 1999.

———. *The United States and the Metric System.* 1999. Webpage. National Institute of Standards and Technology. Available: *http://ts.nist.gov/ts/htdocs/200/202/lc1136a.htm.* December 20, 1999.

Parker, Sybil P., ed. *McGraw-Hill Dictionary of Mathematics.* New York: McGraw-Hill, 1997.

Sources for Math Symbols and Words Pages. 1999. Webpage bibliography. Available: *http://members.aol.com/jeff570/sources.html.* December 20, 1999.

World Wide Web Consortium. *Mathematical Markup Language (MathML™) 1.01 Specification.* July 7, 1999. Recommendation. World Wide Web Consortium. Available: *http://www.w3.org/TR/REC-MathML/.* December 20, 1999.

———. *W3C's Math Home Page.* September 26, 1999. Webpage. World Wide Web Consortium. Available: *http://www.w3.org/Math/.* December 20, 1999.

13.8 Using Quotations, Citations, and References

International Standards Organization. *Excerpts from International Standard ISO 690-2.* 1999. Webpage. International Standards Organization. Available: *http://www.nlc-bnc.ca/iso/tc46sc9/standard/690-2e.htm.* December 20, 1999.

Li, Xia, and Nancy B. Crane. *Electronic Style: A Guide to Citing Electronic Information.* Westport, CT: Meckler, 1993.

United States Copyright Office. *Digital Millennium Copyright Act.* 1998. Portable document file (.pdf) containing the text of the Digital Millennium Copyright Act. United States Copyright Office. Available: *http://lcweb.loc.gov/copyright/legislation/dmca.pdf.* December 20, 1999.

University of Alberta Libraries. *Citation Style Guides for Internet and Electronic Sources.* August 18, 1999. Webpage. University of Alberta Libraries. Available: *http://www.library.ualberta.ca/library_html/help/pathfinders/style/.* December 20, 1999.

13.9 Creating Indexes

American National Standards Institute. *American National Standards Institute.* 1999. American National Standards Institute. Available: *http://www.ansi.org.* December 20, 1999.

American Society of Indexers. *American Society of Indexers.* 1999. Webpage. American Society of Indexers. Available: *http://www.asindexing.* December 20, 1999.

Aquatic Sciences and Fisheries Thesaurus. *AFIS Reference Series No. 6, Rev. 1.* 1986. Webpage thesaurus. E. Fagett, D. W. Privett, and J. R. L. Sears. Available: *http://www.csa2.com/helpV3/ab.html.* December 20, 1999.

Bartlett, John, and Justin Kaplan, eds. *Bartlett's Familiar Quotations.* New York: Little Brown & Co., 1992.

Bonura, Larry S. *The Art of Indexing.* New York: John Wiley & Sons Inc., 1994.

British Standards Institute. *British Standards Institute.* 1999. Webpage. British Standards Institute. Available: *http://www.bsi.org.uk/.* December 20, 1999.

Chapman, Robert L., ed. *Roget's International Thesaurus.* New York: HarperCollins, 1992.

Fetters, Linda K. *A Guide to Indexing Software.* 5th ed. Corpus Christi, TX: FimCo Books, 1995.

———. *Handbook of Indexing Techniques: A Guide for Beginning Indexers.* 2nd ed. Corpus Christi, TX: FimCo Books, 1999.

International Organization for Standardization. *Information and Documentation—Guidelines for the Content, Organization, and Presentation of Indexes.* Geneva, Switzerland: International Organization for Standardization, 1996.

———. *International Organization for Standardization.* 1999. Webpage. International Organization for Standardization. Available: *http://www.iso.ch/*. December 20, 1999.

Lathrop, Lori. *An Indexer's Guide to the Internet.* Medford, NJ: Information Today, Inc., 1999.

Library of Congress Subject Headings. 21st ed. Washington, DC: Library of Congress, 1997.

McKiernan, Gerry. *Beyond Bookmarks: Schemes for Organizing the Web.* December 9, 1999. Webpage. Iowa State University Library. Available: *http://www.public.iastate.edu/~CYBERSTACKS/CTW.htm.* December 20, 1999.

Medical Subject Headings. Washington, DC: United States Government Printing Office, 1995.

Mulvany, Nancy C. *Indexing Books.* Chicago: University of Chicago Press, 1993.

National Information Standards Organization. *Guidelines for the Construction, Format, and Management of Monolingual Thesauri.* Oxon Hill, MD: National Information Standards Organization, 1993.

———. *National Information Standards Organization Home Page.* 1999. Webpage. National Information Standards Organization. Available: *http://www.niso.org/index.html.* December 20, 1999.

O'Connor, Brian C. *Explorations in Indexing and Abstracting: Pointing, Virtue, and Power.* Englewood, CO: Libraries Unlimited, Inc., 1996.

Sears, Minnie Earl, and Joseph Miller, eds. *Sears List of Subject Headings.* 16th ed. New York: H. W. Wilson, 1997.

Towery, Margie. *Indexing Specialties: History.* Medford, NJ: Information Today, Inc., 1998.

Weinberg, Bella Hass. *Can You Recommend a Good Book on Indexing? Collected Reviews on the Organization of Information.* Medford, NJ: Information Today, Inc., 1998.

Wellisch, Hans H. *Indexing from A to Z.* 2nd ed. New York: H. W. Wilson Co., 1996.

13.10 Creating Nontextual Information

Allen, Nancy. "Ethics and Visual Rhetorics: Seeing's Not Believing Anymore." *Technical Communication Quarterly* 5 (1996): 87–105.

Barton, Ben F., and Marthalee S. Barton. "Modes of Power in Technical and Professional Visuals." *Journal of Business and Technical Communication* 7 (1993): 138–62.

Benson, Phillipa J. "Problems in Picturing Text: A Study of Visual/Verbal Problem Solving." *Technical Communication Quarterly* 6 (1997): 141–61.

Bertin, Jacques. Trans. William J. Berg. *Semiology of Graphics.* Madison: University of Wisconsin Press, 1983.

Bosley, Deborah S. "Study of Gender and Its Influence on Visual Design." *Technical Communication* 40 (1993): 543–47.

Dreyfuss, Henry. *Symbol Sourcebook.* New York: McGraw-Hill, 1972.

Duchastel, P. C. "Research on Illustrations in Text." *Educational Communication and Technology* 28 (1980): 283–387.

Fukuoka, Waka, Yukiko Kojima, and Jan H. Spyridakis. "Illustrations in User Manuals: Preference and Effectiveness with Japanese and American Readers." *Technical Communication* 46 (1999): 167–76.

Horton, William. "New Media Literacy." *Technical Communication* 41 (1994): 562–63.

———. "Visual Literacy: Going beyond Words in Technical Communications." *Technical Communication* 40 (1993): 146–48.

Kostelnick, Charles, and David D. Roberts. *Designing Visual Language: Strategies for Professional Communicators.* Boston: Allyn and Bacon, 1998.

Lohse, Gerald, Neff Walker, Kevin Biolsi, and Henry Rueter. "Classifying Graphical Information." *Behaviour and Information Technology* 10.5 (1991): 419–36.

Modley, Rudolph. *Handbook of Pictorial Symbols.* New York: Dover Publications, Inc., 1976.

Murch, Gerald M. "Using Color Effectively: Designing to Human Specifications." *Technical Communication* 32.4 (1985).

Tanner, Beth, and Pete Larson. "Worth a Thousand Words: Choosing and Using Illustrations for Technical Communication." *Technical Communication* 41 (1994): 150–57.

Williams, Thomas R., and Deborah A. Harkus. "Editing Visual Media." *IEEE Transactions on Professional Communication* 41 (1998): 33–47.

13.11 Creating Usable Data Displays

Beninger, James R., and Dorothy L. Robyn. "Quantitative Graphics in Statistics: A Brief History." *The American Statistician* 32.1 (1978): 1–11.

Huff, Darrel. *How to Lie with Statistics.* New York: W. W. Norton & Company, 1954.

Neurath, M. "Isotype." *Intercultural Science* 3 (1974): 127–50.

Tufte, Edward. *Envisioning Information.* Cheshire, CT: Graphics Press, 1997.

———. *The Visual Display of Quantitative Information.* Cheshire, CT: Graphics Press, 1983.

———. *Visual Explanations: Images and Quantities, Evidence and Narrative.* Cheshire, CT: Graphics Press, 1997.

Wainer, Howard. *Visual Revelations.* New York: Springer-Verlag, 1997.

Winn, William. *Encoding and Retrieval of Information in Maps and Diagrams.* Boston: American Educational Research Association, 1990.

13.12 Designing Useful Documents

Bringhurst, Robert. *The Elements of Typographic Style.* 2nd. ed. Vancouver, BC: Hartley & Marks, 1997.

Burt, Cyril. *A Psychological Study of Typography.* Cambridge: Cambridge University Press, 1959.

Hartley, James. *Designing Instructional Text.* New York: Nichols Publishing Company, 1977.

Hurlburt, Allen. *The Grid.* New York: Van Nostrand Reinhold, 1978.

Keyes, Elizabeth. "Typography, Color, and Information Structure." *Technical Communication* 40 (1993): 638–54.

Kostelnick, Charles. "Supra-Textual Design: The Visual Rhetoric of Whole Documents." *Technical Communication Quarterly* 5 (1996): 9–33.

Müller-Brockmann, Josef. *Grid Systems in Graphic Design.* New York: Visual Communication Books, 1985.

Schriver, Karen A. *Dynamics in Document Design: Creating Texts for Readers.* New York: John Wiley & Sons, 1997.

Skillin, Marjorie E., and Robert Malcolm Gay. *Words into Type.* 3rd ed. Upper Saddle River, NJ: Prentice Hall, 1974.

Sullivan, Patricia. "Visual Markers for Navigating Instructional Texts." *Journal of Technical Writing and Communication* 20 (1990): 255–67.

Tinker, M. A. *Bases for Effective Reading.* Minneapolis: University of Minnesota Press, 1965.

———. *Legibility of Print.* Ames: Iowa State University Press, 1963.

Tschichold, Jan. Trans. Hajo Hadeler. *The Form of the Book: Essays on the Morality of Good Design.* Vancouver, BC: Hartley & Marks, 1991.

Williams, Thomas R., and Jan H. Spyridakis. "Visual Discriminability of Headings in Text." *IEEE Transactions on Professional Communication* 35 (1992): 64–70.

Wright, Patricia. "Presenting Technical Information: A Survey of Research Findings." *Instructional Science* (1977).

Zachrisson, B. *Studies in the Legibility of Printed Text.* Stockholm: Almqvist & Wiksell, 1965.

Index

Numerics

24-hour scale 4.59, 11.88–89
100 percent column charts 11.241
100 percent surface charts 11.229

A

A4 bilingual and multilingual documents 12.86–91
Abbreviations. *See also* Acronyms; Symbols
  apostrophes and 4.115–16
  considerations 6.1
  examples 6.4
  explaining 6.24–26
  genus names 6.77
  health and medical terms 6.1, 6.68–79
  indexing 9.31, 9.102–04
  initial capital letters in 5.49
  multilingual terminology database 2.91
  organization names 6.1, 6.48–68
  parentheses and 4.128
  periods and 4.71–74
  as professional shorthand 6.1, 6.2–39
  punctuation 6.5, 6.33–39
  scientific terms 6.1, 6.80–124
  selection considerations 6.19–26
  in tables 11.104
  telecommunications 6.169
About This Manual section, design decisions 12.78
Absolute modifiers 3.30
Absolute phrases 3.4
Abstract nouns 3.8
Abstracts 12.57–58, 12.75
Academic degrees, capitalization 5.50
Academic organizations 6.41–45
Acknowledgments, design decisions 12.74
Acronyms. *See also* Abbreviations
  beginning sentences with 6.49
  considerations 6.9
  examples 6.11–13
  indexing 9.31, 9.102–4
  indirect articles and 6.14
  multilingual terminology database 2.91
  parentheses and 4.129
  telecommunications 6.169

Action verbs 6.156
Active electronic components 6.161–62
Active voice
Controlled English 2.7
topical shifts 3.50
Adjectives. *See also* Adverbs
avoiding verb suffixes 2.109
capitalization 5.47
common problems 3.31–32
compound 2.82
compound terms serving as 4.66
expanding into relative clauses 2.87
foreign phrase used as 4.97
for generic descriptions of the armed forces 6.60
-ing words 2.68
multiple common meanings 2.99
as nouns, avoiding 2.110
predicate adjectives 4.98–99
in series 4.29–30
unit modifier containing 4.85–87
Adverbs. *See also* Adjectives
avoiding adverbial interrupters 2.42–44
common problems 3.31–32
conjunctive 2.44, 3.49
multiple common meanings 2.99
transitional 4.47
Agreement of pronouns 3.23
Aircraft terminology 6.63
Algebraic symbols 7.50, 7.53
All-capital abbreviations 4.116
All-capital text, design decisions 12.37
"all," Global English guidelines 2.21
Alphabetic indexes 9.27–31, 9.107
Ambiguity
avoiding 2.38
collective nouns 3.26
present participles 2.64
pronoun references 3.17
"scope of conjunction" problems 2.71
"scope of modification" problems 2.72
sentence structure problems 2.15
"stacked nouns" 2.81, 3.15–16
American Chemical Society, abbreviations 6.36
American Society of Mechanical Engineers, abbreviations 6.35
American spelling, compared to British spelling 5.65. *See also* English language; Spelling
American Standard Code for Information Interchange (ASCII) 6.171, 9.33–34, 9.66
Amplification
colon and 4.53–55
dash and 4.62
em dash and 4.63
Analytic geometry, symbols 7.53
Anatomical terms 6.72
"and" article
compound pronouns and 3.18–20
Global English 2.71–76
Animation, software characteristics 11.24
Annotations, in charts 11.175–77
Antecedents
compound pronoun 3.18–20
pronouns and 2.20
singular pronouns and 3.24
Apostrophe 4.3, 4.108–16
contractions and 4.114
definition 4.108
physics terms 6.118
plurals and 4.115–16
plurals of initialisms 6.38–39
in possessive nouns 3.10
possessives and 4.109–13
Appendixes 1.78, 10.60, 12.80–82
Application packages 6.135–36
Appositives
abbreviations and 4.115–16
characteristics 3.4
commas and 4.22–24
definition 3.12
restrictive 8.79
Arabic numerals, in tables 11.84
Archived sources 12.99

Area designation, for charts 11.139–41
Armament, types of 6.64
Armed forces, generic descriptions of 6.60
Articles
  audience considerations 1.43
  communication characteristics 1.37
  design decisions 12.54–62
  indirect 6.14
  information classes and 12.8
  specialized vocabulary 1.44
Articles (*a*, *an*, *the*) 2.53–55
"*assign*" verb 2.84
Association names 6.57
Astronomical naming conventions 6.100–1
Atomic particles 6.121–25
Atomic weight, abbreviation 6.35
Audience analysis 1.1–10. *See also* Documents; Genres; Media selection
  articles 1.43
  booklets 1.46
  brochures 1.47
  citation methods and 8.87–88
  classifying documents and 12.109
  conceptual documents 1.23
  document design and 12.114
  formal 1.2–3
  genre considerations 1.38
  indexing and 9.8–10
  informal 1.2, 1.4
  job aid documents 1.32
  marketing documents 1.20
  media considerations 1.39, 1.44
  newsletters 1.49
  procedural documents 1.26
  referential documents 1.35
  source materials 1.126
  tutorials 1.29
  typeface considerations 12.27
  usage patterns 1.131
  user guides/operation manuals 1.73
Audience characteristics 1.5. *See also* Non-native audiences
Audience objectives 1.6–9
Audience profile 1.10, 1.76
Audience training 1.14–16
Author-date citation 8.95–100
Authors
  audience assumptions 1.123
  indexers and 9.12

B

Back cover, design decisions 12.85
Band surface charts 11.226–28
Base line, in charts 11.173
Bayer method for "naked eye" stars 6.93
Bibliographical notes 8.121–25. *See also* Citations
Bibliography software applications 8.90
Bilingual documents 12.86–91
Binding, design decisions 12.125–33
Biological terminology 6.102–3
Block quotations
  capitalization 8.70–71
  definition 8.19
  omitting parts of 8.54–58
  setting in text 8.40–45
Body text 12.32, 12.60
Boilerplate (standard) paragraphs 3.56, 3.57
Boldface
  correct usage 7.60
  in data displays 11.5
  in indexes 9.49
  type sizes and 12.37
Bond paper 12.124
Booklets
  audience considerations 1.46
  communication characteristics 1.37
  information classes and 12.8
  length 1.45
  size considerations 12.12
Borders
  for charts 11.181
  for tables 11.68
Braces, typographic details 7.79

Brackets
breaking equations and 7.73
clarifying notes in 8.53
definition 4.134
editorial interjections and 4.135
for endnote numbers 8.112
references and 4.129
typographic details 7.79
Brand names, capitalization 6.75
Brightness, of printing paper 12.122
British spelling, compared to American spelling 5.65. *See also* English language; Spelling
Brochures
audience considerations 1.47
communication characteristics 1.37
design decisions 12.52–53
information classes and 12.8
length 1.48
size considerations 12.12
Business research reports
purpose 1.91
sections 1.92

C

Calculus, symbols 7.53
Callouts, for illustrations 12.32, 10.68–69
Camera-ready pages 12.120–21
Capitalization. *See also* Spelling
all-cap abbreviations 4.116
amplification and 4.55
beginning sentence 5.39–40
column headings 11.32
computer documentation 5.56–57
copyright 5.52
in data displays 11.6
data names 6.150
drug terminology 6.74–75
geology terms 6.116
government and legislative names 6.55–56
in indexes 9.50
infectious organisms 6.76
listings 5.41
lunar features 6.88
organization names 6.40
parentheses and 5.42
physics terms 6.118
planet features 6.86
planet names 6.85
proper nouns 5.47–51
quotations 5.43, 8.66–71
registered names 5.52
in row headings 11.57
salutations and closings 5.44
small caps 8.98
software development documentation 6.148
trade names 5.52
trademarks 5.52
Captioning illustrations 10.62–67, 12.32
Carbon copy list 1.54
caret marks, indicating quotations 4.122
Catalogs, astronomical naming conventions 6.100–1
Causative construction, "*get*" 2.32
Cautions 12.43–44
"*-ce*" word ending 5.31
"*-cede*" word ending 5.38
"*-ceed*" word ending 5.38
Chapters, design decisions 12.79
Charts 11.215–17. *See also* Data displays; Diagrams; Graphs; Illustrations; Tables
abbreviations in 11.185
annotations 11.175–77
area or volume designation 11.139–41
borders 11.181
characteristics 11.121–32
components 11.133, 11.134
correct usage 11.118–20
curves in 11.145, 11.149
data values on 11.135–41, 11.174
design considerations 11.132
grid lines 11.172
inserts 11.178–79
integrated with text 11.11

keys and legends 11.182
lines in 11.145–48, 11.184
plotting data 11.138
purpose 11.117
relationships in data 11.3
scales for 11.150–71
selection considerations
100 percent column chart 11.241
100 percent surface charts 11.229
band surface charts 11.226–28
column/bar 11.231–39, 11.248–57
cumulative curve 11.221–22
curve charts 11.218–19, 11.223–25
frequency distribution 11.220
high-low 11.240
histogram 11.242
matrix charts 11.116
net-difference surface charts 11.230
paired-bar 11.246
pie charts 11.195, 11.196, 11.258–66
scale charts 11.205–7
scatter charts 11.208–10
separate segment charts 11.267
sliding-bar 11.247
surface charts 11.223–25
triangular 11.211–14
special effects 11.186–99
textual features 11.183–84
time lines 11.243–45
types of 11.204
Chemistry terminology 6.104–13, 7.43, 7.52
Circuit technologies 6.158
Circular charts 11.215–17
Citations. *See also* Copyright; Quotations; References
considerations 8.86–87
electronic sources 8.126–31
ethical considerations 8.1, 8.14
footnotes 8.101–6
guidelines 8.12–16
in-text 8.95–100
legal guidelines 8.4–6
Classified index 9.28–29
Classifying documents 12.108–12
Classroom tutorials 1.69
Clauses
adverbial interrupters, avoiding 2.42
conditional 2.85, 2.86
dangling participles and 3.35
dependent
characteristics 3.5
common problems 3.33
compound predicates in 4.7
as fragments 3.42
if 2.86
independent
characteristics 3.5
common problems 3.33
introductory 4.9–10, 4.11
joined by conjunctions 4.45–46
joined without conjunctions 4.44
logical relationship 2.14, 2.43
nonrestrictive 4.13
relative 2.25, 2.60–62, 2.64, 2.87
restrictive 4.13–14
in series 4.19, 4.51–52
types of 3.5
Clock time 7.8
Closing quotation marks 4.119–21
Coated paper 12.124
Cognitive performance 1.132
Collecting subject information
audience considerations 1.123
interviewing sources 1.122, 1.127–29
types of 1.121
written source materials 1.121, 1.124–26
Collective nouns
as pronoun antecedents 3.20
verb agreement with 3.25
College colors, capitalization 5.50
Colloquialisms, avoiding 2.102
Colon 4.3, 4.50–66, 4.127–37
amplification and 4.53–55
complete sentence following 5.40
definition 4.50

Colon *(continued)*
  elements in a series and 4.51–52
  in indexes 9.100
  long quotation and 4.56
  in mathematical equations 7.83
  ratios and 4.57
  time of day and 4.58–60
Color combinations
  in charts 11.187–93
  design decisions 12.37
  graphic specifications 12.105
Color, of printing paper 12.122
Column/bar charts 11.231–39, 11.248–57
Column headings, in tables 11.29–50
Columns, in indexes 9.43
Comb binding 12.131
Comets 6.89
Commas 4.3, 4.4–42
  adjectives in series and 4.29–30
  appositions and 4.22–24
  in company names 6.51
  compound predicates and 4.6–8
  in compound sentences 4.5
  contrasting phrases and 4.25
  dates and 4.33–36
  definition 4.4
  deliberate omissions and 4.26–27
  direct quotations and 4.31–32
  ending quotations 8.81
  in indexes 9.52
  introducing quotations 8.79
  introductory expressions after 4.11–12
  with introductory phrases or clauses 4.9–10
  items in series 4.16–18
  in large numbers 7.14
  numbers and 4.38–41
  phrases and clauses in series 4.19
  in place of semicolons 4.46
  within quotation marks 8.80
  reference numbers and 4.42
  restrictive phrases or clauses and 4.13–14
  similar or identical words and 4.28
  specifying names with 4.36–37
  transitional words or phrases and 4.20
  transposed terms and 4.21
Commercial documents 1.43
Commercial magazines, characteristics 12.12
Commercial organizations 6.46–54, 9.51
Commercial software 6.135–36
Commercial websites 1.117
Communication media characteristics 1.37. *See also* Audience analysis; Information communication characteristics
Complete sentences
  introduced by colons 4.55
  introducing block quotations 8.55
  in lists 3.47, 4.70
Complex sentences 3.6
Component edits 3.39
Compound adjectives 2.82
Compound/complex sentences 3.6
Compound numbers, spelled out 7.22
Compound predicates 4.6–8
Compound pronoun antecedents 3.18–20
Compound sentences
  characteristics 3.6
  commas in 4.5
  coordinating conjunctions in 4.12
  semicolons and 4.44
Compound subjects 3.28–29
Compound technical terms 4.103
Compound terms, serving as adjectives 4.66
Compound verbs 4.99
Computer-assisted indexing
  considerations 9.59–60
  high-end publishing packages 9.59, 9.74–75
  markup languages 9.59, 9.71–73
  recommendations 9.76–81
  stand-alone programs 9.59, 9.67–70
  with word processors 9.59, 9.61–62

Computer screens
  resolution 12.13
  "safe" typefaces 12.28
  white space and 12.20
Computer terminology
  commands 4.124, 6.140–41
  hardware 6.132–33
  mainframe computer software 6.138
  messages 6.143
  personal computer software 6.137
  procedures 6.154–56
  programming languages 6.151–55
  software development documentation 6.144–50
  software terminology 6.134–40
  trends 6.129–31
Conceptual documents
  audience considerations 1.23
  purpose 1.18, 1.22
  textual features 1.24
Conclusion, for articles 12.61
Concrete nouns 3.8
Conditional clauses 2.85
Conditional tense modal verbs 2.35
Conjunctions
  avoiding verb suffixes 2.109
  clauses joined by 4.45–46
  clauses joined without 4.44
  commas before 4.8
  coordinating 3.49, 4.5
  introductory expressions after 4.11–12
  joining clauses 4.45–46
  role in sentences 3.4
  subordinating 2.85
  temporal 2.67
  triage edits 3.40
Conjunctive adverbs 2.44, 3.49
Connection maps 11.289
Connectives, triage edits 3.40. *See also* Conjunctions
Consecutive words, abbreviating 6.22
Consistency
  abbreviations 6.6
  editing for 3.48–59
  hyphenation 4.87
  technical manuals 12.63
  terminology 6.154
Consonant endings 5.34–37
Constellations 6.91
Context-sensitive help systems 1.101
Continued tables 11.96–100
Contractions 4.114
Contrasting phrases 4.25
Controlled English 2.1, 2.4–9
  cost-effectiveness 2.5
  developing 2.6
  limitations 2.4, 2.7, 2.92
  options 2.9
  Simplified English 2.8
Conventional word combinations 2.96
Conversations, quoting 8.27
Coordinating conjunctions
  in compound sentences 4.5
  joining clauses 4.45–46
  logical transitions 3.49
Copyright 5.52, 5.54. *See also* Citations
Copyright page, design decisions 12.73
"Core semantic meaning" 2.92
Corporate names 5.53
Correspondence
  carbon copy list 1.54
  communication characteristics 1.37
  components 1.51, 1.53
  cover letters 1.55
  quotations 8.21, 8.24, 8.31–32
  reviewing 1.53
  subject line 1.52
Cost, of printing paper 12.122
CRC Handbook of Chemistry and Physics 7.43, 7.52
Cross-platform compatibility 1.41
Cross-references 9.48, 9.96–99
Culture-specific ideas, avoiding 2.106
Cumulative curve charts 11.221–22
Currency symbols, in tables 11.86
Curve charts 11.218–19, 11.223–25
Curves, in charts 11.145, 11.149
Cut-in heads 11.43–47

D

Dangling modifiers 3.36
Dangling participles 3.35
Dash 4.3, 4.50–66, 4.127–37. *See also* Em dash; En dash; Hyphen; Slash
Data displays. *See also* Charts; Diagrams; Illustrations; Tables
  characteristics 11.4–21
  creating 11.22–25
  definition 11.1, 11.2
  design considerations 11.16–17
  as in-text notes 8.97–100
  indexing 9.57
Data names 6.151
Data plotting 11.162–64
Data values, in charts 11.135–41, 11.174, 11.238. *See also* Values
Dates
  capitalization 5.50
  commas and 4.33–36
  guidelines 7.9–10
  in tables 11.87
Decimal marker 7.13
Decimals
  digits to the right of 4.39
  international considerations 4.41
  in scientific notation 7.28
  in tables 11.85
  zeros and 7.4
Decision tables 11.111–12
Decked heads 11.39–42
Deliberate omissions 4.26–27
Delivery methods 1.38–44
Dependency diagrams 11.290
Dependent clauses
  characteristics 3.5
  common problems 3.33
  compound predicates in 4.7
  as fragments 3.42
Deviation column/bar charts 11.255–56
Device states, telecommunication devices 6.171–72
Diagrams 11.268–90. *See also* Charts; Data displays; Graphs; Illustrations; Tables
  components
    links 11.276–80
    symbols 11.272–75
    types of 11.271
  integrated with text 11.11
  purpose 11.268–70
  relationships in data 11.3
  types of 11.281–90
Dialogue, quoting 8.27–28
Diction 5.60
Dictionaries. *See also* English language; Spelling
  considerations 5.1
  indexing and 9.13
  local 6.6–7
  in spelling checkers 5.6, 5.8
The Digital Millennium Copyright Act 8.2
Digits. *See* Numbers
Direct object, with intransitive verb 2.50
Direct questions 4.78
Direct quotations
  commas and 4.31–32
  definition 8.17
  guidelines 8.21, 8.22–23
Displaying illustrations 10.14–17
Distance tables 11.114–15
Distributing documents. *See* Delivery methods
Divided column/bar charts 11.252–54
Document design
  considerations 12.3–4
  graphic production
    camera-ready pages 12.120–21
    considerations 12.119
    paper selection 12.122–24
  large document sets
    accessibility 12.113–18
    categorizing information 12.108–12
    considerations 12.107
  page and screen
    considerations 12.39
    design decisions 12.38
    formal format elements 12.10

formatting 12.9
grids 12.48–50
leading or line spacing 12.33–34
line width 12.35–37
lists 12.46
margins 12.23–25
page numbering 12.39–40
revision notices 12.47
running headers and footers 12.41–44
size considerations 12.11–17, 12.38–50
source notes and references 12.45
typefaces 12.26–30
white space 12.18–22
planning 12.1
purpose 12.2
specific information types
A4 bilingual and multilingual documents 12.86–91
articles 12.54–62
brochures and newsletters 12.52–53
on-screen information 12.92–106
technical manuals 12.63–85
technology and 12.5–8
Documents. *See also* Manuals; *specific types of documents*, e.g. Articles
binding 12.125–33
classifying by type 1.17
communication characteristics 1.37
computer documentation 5.56–57
conversion problems 4.2
delivery methods 1.38
format incompatibilities 1.41
genres 1.12, 1.17, 1.18
paper size 12.15
pattern of use 1.15
planning
activities 1.120–39
parallelism 3.59
structure 1.119
purpose and structure 1.13
software considerations 1.42
source materials 8.92–94
textual features 1.16
typology 1.18, 1.37–38
Dot charts 11.208–10
Dot leaders 11.55–56
Dot pitch 12.13
Double hyphen 4.61
Double-posting index entries 9.89
Drop shadows, in charts 11.197
Drug terminology 6.74–75
Durability, of printing paper 12.122

E

EDGAR Database 6.54
Editing. *See also* Grammar; Spelling
brackets for editorial interjections 4.135
ensuring consistency 3.48–59
illustrations 10.40
index entries 9.115
levels of 3.39
spelling checkers and 5.3–8
static information 12.102
Educational websites 1.117
Electron shells 6.124–25
Electronic components
active 6.161–62
passive 6.159–60
Electronic documents. *See also* Help systems
circuit technologies 6.158
delivery considerations 1.38
information classes and 12.8
information design 12.14
purpose 1.116
Electronic mail (e-mail)
document delivery considerations 1.38
informality 1.60
purpose 1.59
Electronics terminology
integrated circuits 6.167
passive electronic components 6.159–60
trends 6.157

Elementary geometry, symbols 7.53
Elements, horizontal spacing 7.61, 7.62–66
Ellipses
in mathematical expressions 4.77
in quotations 4.76, 8.63–65
Em dash
in indexes 9.52
other punctuation and 4.63–64
word-processing software and 4.2
Embedded indexing 9.60, 9.63–66, 9.74, 9.79
Emoticons 4.138
Emotional statements 4.82
Emphasis 5.40
En dash
in indexes 9.52, 9.100
range of numbers and 4.42, 4.65, 11.90
two-noun combinations and 4.66
word-processing software and 4.2
En space, separating trigonometric and hyperbolic functions 7.63
Encyclopedic information 1.80
Endnotes 8.107–16. *See also* Footnotes
Engineering, metric conversion policies 7.41
English language. *See also* Foreign languages; Spelling; Translation
flexibility 2.45
homonyms 5.58–61
multilingual terminology database 2.91
sentence types 3.6
spelling variations 5.62–64
verb complements 2.46–48, 13.2
Enumeration, en dash 4.63
Epigraphs 8.21, 8.24, 8.25–26
Equals sign, breaking equations 7.72, 7.73
Equations
breaking 7.72–76
colon in 7.83
placement 7.67–71
punctuating 4.139, 7.82–84
Equipment
hardware terminology 6.132–33
medical 6.78
surveillance 6.68
test results 5.54
Equivalent expressions 2.78
Error messages 6.143
Error recovery information 1.77
European Association Aerospace Industries 2.8
Exclamation mark 4.3, 4.81–82, 4.121
Exponents 7.29
Expressions, equivalent 2.78
Extranet 1.118

F

Facsimile documents, delivery considerations 1.38
Feasibility studies, sections 1.96
Field (table body) 11.58–61
Fields 11.43–47
Figurative language, avoiding 2.103
Figures 8.97, 12.77. *See also* Illustrations
File formats
compatibility considerations 1.42, 11.24
graphic specifications 12.105
Finish, of printing paper 12.122
"First aid" edits 3.40
Floppy disks 1.38
Flowcharts 11.286–88
Focus groups 1.137
Footnotes. *See also* Endnotes
guidelines 8.101–6
indexing 9.100
quotations 8.21, 8.24, 8.33
Foreign audiences. *See* International audiences; Non-native audiences
Foreign terms. *See also* International audiences; Spelling; Translation
adjectives 4.97
in indexes 9.105
phrases 5.67
plurals of 5.14–18

Formal audience analysis 1.2–3
Formal introduction, of quotations
  8.43–44
Formal outlines 1.135–36
Formal titles 5.50
Formal writing
  foreign terms and 5.16
  spelling 5.64
Formatting
  abbreviations 6.27–28
  indexes 9.40–53
  page and screen 12.9
  technical manuals 12.65–85
Formulas
  chemical reactions 6.113
  in tables 11.91
Four-dot ellipsis 8.64
Fractions
  mixed numbers and 7.24
  spelled out 7.19, 7.23
  writing 7.11–12
Fragments
  dependent clauses as 3.42
  in lists 3.45–47
  phrases as 3.43
Frequency distribution charts 11.220
Frequency of use, design decisions
  12.110
Front cover, design decisions 12.68
Front matter, design decisions
  12.69–70
Fronting modifiers 2.43
Fronting prepositional phrases 2.27
Fronting sentence constituents 2.28
FTP (file transfer protocol) 1.40
Function words, syntactic cues
  2.56–58

G

Galaxies 6.99–101
"-*ge*" word ending 5.31
Generic names 6.74
Genres 1.12, 1.17, 1.18. *See also*
  Audience analysis; Manuals
Genus names 6.77
Geographical location 4.36–37
Geology terminology 6.114–16
Geometric scale, in charts 11.170–71
Geometry, symbols 7.53
Gerunds 2.63–70, 3.13, 6.156
"*get*" construction
  causative 2.32
  passive 2.31
GIF (Graphics Interchange Format)
  12.105
"*give*" verb 2.84
"*given*"/"*given that*" construction 2.33
Global English 2.1, 2.10–111
  guidelines, types of 2.11
  parts of speech 2.92
  qualities 2.10
  sentence structure guidelines 2.8,
    2.9, 2.12–55
  syntactic cues procedure guidelines
    2.56–76
  terminology guidelines 2.88–111
Glossary, design decisions 12.83
Government names 6.55–56
Grammar. *See also* Editing
  basics 3.1, 3.3–6
  common problems 3.1, 3.7–37
  Global English 2.30–37
  goals 3.1
  grammatically or semantically
    incomplete words 2.111
  parts of speech 3.4
  revision strategies 3.1, 3.38–59
  style guides 3.2, 3.38
Grammar checkers 4.2
Graphic specification, on-screen
  information 12.105–6
Graphical user interfaces 6.145
Graphics, component edits 3.39
Graphs. *See also* Charts; Diagrams;
  Illustrations; Tables
  distortion in 11.127
  numbers in 7.36–37
  plotting data 7.39
  in tables 11.92
Greek letters
  correct usage 7.59

Greek letters *(continued)*
identifier with constellation names 6.93
indexing 9.113
Grid lines, in charts 11.172
Grid linkage, in diagrams 11.280
Grids, page and screen 12.48–50
Grouped column/bar charts 11.248–51
Guideposts 1.131
Guides. *See* Manuals

H

Half-title page 12.71
Hands, in illustrations 10.50
Hands-on evaluation 1.130
Hanging indent style, in indexes 9.42
Hardware terminology 6.132–33
Harvard observatory system, star names 6.96
"*have*" causative construction 2.31
Head-stub-field separators, in tables 11.69
Headings
in index entries 9.85–92
type sizes, in documents 12.32
Health terminology. *See* Medical terminology
Help systems
accessing 1.99
communication characteristics 1.37
complex 1.103
context-sensitive 1.101
layers 1.102
navigation 1.100
purpose 1.97
space considerations 1.98
Heuristics 1.137
High-low charts 11.240
Highlighting
illustrations 10.51–55
techniques 12.37
Histogram charts 11.242
Holidays, capitalization 5.50
Home page 1.116, 12.97
Homonyms 5.5, 5.58–61
Honorary titles, capitalization 5.50
Horizontal scale, for charts 11.156
"House" style 3.39
Human translators, ambiguity problems 2.39
Humanities, style guides 8.88
Humor, avoiding 2.105
Hybrid newsletters 1.49
Hyperbolic functions, symbols 7.53
Hyperlinks 1.116
Hyphen 4.3, 4.83–105. *See also* Dash; Slash
in compound adjectives 2.82
definition 4.83–84
foreign phrase used as adjective 4.97
in indexes 9.100
in noun strings 2.19, 2.81, 4.83
with numbers 7.20–26
numbers and 4.102
predicate adjectives 4.98–99
prefixes and suffixes 4.104–5
quantum numbers 6.125
self- compounds, 4.100
single letter and 4.96
suspended 4.101
technical terms and 4.103
unit modifier
containing a proper name 4.93
containing adjectives and 4.85–87
containing adverb and hyphen 4.90–92
containing number and unit of measure 4.88–89
longer than two words 4.94–95
word division guidelines 5.25–27

I

Icons
cultural considerations 10.35–36
iconic characters 4.138
in illustrations 10.28
Identical words 4.28

Idiomatic two- or three-word verbs 2.100
"if" clause 2.85, 2.86
Illustrations. *See also* Data displays; Diagrams; Tables
  acquiring 10.4, 10.5, 10.6–11
  color considerations 10.55
  component edits 3.39
  considerations 10.4, 10.7, 10.42
  constructing 10.37–41
  cultural considerations 10.50
  em dash and 4.62
  hands in 10.50
  highlighting 10.51–55
  in-text notes 8.97–100
  indexing 9.57
  infographics 10.26
  labeling 10.61–74
  line drawings 10.22–25
  Minimum Word Strategy 2.3
  numbering 10.56–60
  photographs 10.19
  positioning 10.46–48
  preceded by a colon 4.50
  printing or displaying 10.5, 10.14–17
  processing 10.5, 10.12
  renderings 10.20–21
  resolution 10.8–11
  selection considerations 10.18
  software characteristics 11.24
  symbols 10.27
  type legibility for text 10.43–45
  type sizes and 12.32
  typefaces 12.27
  views 10.49
  white space and 12.18
Images. *See* Illustrations
Imperative sentences 3.21
Imperative verbs 6.155
Implied *you* 3.21
"in order to" construction 2.40
In-text citations 8.95–100
"in that" construction 2.34
Inanimate objects 2.51
Incomplete sentence 8.60
Indented indexes 9.41, 9.42
Independent clauses 3.5
Index
  acronyms 9.31, 9.102–4
  definition 9.5
  design decisions 12.84
  document design and 12.116
  example 9.53
  formatting 9.40–53
  goals 9.6
  master index 12.116, 12.118
  planning 9.7–25
  type choices 9.47–51
  user guides/operation manuals 1.79
Index cards, preparing nonlinear outlines 1.138
Index lines, in charts 11.173
Indexers
  access to source files 9.75–76
  characteristics 9.11–14
Indexing
  computer-assisted
    considerations 9.59–60
    high-end publishing packages 9.59, 9.74–75
    markup languages 9.59, 9.71–73
    recommendations 9.76–81
    stand-alone programs 9.59, 9.67–70
    with word processors 9.59, 9.61–62
  key terms 9.4
  phases of 9.1–2
  preparation time 9.18–20
  procedures
    considerations 9.54
    iterative steps 9.55
    selecting entries 9.56–58
    writing and editing entries 9.82–115
  sorting methods 9.26–39
  special considerations 9.102–4
  standards 9.3
  style guides 9.14, 9.21–25
Indirect articles 6.14

Indirect questions, punctuating 4.80
Indirect quotations
definition 8.18
guidelines 8.35–37
introducing 8.45
Infectious organisms 6.76–77
Infinitive phrases
ambiguity problems 2.39
Global English 2.26, 2.63
Infinitives 3.14
Infographics 10.26
Informal audience analysis 1.2, 1.4
Informal introduction, of quotations 8.42
Informal language, avoiding 2.103
Informal outlines 1.137
Information analysis 1.11–13
Information characteristics. *See also* Audience analysis
document typology 1.18
identifying 1.11–13
retrievability 11.14–15, 11.18–21
Information classes 12.8, 12.94–95
Information structure, parallelism 3.58
Informational newsletters 1.49
Informational websites 1.117
-ing words 2.63–70
Initial capital letters
in abbreviations 5.49
in data displays 11.6
in indexes 9.50
Initialisms. *See also* Abbreviations; Acronyms
beginning sentences with 6.49
considerations 6.9
examples 6.3, 6.10
multilingual terminology database 2.91
omitting periods in 4.73
plurals of 6.38–39
Inserts, in charts 11.178–79
Installation instructions, parallelism 3.55
Integrated circuits 6.167
Intensifying modifiers 3.30
Interjections, role in sentences 3.4
Internal newsletters 1.49
International audiences. *See also* Non-native audiences; Translation
A4 bilingual and multilingual documents 12.86–91
illustrations for 10.29–36
imperative sentences, cultural implications 3.21
large numbers and 7.15
quotations 8.21, 8.24, 8.34
terminology guidelines 2.88
International System of Units 7.41, 7.44
Internet
accessing 1.118
citations 8.128
Interviewing sources
guidelines 1.129
hands-on evaluation 1.130
preparation 1.128
purpose 1.122, 1.127
Intranet, accessing 1.118
Introduction, for articles 12.59
Introductory notes, in indexes 9.114
Inverted question mark 4.2, 4.79
Inverted sentences, Global English guidelines 2.37
Irony, avoiding 2.105
"*-ise*" word ending 5.36
"*it*" pronoun
common problems 3.17
Global English guidelines 2.20
Italic type
correct usage 7.58
in data displays 11.5
design decisions 12.27
in indexes 9.49
quantum numbers 6.123
"*-ize*" word ending 5.36, 5.37

J

Job aid documents
action-oriented approach 1.87
purpose 1.18

task cards 1.88
task information 1.86
types of 1.31
JPEG (Joint Photographic Experts Group) 12.105
Justification
margins 12.25
row headings 11.53

K

Keys, for charts 11.182
Keyword list 9.21–25

L

Labels
for links in diagrams 11.279
for scales in charts 11.154–56
Large numbers 7.13–15
Latin names
abbreviated 4.72
in indexes 9.106
medical terms 6.69, 6.71
novas 6.98
stars 6.93
Leaders (dot leaders) 11.55–56
Leading 12.22, 12.33–34, 12.36
Legal issues, component edits 3.39
Legends, for charts 11.182
Legislative names 6.55–56
Letter-by-letter sorting 9.36, 9.70
Letters 1.55–56, 5.44–45
Line-by-line indexes. *See* Indented indexes
Line drawings 10.22–25
Line spacing 12.33–34
Line width 12.35–37
Linear scales, in charts 11.166
Lines, in charts 11.145–48, 11.184
Linked sources 12.99
Links, in diagrams 11.276–80
List of figures, design decisions 12.77
Lists
in appendixes 1.78
definitions of variables 7.84
design decisions 12.46
fragments used in 3.45–47
keywords 9.21–25
numbered
for endnotes 8.111
parentheses and 4.130–31
reference lists
bibliography 8.121–25
guidelines 8.117–20
vertical 4.68–70
Local dictionaries 6.6–7
Locators, in indexes 9.49, 9.91–92, 9.100
Logarithmic scale, in charts 11.167–69
Logic, symbols 7.53
Logical transitions, between grammatical structures 3.49
Long column headings 11.34
Long noun phrases 2.17–18, 2.81
Long quotations 4.56
Long sentences
Global English 2.14
parallelism 3.54
Long tables 11.102
Look-up-a-value tables 11.110
Loose-leaf ring binding 12.129
Lowercase letters
apostrophes and 4.115
chemical compounds 6.108, 6.110
in data displays 11.6
drug terminology 6.74
for elements and vitamins 5.49
geological features 6.113
listings 5.41
measurements 5.48
medical conditions 6.74
minerals 6.115
names of atomic and nuclear particles 6.122
naval 5.50
naval vessels 6.66
plural form of military terminology 6.63
software development terms 6.147

M

Machine translation systems
  ambiguity problems 2.39
  intransitive verbs 2.50
  syntactic cues 2.56
Magazines, characteristics 12.12
Mainframe computer software 6.138
Male pronoun 3.24
Mantissa, in scientific notation 7.30
Manuals. *See also specific types of manuals*, e.g. Technical manuals
  communication characteristics 1.37
  information classes and 12.8
  information needs 1.63
  modular 1.66
  organizing information 1.64, 1.71
  purpose 1.61
  size considerations 1.65
  types of 1.62
Margins 12.23–25
Marketing documents
  audience considerations 1.20
  purpose 1.18
  textual features 1.21
  types of 1.19
Master glossary 12.117–18
Master index 12.116, 12.118
Mathematical expressions. *See also* Symbols
  ellipsis in 4.77
  numerals in 7.3
  physical arrangement 7.61
  punctuating 7.82–84
  setting 7.55
  SI units and 7.47
  typographic details 7.77
Mathematical symbols 6.17, 7.53
Mathematical typography 7.54–60
Mathematics, style guides 8.88
Matrix charts 11.116
Measurements. *See also* Symbols
  derived from proper names 5.48
  expressing as numerals 7.1
  significant digits in 7.38
  in tables 11.83
  traditional units 7.48
  units of 7.41–48
Mechanical binding 12.128–31
Media selection considerations 1.39–42
Medical terminology 6.69–80
  abbreviations 6.1, 6.68–79
  anatomical terms 6.72
  considerations 6.69–71
  diseases and conditions 6.74
  drug terminology 6.74–75
  equipment 6.78
  infectious organisms 6.76–77
  procedures and treatments 6.79
  style guides 8.88
  vitamins and minerals 6.80
Memoranda 1.57–58
Meteors 6.90
Metric Conversion Act (1975) 7.41
Military terminology 6.58–67
Minerals 6.80, 6.115
Minimum Word Strategy 2.1, 2.3
Minus sign
  breaking equations and 7.73
  in tables 11.86
  typographic details 7.77
Misplaced modifiers 3.33–34
Misspellings 5.66. *See also* Spelling
Mixed numbers 7.11–12, 7.24
Mnemonic forms, military terminology 6.62
Modal verbs
  conditional tense 2.35
  implications 2.41
Modifiers. *See also* Adjectives; Adverbs
  common problems 3.8, 3.30–59
  dangling 3.36
  misplaced 3.33–34
  possessive nouns and 7.26
  role in sentences 3.4
  squinting 3.37
  triage edits 3.40
  unit modifiers
    containing a proper name 4.93
    containing adjectives and hyphen 4.85–87

containing adverb and hyphen 4.90–92
containing number and unit of measure 4.88–89
hyphens in 7.20, 7.25
longer than two words 4.94–95
Modular modules 1.66
Month names 5.50
Mood 3.50
Moons 6.87–88
Multicolumn side-by-side publications, multilingual documents 12.91
Multilingual documents 12.86–91
Multilingual terminology database 2.89–91
Multimedia dynamic information 12.103–4
Multiple common meanings 2.98
Multiplication sign 7.73
Multipurpose reference manuals 1.82
"*must have been*" construction 2.41

N

"Naked eye" stars, Bayer method 6.93
Naming conventions, astronomical 6.100–1
National Aeronautic and Space Administration (NASA), abbreviations 6.35
National flags, capitalization 5.50
National Institute for Standards and Technology (NIST), *Guide for the Use of the International System of Units* 7.43, 7.44
Natural sciences, style guides 8.88
naval, lowercase 5.50
Naval vessels 6.65–67
Navigation
help systems 1.100
websites 1.113
"*need not*" construction 2.36
"*need*" plus infinitive, ambiguity problems 2.39
Net-difference surface charts 11.230
Newsletters
communication characteristics 1.37
design decisions 12.52–53
information classes and 12.8
purpose 1.49
size considerations 1.50, 12.12
Non-native audiences 2.39, 2.71–72. *See also* Controlled English; Global English; International audiences; Minimum Word Strategy
Nonlinear outlines 1.138
Nonrestrictive clauses 4.13
Nonrestrictive phrases 4.13
Nonstandard English 2.45
Nonstandard quotations 8.21, 8.24. *See also* Correspondence; Epigraphs; Footnotes; Foreign languages
Nontechnical terminology 2.91
Nonzero digits 7.38
"*nor*" conjunction
joining compound antecedants 3.19
joining compound subjects 3.29
"*not*," Global English guidelines 2.21, 2.22
Notes
for charts 8.97, 11.180
design decisions 12.43–44
for illustrations 8.97, 10.72–74
in tables 11.31, 11.71–77
Noun phrases
characteristics 3.4
as fragments 3.45–47
hyphenation 2.19, 2.81, 4.83
in illustration captions 10.65
joined by *and* 2.73
joined by *or* 2.79
in lists 3.46
long 2.17–18, 2.81
translation considerations 2.91
Nouns
abstract 3.8
adjectives as, avoiding 2.110
avoiding verb suffixes 2.109
collective 3.20

Nouns *(continued)*
common problems 3.7, 3.8–16
concrete 3.8
correct usage 2.107–8
in illustration captions 10.65
as index entries 9.83–84
modified 2.64
multiple common meanings 2.99
plural
apostrophes and 4.115–16
collective 3.20
compared to possessive nouns 3.9–11
correct usage 2.52
foreign terms 5.14–18
forms of 5.9–13
as index entries 9.84
joined by compound antecedents 3.19
possessive 4.110
singular/plural dilemma 5.19–22
possessive 4.109–13, 7.26
proper 2.91
relationships 2.52
role in sentences 3.4
singular
collective 3.20
correct usage 2.52
as index entries 9.84
joined by compound antecedents 3.19
possessive 4.109
singular/plural dilemma 5.19–22
verb agreement with 3.26
"stacked nouns" 2.81, 3.15–16
suspended compounds and 4.101
triage edits 3.40
two-noun combinations, with en dash 4.66
Novas 6.98
Nuclear particles 6.121–25
Number theory, symbols 7.53
Numbered lists 4.130–31
Numbers. *See also* Decimals; Measurements; Values
apostrophes and 4.115
clock time 7.8
commas and 4.38–41
converting to text 6.30
in data displays 11.19
dates and 7.9–10
for endnotes 8.109, 8.110, 8.112
for footnotes 8.103
fractions and mixed numbers 7.11–12
for front matter pages 12.70
in graphs 7.36–37
with hyphens 7.20–26
hyphens and 4.102
for illustrations 10.56–60
indexing 9.109–11
large 7.13–15
mathematical expressions 4.77, 7.3
measurements 5.48, 7.1, 7.38, 7.41–48
numerals and 7.1
orders of magnitude 7.40
page and screen 12.40
percentages 7.5
powers of ten 7.31–36
proportios and 7.6
quantum 6.123
range of 4.65
reference 4.42
scientific notation 7.27–30
significant digits in 7.38
spelled out 7.16–19
for table columns 11.38
in tables 7.36–37, 11.82–91, 11.99, 11.105
time spans 7.7
unit modifier containing number and unit of measure 4.88–89
Numeric value 2.52

O

Oblique views, in charts 11.195
Office automation products, characteristics 12.12
Omissions
deliberate 4.26–27
within quotations 4.76

Omnibus Trade and Competitiveness Act (1986) 7.41
On-screen information
  considerations 12.92–93
  graphic specifications 12.105–6
  home page 12.97
  information classes and 12.94–95
  multimedia dynamic information 12.103–4
  section pages 12.98
  splash page 12.96
One hundred percent column chart 11.241
One hundred percent surface charts 11.229
One-word modifiers 2.21
Online spelling checkers 5.8
"*only*," Global English guidelines 2.21
Opacity, of printing paper 12.122
Operator manuals
  appendixes 1.78
  conceptual information 1.76
  index 1.79
  purpose 1.73
  task information 1.74
"*or*" conjunction
  indicating synonyms 2.77–80
  joining compound antecedants 3.19
  joining compound subjects 3.29
Oral sources, quotations 8.21, 8.24, 8.27–30
Orders of magnitude 7.40
Organization charts 11.281–85
Organization names. *See also* Proper nouns
  abbreviations 6.1, 6.48–67
  academic 6.41–45
  capitalization 6.40
  commercial 6.46–54
  government and legislative 6.55–56
  societies and associations 6.57
Organizational methods, selection considerations 1.131, 1.139
Organizations, capitalization 5.50
Outlines
  example 1.136, 1.138
  formal 1.135
  informal 1.137
  nonlinear 1.138
  parallelism 3.59
  purpose 1.134
  types of 1.136
Oversize tables 11.102–3
Overview design 12.115

P

Page and screen grids 12.48–50
Page numbering 12.39–40
Paired-bar charts 11.246
Paragraph format, indexes 9.41
Paragraphs
  boilerplate 3.56
  topical shifts 3.50
Parallelism, common problems 3.51–59
Parentheses 4.3, 4.50–66, 4.127–37
  abbreviations and 4.128
  breaking equations and 7.73
  definition 4.127
  for endnotes 8.110
  in indexes 9.52
  indicating equivalent expressions 2.78
  numbered lists and 4.130–31
  parentheses within 4.136–37
  punctuating parenthetical comments 4.132–33
  references and 4.129
  sentences enclosed in 5.42
  typographic details 7.79
Parenthetical definitions 3.12
Participial phrases
  common problems 3.35
  introduced by an *-ing* word 2.68
  as subjects 4.15
Participles
  dangling 3.35
  past 2.60–62
  present 2.63–70

Parts of speech
  Global English 2.92
  grammatical functions 3.4
Passive electronic components 6.160–61
Passive voice
  Controlled English 2.9
  dangling modifiers and 3.36
  dangling participles and 3.35
  Global English 2.31
  topical shifts 3.50
Past participles 2.60–62
Patterns, in charts 11.186
Percentages 7.5, 11.86
Perfect binding 12.132–33
Period 4.3, 4.67–82
  in abbreviations 6.33–34
  abbreviations and 4.71–74
  abbreviations with 4.115
  definition 4.67
  eliminating in abbreviations 6.5
  ending quotations 8.81
  between the hour and minute, British usage 4.58
  within quotation marks 8.80
  raised 4.75
  vertical lists and 4.68–70
Person, consistency issues 3.50
Personal computer software 6.137
Phenomena 6.120
Photographs 10.19
Phrases
  amplifying 4.54
  contrasting 4.25
  foreign phrase used as adjective 4.97
  as fragments 3.43
  introductory 4.9–11
  nonrestrictive 4.13
  parallelism 3.53
  participial
    common problems 3.35
    introduced by an *-ing* word 2.68
    as subjects 4.15
  replacing with single words 2.101
  restrictive 4.13–14
  in series 4.19, 4.51–52
  special effects 4.123
  transitional 3.49, 4.20
  types of 3.4
Physical performance 1.132
Physics terminology
  guidelines 6.117–25
  style guides 7.43, 8.88
  symbols for quantities 7.52
Pictographic column/bar charts 11.257
Pictorial elements, in charts 11.198. *See also* Illustrations
Pictorial pie charts 11.265–66
Pie charts 11.195, 11.196, 11.258–66
Pixels 12.13, 12.106
Planets 6.84–86
Plastic comb binding 12.129
Plotting data 11.138, 11.162–64
Plus sign
  breaking equations and 7.73
  in tables 11.86
  typographic details 7.77
Point of view 3.50
Points (type sizes) 12.31–32
Popular press documents 1.43
Possessive nouns
  apostrophes and 4.109–13
  compared to plural nouns 3.9–11
  modifiers and 7.26
Powers of ten 7.31–36
Predicate adjectives 4.98–99
Predicate nominative 8.79
Preface, design decisions 12.75
Preferred spellings 5.2
Prefixes
  chemical compounds 6.109
  hyphens and 4.104–5
  Système Internationale (SI) 11.161
  word division guidelines 5.27
Prepositional phrases
  characteristics 3.4
  Global English 2.23–24, 2.27
Prepositions
  in index entries 9.90
  multiple common meanings 2.99

role in sentences 3.4
standardizing 2.97
triage edits 3.40
Present participles 2.63–70
Press size 12.16
Print documents, delivery considerations 1.38
Printer quotation marks 4.2
Printing illustrations 10.14–17, 10.41
Printing paper 12.122–24
Procedural documents
audience considerations 1.26
parallelism 3.55
purpose 1.18, 1.25
size considerations 12.12
textual features 1.27
Product knowledge 1.122
Product names, 5.53, 6.139
Professional journal
media considerations 1.43
size considerations 12.12
Programming language-based help 1.103
Progress status reports 1.93
Promotional newsletters 1.49
Pronouns
agreement of 3.23
antecedents 3.24
common problems 3.7, 3.17–24
consistency 3.50
references to 3.17
role in sentences 3.4
second-person 3.21
sexist language and 3.24
translation considerations 2.20
Proofreading indexes 9.115
Proper names
indexing 9.108
possessive 4.112
specifying with commas 4.36–37
Proper nouns. *See also* Organization names
capitalization 5.39, 5.47–51
geology terms 6.116
multilingual terminology database 2.91
software development documentation 6.149
unit modifier containing 4.93
Proportions, numeric experssion 7.6
Proposals 1.94, 1.95
Psychology, style guides 8.88
Public relations, newsletters 1.49
Publishing, style guides 8.88, 9.21
Punctuation. *See also specific types of punctuation*, e.g. Commas
in abbreviations 6.5, 6.33–39
component edits 3.39
for endnotes 8.110
flexibility 4.1
in indexes 9.52
mathematical expressions 7.82–84
preceding omission in multisentence quotations 8.65
in quotations 8.72–85
table text 11.80
Punctuation marks, typeset-quality 4.2

Q

Qualifier, with electronics terms 6.160
Qualitative information, in charts 11.118
Quantities, representing with symbols 7.49
Quantum numbers 6.123
Question and answer format, quotations 8.30
Question mark 4.3, 4.67–82
within closing quote 4.121
definition 4.78
inverted 4.2, 4.79
Quotation marks 4.3, 4.117–26
in block quotations 8.76–78
correct usage 4.118
definition 4.117
guidelines 8.73–75
placement, with other punctuation 4.119–22
specialized usage 4.123–25

Quotations
  adding material to 8.48–53
  capitalization 5.43, 8.66–71
  citing 8.12–16
  copyright considerations 8.2
  correct usage 8.8–11
  definition 8.7
  direct
    commas and 4.31–32
    definition 8.17
    guidelines 8.21, 8.22–23
    types of 8.19–21
  ethical considerations 8.1, 8.14
  indirect
    definition 8.17
    guidelines 8.35–37
    introducing 8.45
  legal guidelines 8.4–6
  long 4.56
  omitting parts of 4.76, 8.54–58
  punctuating 8.72–85
  revising, guidelines 8.46–47
  setting in text 8.38–45
  types of 8.17–37

R

Races, capitalization 5.50
Ragged-right margins 12.24
Raised periods 4.75
Ranges, in tables 11.90
Ratios
  colon and 4.57
  expressing as numerals 7.6
  heading sizes 12.32
  white space 12.20–21
Readability
  Global English 2.12
  static information 12.101
Reader access, design decisions 12.66–67
Reference numbers 4.42
Reference of base line, in charts 11.173
References
  for articles 12.62
  brackets and 4.129
  to charts 11.200–3
  considerations 8.86–87
  design decisions 12.45
  to tables 8.97, 11.101
Referential documents
  audience considerations 1.35
  multipurpose 1.82
  purpose 1.18, 1.80, 1.81
  size considerations 12.12
  textual features 1.35
  types of 1.34
Refresh rate, of computer screens 12.13
Regional websites 1.117
Registered names
  capitalization 5.52, 5.53
  commercial software 6.135–36
Relative clauses
  clarity 2.25
  editing 2.64
  expanding adjectives into 2.87
  expanding past participles into 2.60–62
Renderings 10.20–21
Reports
  communication characteristics 1.37
  purpose 1.89
  test results 5.54
  types of 1.90–96
"*require*" plus infinitive 2.39
Research reports
  business 1.91–92
  scientific 1.90
  tracking materials 8.90–91
Reserved words 6.153
Resolution
  computer screens 12.13, 12.20
  illustrations 10.8–11
Restricting modifiers 3.30
Restrictive appositives 8.79
Restrictive clauses 4.13–14
Restrictive phrases 4.13–14
Reverse type 12.37
Revision notices 12.47
Roman numerals, for front matter pages 12.70

Roman type 7.57
Row headings, tables 11.51–57, 11.97
Ruled boxes 12.44
Ruled tables 8.97, 11.62–70, 11.98
Run-in quotations
  capitalization 8.68–69
  definition 8.19
  omitting parts of 8.59–65
Run-on sentences 3.44
Running headers and footers 12.41–44

S

Sans serif typefaces 12.27, 12.28
Satire, avoiding 2.105
Scale charts 11.205–7
Scales, for charts 11.150–71
Scatter charts 11.208–10
Scientific notation 7.27–30
Scientific research reports 1.90
Scientific terminology 6.81–125
  abbreviations 6.5–39
  astronomical terms 6.83–101
  biological 6.102–3
  chemical 6.104–13
  considerations 6.81–82
"Scope of conjunction" problem 2.71
"Scope of modification" problem 2.72
Second-person pronouns 3.21
Section divisions, design decisions 12.79
Section pages, on-screen information 12.98
"*-sede*" word ending 5.38
See also reference 9.96, 9.98
See reference 9.96–97
Self- compounds, hyphenation 4.100
Semantics
  "core semantic meaning" 2.92
  Global English 2.29, 2.51
  grammatically or semantically incomplete words 2.111
Semicolon 4.3, 4.43–49
  connecting clauses joined without conjunctions 4.44
  definition 4.43
  elements in a series and 4.49
  in indexes 9.52
  transitional adverbs and 4.47
  transitional terms and 4.48
Sentence constituents, Global English guidelines 2.28
Sentence structures. *See also* Capitalization; Punctuation
  abbreviated terms and 6.29–30
  common problems 3.41–47
  complete sentences
    introduced by colons 4.55
    introducing block quotations 8.55
    in lists 3.47, 4.70
  complex sentences 3.6
  compound sentences
    characteristics 3.6
    commas in 4.5
    coordinating conjunctions in 4.12
    semicolons and 4.44, 4.45–46
  Controlled English 2.7, 2.9
  Global English 2.12–55
  imperative sentences 3.21
  incomplete sentences 8.60
  logical transitions 3.49
  parallelism 3.52, 3.54
  parenthetical comments and 4.132–33
  parts of speech 2.92, 3.4
  "scope of conjunction" problem 2.71
  sentence types 3.6
  sentences enclosed in parentheses 5.42
  simple sentences 3.6
Separate joined publications technique, multilingual documents 12.89
Separate publications technique, multilingual documents 12.88
Separate segment charts 11.267
Sequence linkages, in diagrams 11.280

Series
  adjectives in 4.29–30
  colon and 4.51–52
  hyphenated compounds in 7.21
  items in 4.16–18
  phrases or clauses in 4.19
  semicolons and 4.49
  symbols used in 11.143
Serifed typefaces 12.27
Set theory, symbols 7.53
Sexist language 3.24, 3.39
Shading, in charts 11.186
Short sentences, Global English 2.13
"*should,*" conditional tense modal verb 2.35
Sic, definition 8.51
Side-by-side separators, for tables 11.70
Silent -*e*, and suffixes 5.28–30
Similar words 4.28
Similies 4.138
Simple sentences 3.6
Simplified English 2.8
Single hyphen 4.61
Single letters, and hyphens 4.96
Single words
  abbreviating 6.21
  replacing phrases 2.101
Singular pronouns, antecedents 3.24
Size
  data displays 11.8
  manuals 1.65
  page and screen 12.11–17, 12.38–50
  page grids 12.50
  typefaces 12.31–32
Slang, avoiding 2.102
Slash. *See also* Dash; Hyphen
  in abbreviated forms 6.37
  in citations with URLs 8.131
  correct usage 4.106–7
Sliding-bar charts 11.247
Small caps, introducing references 8.98
So-called expressions 4.126
Social sciences, style guides 8.88
Society names 6.57
Soft endings ("-*ce*," "-*ge*") 5.31
Software
  creating data displays 11.23, 11.25
  cross-version compatibility 1.41
  file formats 1.42
  mainframe 6.138
  personal computer 6.137
  supportive database programs 8.90–91
  terminology 6.134–40
Software documentation, specialized terminology 6.144–50
Solidus 6.149. *See also* Slash
Sorting methods, indexing 9.26–39
Spaces
  in large numbers 7.14
  between table rows 11.81
Spanish, direct questions 4.79
Spanners, table columns 11.39–42
Special characters
  indexing 9.37–39, 9.104, 9.112–13
  in tables 11.74
Special terms, enclosing with quotation marks 4.125
Specialized vocabulary
  media considerations 1.44
  multilingual terminology database 2.89–91
  thesauri 5.60
Speeches, quoting 8.27
Spelling. *See also* Capitalization; Editing; English language; Foreign languages
  abbreviated terms and 6.23
  American spelling compared to British spelling 5.65
  component edits 3.39
  considerations 5.1
  misspellings in quotations 8.49, 8.51
  preferred 5.2
  prefixes 5.27
  singular/plural dilemma 5.19–22
  suffixes 5.27
  variations 5.62–64
Spelling checkers 5.3–8

Spiral binding 12.129, 12.130
Splash page 12.96
Spreadsheets, software 11.24
Square brackets, for endnotes 8.110
Squinting modifiers 3.37
"Stacked nouns" 2.81, 3.15–16
Stand-alone indexing programs
  9.67–70
Stand-alone tutorials 1.68
Standard Generalized Markup
  Language (SGML) 9.71
Standardizing prepositions 2.97
Standardizing terminology 2.94–110
Standards
  citations for electronic sources
    8.131
  indexing 9.3
  military 6.59–61
  National Institute for Standards and
    Technology (NIST) 7.43, 7.44
  technical manuals 12.64
  websites 1.115
Stars 6.92–98
Statistics
  software characteristics 11.24
  symbols 7.53
Stock exchange listings 6.54
Stub head, column heading 11.48–50
Style, component edits 3.39
Style guides
  for abbreviated terms 6.5–39
  citation variations 8.96
  grammar 3.38
  indexing 9.14, 9.21–25
  recommended 3.2
  for specific professions 8.88
Subheadings
  index entries 9.85–92
  table columns 11.36–37
Subject and verb agreement 3.50
Subject complements, verb agreement
  3.27
Subject indexes, compared to
  alphabetic indexes 9.27–31
Subject line, in correspondence 1.52
Subordinating conjunctions 2.85
Subscripts, typographic details 7.80
Subtitling illustrations 10.68–69
Suffixes
  hyphens and 4.104–5
  spelling considerations 5.28–38
  word division guidelines 5.27
Summarizing data, in tables 11.106–8
Superscripts
  for endnote numbers 8.111
  typographic details 7.80–81
Surface band charts 11.226–28
Surface charts 11.223–25
Surveillance terms 6.68
Suspended hyphen 4.101
Symbols. *See also* Abbreviations;
    Mathematical expressions;
    Measurements
  arbitrary changes in 6.18
  astronomical terms 6.83–101
  biological terms 6.102–3
  for charts 11.142–44
  chemical 6.104–7, 6.111–12
  considerations 6.9
  cultural considerations 10.35–36
  for currency 11.86
  for data points 11.137
  definition 6.15–17
  in diagrams 11.272–75
  in illustrations 10.27
  indexing 9.112–13
  mathematical 6.17, 7.54–60
  media considerations 1.44
  in pictorial pie charts 11.265
  punctuation 6.33
  scientific 6.1, 6.80–124
  selection considerations 7.49–53,
    7.56
  in tables 11.74
  trademarks 5.53
Synonyms
  avoiding 2.95
  as index entries 9.93–95
  indicating with *or* 2.77
  multiple common meanings 2.98
Syntax, in Global English 2.29,
  2.56–76

System configuration 12.112
System information, user reference manuals 1.83
System messages 6.143
System responses 1.77
Système Internationale (SI) 7.41, 7.44, 11.161

T

Table of contents 12.67, 12.76
Tables 11.26–116. *See also* Charts; Data displays; Diagrams; Graphs; Illustrations
  abbreviations in 11.104
  in appendixes 1.78
  complex
    continued tables 11.96–100
    decision tables 11.111–12
    distance tables 11.114–15
    look-up-a-value 11.110
    matrix charts 11.116
    oversize 11.102–3
    summarizing data 11.106–8
    types of 11.95, 11.109–16
  components
    column headings 11.29–50
    field (body) 11.58–61
    field spanners 11.43–47
    row headings 11.51–61
    rules 11.62–70, 11.98
    spanners 11.39–42
    stub head 11.48–50
  contents
    graphics 11.92
    numbers 11.82–91
    text 11.80–81
    values 11.93–94
  correct usage 11.27
  integrated with text 11.11
  numbers in 7.36–37, 11.99, 11.105
  relationships in data 11.3
  ruled 8.97
  types of 11.79
Tag languages, software characteristics 11.24
Task information
  activities 1.132
  job aids 1.86
  ordering 1.133
  user guides/operation manuals 1.74
  user reference manuals 1.85
Task orientation 12.111
Technical letters 1.56
Technical manuals
  consistency 12.63
  design decisions 12.63–85
  format issues 12.65–85
  standards 12.64
Technical terminology
  abbreviations 6.1, 6.125–71
  computer terms 6.129–31
  considerations 6.126–28
  electronics 6.157–67
  hyphenation 4.103
Telecommunications terminology 6.168–72
Telegraphic writing style, avoiding 2.58
Template, page grid 12.50
Temporal conjunction 2.67
Terminology. *See also* Audience analysis; Technical terminology; Translation
  Global English 2.88–111
  international audiences 2.88
Text
  complete reference in 8.96
  converting to numbers 6.30
  document body 12.32
  integrated with data displays 11.9–13
  static information 12.100–2
  in tables 11.33, 11.80–81
  white space and 12.18
Textual features
  for charts 11.183–84
  classifying documents 1.16
  conceptual documents 1.24
  in data displays 11.5
  fractions 7.11
  in illustrations 10.43–45

job aid documents 1.33
marketing documents 1.21
procedural documents 1.27
referential documents 1.36
tutorials 1.30
Textured paper 12.124
Textures, in charts 11.186
"*that*" pronoun 2.83
"*the*" article 2.53–55
"*them*" pronoun, Global English guidelines 2.20
Thesauri 5.60. *See also* Dictionaries
"*this*" pronoun, common problems 3.17
Three-dimensional effects, in charts 11.194–97
Time lines 11.243–45
Time of day
colon and 4.58–60
en dash and 4.65
expressions 4.113
plotting in charts 11.138
in tables 11.88–89
Time spans 7.7
Title, for articles 12.55–56
Title page, design decisions 12.71–72
Tone 3.50
Topical shifts 3.50
Trade directories 6.53
Trade journals, media considerations 1.43
Trade names
capitalization 5.52
indexing 9.51
medical terminology and 6.75
Trademarks 5.52, 5.53
Trailing zeros 7.38
Training manuals 1.71
Transitional adverbs 4.47
Transitional words and phrases 3.49, 4.20, 4.48
Transitive verbs 2.49–50
Translation. *See also* Foreign terms; International audiences; Non-native audiences
alternative 2.3
ambiguity problems 2.15, 2.39
cultural considerations 10.35–36, 10.50
large numbers and 7.15
multilingual terminology database 2.89–91
pronouns 2.20
"stacked nouns" problem 2.81, 3.15–16
Transposed terms 4.21
Tree structure
diagram technique 11.280
websites 1.110, 1.138, 11.280
Triage edits 3.40
Triangular charts 11.211–14
Tribes, capitalization 5.50
Trigonometric functions
en space and 7.63
symbols 7.53
Tutorials
audience considerations 1.28–29
characteristics 1.67
classroom 1.69
purpose 1.18
recovery information 1.72
sequence of procedures 1.70
size considerations 12.12
stand-alone 1.68
textual features 1.30
training manuals 1.71
Two-column side-by-side publications, multilingual documents 12.90
Two-noun combinations 4.66
Type sizes
design decisions 12.31–32
highlighting techniques and 12.37
interactions with line width and leading 12.36
Typefaces 12.26–30
definition 12.26
design decisions 12.27
for displayed information 12.28–29
running headers and footers 12.42
selection considerations 12.27, 12.30
Typography. *See* Textual features

U

Uncoated paper 12.124
Unconventional English 2.45
Unconventional verb complements 2.48
Underlining
 in data displays 11.5
 type sizes and 12.37
Unit modifier
 containing a proper name 4.93
 containing adjectives and hyphen 4.85–87
 containing adverb and hyphen 4.90–92
 containing number and unit of measure 4.88–89
 hyphens in 7.20, 7.25
 longer than two words 4.94–95
Unit symbols 11.144
United Nations mandate 6.60
United States Securities and Exchange Commission (SEC) 6.54
Uppercase letters 5.41
URLs (universal resource locations)
 citations 8.131
 slashes in 4.107
U.S. Patent and Trademark Office 5.53
Usability tests 1.137
User actions 6.146
User guides
 appendixes 1.78
 conceptual information 1.76
 index 1.79
 purpose 1.73
 recovery information 1.77
 task information 1.74
User-interface design 1.137
User reference manuals
 purpose 1.84
 system information 1.83
 task-oriented 1.85

V

Values. *See also* Measurements; Numbers
 in charts 11.135–41, 11.151, 11.153, 11.172, 11.174
 in tables 11.93–94
Variables, list of definitions 7.84
Vehicle weight 7.51
Verb complements 2.46–48, 13.2
Verb phrases
 characteristics 3.4
 Global English 2.16, 2.27
 joined by *and* 2.73
 joined by *or* 2.80
Verb suffixes, avoiding 2.109
Verb tenses, Controlled English 2.7
Verbals, characteristics 3.4
Verbs
 action 6.156
 agreement with collective noun subjects 3.25
 agreement with compound subjects 3.28–29
 agreement with subject complements 3.27
 common problems 3.7, 3.25–29
 compound 4.99
 correct usage 2.107–8
 editing 2.84, 3.40
 idiomatic 2.100
 imperative 6.155
 modal, conditional tense 2.35
 multiple common meanings 2.99
 role in sentences 3.4
 specific forms 2.83
 subject and verb agreement 3.50
 tense 2.16, 3.50
 transitive 2.49–50
 triage edits 3.40
 user actions 6.146
Version numbers, in product names 6.139
Vertical lists 4.68–70
Vertical scale, for charts 11.152

Vitamins 6.80
Voice. *See* Active voice; Passive voice
Volume designation, for charts
  11.139–41

W

Warnings 1.77, 12.43–44
Websites
  accessing 1.118
  communication characteristics 1.37
  components 1.110, 1.112
  document delivery considerations
    1.38
  elements 1.116
  navigation 1.113
  purpose 1.109
  special effects 1.114
  standards 1.115
  types of 1.117
  updating 12.92
  writing style 1.111
Weight, of printing paper 12.122
White space 12.18–22
Wide tables 11.103
Wide World Web, citations 8.131
Wildcard characters 2.91
Wire-o binding 12.129, 12.130
Wire stitching 12.126–27
Wizards
  audience considerations 1.107
  communication characteristics 1.37
  purpose 1.104, 1.106, 1.108
  required responses 1.105
Word-by-word sorting 9.35, 9.39, 9.70
Word combinations, conventional
  2.96
Word division, guidelines 5.25–27
Word endings
  "*-ce*," "*-ge*" 5.31
  "*-ceed*," "*-sede*," and "*-cede*" rule 5.37
  consonent endings 5.34–37
  English language and 5.60
  "-ise" 5.36
  "-ize" 5.36, 5.37
  "*-y*" ending 5.32–33
Word-processing software, typeset-
  quality punctuation marks 4.2
Word processors
  alphabetized lists and 9.61–62
  embedded indexing and 9.60,
    9.63–66
  software characteristics 11.24
Words per line 12.35

Y

"*-y*" ending 5.32–33
Years 7.10
"*You*," implied 3.21

Z

Zeros, and decimals 7.4, 7.38